Water Management, Partnerships, Rights, and Market Trends

An Overview for Army Installation Managers

Beth E. Lachman, Susan A. Resetar, Nidhi Kalra, Agnes Gereben Schaefer, Aimee E. Curtright

Prepared for the United States Army

For more information on this publication, visit www.rand.org/t/rr933

Library of Congress Cataloging-in-Publication Data is available for this publication.

ISBN 978-0-8330-9046-1

Published by the RAND Corporation, Santa Monica, Calif.

© Copyright 2016 RAND Corporation

RAND® is a registered trademark.

Preface

This project, "Water Management, Partnerships, Rights, and Market Trends: An Overview for Army Installation Managers," responds to an Army request to assess existing market mechanisms and partnership opportunities that Army installations could use to take advantage of industry and other investments in installations' water systems. The Army also sought recommendations for how current policies and activities could be adjusted to help use these mechanisms and opportunities. The project had three main tasks:

- Review the literature on market mechanisms and partnership opportunities for leveraging investments in installations' drinking and wastewater systems.
- Conduct Army installation case studies of the drivers and barriers to using market mechanisms and partnership opportunities to help fund installations' water activities.
- Develop recommendations about how policies and activities can be adjusted so installations can more effectively take advantage of market mechanisms and partnership opportunities.

As the title implies, this report is intended to be a resource for those involved in developing policies and practices for installation water supply, conservation, and use. This report also responds to another Army project request, "A Guide to Army Installation Water Rights: How to Establish, Maintain, Document, and Enhance Your Water Rights," on developing a set of tools for Army installation and garrison commanders and water managers on installation water rights by serving as a reference document on water management, partnerships, rights, and market trends. It will be of interest to those involved in these issues at Headquarters Department of the Army and at installations across the United States. This report should also be of interest to the other Services within the Department of Defense (DoD) and some other federal land managers who have an interest in these water-management issues.

This research was sponsored by the Assistant Secretary of the Army (Installations, Energy & Environment) and conducted within the RAND Arroyo Center's Strategy, Doctrine, and Resources Program. RAND Arroyo Center, part of the RAND Corporation, is a federally funded research and development center sponsored by the United States Army.

The Project Unique Identification Codes (PUICs) for the projects that produced this document are RAN136472 and RAN126132.

Executive Summary

This study assesses existing water market mechanisms (such as water banking and auctions) and partnership opportunities that Army installations can potentially use to improve installation water programs and their investments in drinking and wastewater systems. Because such mechanisms and opportunities depend on water-management practices and water rights, the study also provides an overview of these areas. In addition, the report provides detailed case studies of these issues within Colorado and Fort Carson and within Arizona and Fort Huachuca. Lastly, the report provides recommendations about how Army policies and activities can be adjusted to improve installations' water security, programs, and infrastructure investments.

Contents

Figures

Tables

Summary

"Water, water everywhere, nor any drop to drink."

The well-known line from Samuel Taylor Coleridge's "Rime of the Ancient Mariner" does not yet describe the situation in the United States. However, cities, counties, states, and the federal government (including Army installations) all are confronting issues associated with the supply, transport, and distribution of water. Because water is essential, the supply finite, the demand growing, and many countervailing interests are involved, the problem is, not surprisingly, complex.

Part of the complexity stems from the history of water supply and use in the United States. As the country grew and the population moved westward, considerable effort went into developing reliable sources of water for community, agriculture, mining, and other development purposes, which the federal government often subsidized. These water projects involved building an immense water infrastructure in the form of dams, reservoirs, communal wells, conveyance channels, tunnels, and aqueducts to gather, store, move, and use water. Complex intra- and inter-state agreements also were negotiated for managing and sharing water resources. By the middle of the 20th century, this water infrastructure largely had been built, and most water sources already were claimed and allocated for different community, agriculture, industry, and other needs. While water infrastructure replacements and refinements will continue, the day of building large infrastructure projects to move water from one place to another is over.

Water supply is finite, but the demand is not. Indeed, demand for water continues to grow. This occurs in part because the population and economy continue to grow, placing more pressures on this limited resource. In fact, in many parts of the country, lakes, rivers, and other surface water sources and groundwater sources are now over-allocated, causing competition for the rights and access to water. Farmers, ranchers, cities, towns, oil and gas companies, other industries, environmental groups, and other entities compete with military installations and other federal landowners for water. All of them find it more and more challenging to share water resources, especially during droughts. The long-term access to reliable sources of water is uncertain.

In a true market situation, when supply is constant and demand increases, the normal response is that prices go up. However, water is quite different, and this adds another level of complexity. Because water is widely seen as a public good, it also is seen as something that should not cost very much. Thus, local, state, and federal government agencies find themselves supplying an increasing demand without being able to use traditional market techniques to control that demand.

Another complicating factor is that as water has become scarcer, more localities and entities are turning to litigation to preserve their water rights, while others are suing to obtain access. The laws and regulations concerning water have been developed over centuries, and, typically, compromises over water rights and allocation take years, or even decades, to resolve legally. In some cases, the disputes go all the way to the Supreme Court. In addition, under most circumstances, determining water rights is the province of states, though local and federal governments also may have a role in some locales. Given these circumstances, the laws and policies can differ substantially from one location to another, further complicating the situation. For instance, because water also moves across state boundaries, litigants typically must deal with varying laws and policies.

This Report

This report represents a response to a request from Assistant Secretary of the Army (Installations, Energy & Environment), Headquarters, U.S. Army. RAND Arroyo Center was asked to assess existing market mechanisms and partnership[1] opportunities that Army installations can use to take advantage of industry, utility, and other investments in their water systems and to develop recommendations about how current policies and activities can be adjusted to facilitate the use of these mechanisms and opportunities. The main tasks of the project involved a review of the extensive literature on market mechanisms and partnership opportunities; case studies of Army installations of what fosters and hinders market mechanisms and partnerships to help fund installation water activities; and developing a set of recommendations to adjust policies and activities so installations can better take advantage of such mechanisms and partnerships.

This report is written to be a resource for installation water managers with different interests and knowledge. It is not designed to be read cover to cover (although it can be), but rather by individual chapters or topics. At the end of the introduction, we provide a guide on how to read the report to help readers identify the information of most interest to them.

What We Found and What We Recommend

Overall, we find that the traditional way of solving water problems—by increasing access to surface water and groundwater supplies—is no longer viable. These water sources are mostly allocated, and in many cases over-allocated. Today, communities and water management agencies seek to conserve and manage existing water resources more effectively. Many cities and towns have implemented water conservation programs, which have helped decrease per capita use of water. To address unmet needs, water managers are now investing in and using non-

---

[1] A partnership is when two or more organizations agree to work together for mutual benefits and invest in the partnership relationship by sharing responsibilities, resources, and risks.

traditional sources, such as using reclaimed water (treated wastewater), better managing and using stormwater runoff, and even starting to use some desalinated water. Experimentation with water market mechanisms also has been employed to reallocate existing supplies to meet growing demands. Water management today also faces some key challenges, including aging infrastructure, water quality concerns, depleting groundwater aquifers, pressures of population growth, climate change effects on water availability, and continued public demands for low-cost water. These trends put added pressure on effective water management and on the goals for long-term water sustainability.

More detailed findings and recommendations appear below.

Water Market Findings and Recommendations

Wastewater is becoming a valuable asset. A significant fundamental finding is that effluent from wastewater treatment plants has become valuable. It is a new and useful water source and likely will become an important commodity in water market deals in the future. Water recycling, in which treated effluent is being used for many purposes, is becoming more common, especially in parts of the country where traditional water sources are over-appropriated. Reclaimed water is used for landscape and crop irrigation, stream and wetlands enhancement, industrial processes, fountains, decorative ponds, recreational lakes, and toilet flushing. Military installations, such as Fort Carson and Fort Huachuca, have taken advantage of treated effluent to irrigate golf courses. Processed effluent can also help to protect groundwater supplies from seawater intrusion along the coasts and help to recharge aquifers, as Fort Huachuca has done in Arizona. The most successful water auction in the United States, in Prescott Valley, Arizona, was for groundwater credits derived from the city's effluent. Effluent will become an important commodity in more such water market deals in the future.

Reclaimed water has a number of advantages, including reducing reliance on increasingly scarce and expensive potable water sources, reducing discharges of wastewater into oceans and rivers, and providing a consistent local drought-resistance water supply. Effluent also has some disadvantages, most notably the cost for treatment and delivery. Saline content, nutrients, and other chemicals in the water are other potential problems.

Rainwater and stormwater runoff also are becoming valuable alternative water sources, with many of the same types of uses, including irrigation and flushing toilets. Cisterns, rain gardens, and rain barrels can take advantage of this water source but are not a drought-resistance water supply. Thus, capturing rainwater and stormwater runoff is not as valuable from a water market perspective, and many state regulations limit co-mingling stormwater runoff with wastewater. In addition, because of state surface water rights laws, some states restrict the capture and use of rainwater and stormwater because of the potential to affect the water supply of downstream users.

Water markets provide future opportunities for installations. Water market mechanisms are still in their infancy. Many innovative experiments are under way, and markets are evolving.

These mechanisms will continue to evolve and grow, especially because water is becoming a more scarce and valuable commodity. Army installations have limited opportunities today to use water markets but are likely to have more opportunities in the future as the mechanisms increase and develop. Future participation in these markets could help provide long-term water security and financial benefits to Army installations. However, installations' ability to take advantage of water market mechanisms depends on their understanding of the mechanisms, having the expertise to participate in them, and having the water rights to the traditional and alternative (i.e., effluent) water sources.

Recommendations

The Army should:

- Monitor water-market trends and how water-market mechanisms are evolving over time.
- Monitor state and local government policies and issues that pertain to water allocation, especially in locations near major Army installations.
- Ensure that installations understand and track what is happening with local and regional water policy.
- Ensure that installation managers understand the regional and local water use patterns, sources, and infrastructure.
- Ensure that they also consider the trends in the use of water beyond the installation fence line to potentially take advantage of future water-market mechanisms to acquire or market installation water.
- Invest more Army headquarters staff expertise into water issues because of the increasing local shortages of this resource, future threats to water access at installations, and water's critical role in the Army's performance of its installation missions.
- Include the long-term value of water and water-market mechanisms in installation strategic planning.
- Ensure that installations maintain all their existing water rights, including traditional surface and groundwater rights and new alternative water sources, such as wastewater and stormwater runoff.

Findings and Recommendations About Installation Water Rights

Water rights are becoming more important. As a region, state, or locality faces shortages of water, more tension and challenges can emerge over legal water rights. States and local governments are looking to ensure that all water sources, even small ones, are appropriately allocated. Municipal and private interests are more likely to fight to maintain their own water rights and challenge other water rights holders' claims. These battles increasingly will unfold in court.

To participate in any water market activity, an installation or other organization must have clear water rights to its water sources. We found that many installations probably do not fully understand their water rights. This lack of understanding occurs because water is relatively cheap; calculating its future value is difficult; installation water systems have been privatized;

the issues involved are complicated and can trace back many decades or longer; and installation personnel often lack expertise in water's technical and legal dimensions.

Some installations are at risk of losing some water rights. Large installations often have many types of water rights. As water becomes scarcer, state and local governments are looking to tap into any source they can, even small ones. Failure to maintain or submit proper documentation on their water rights, such as not reporting a water source's beneficial use, can place an installation at risk of losing rights. Water rights need to be considered for all water sources, including surface water, groundwater, effluent from installation sewer systems, and stormwater runoff.

State and local governments play a key role in planning and managing water sources and rights. State and local government water managers actively are managing water resources to achieve economic, environmental, and social goals and priorities, largely based on input from water users. However, states have the authority to grant and withdraw water rights. It is important to distinguish between paper and real water rights. An installation may have paper water rights, but problems can ensue if it does not have physical access to the water. If the state or local agency that manages water allocation does not deal with those who violate rights, then actual physical access to water may be lost.

Recommendations

- The Army needs to support installation personnel to ensure that it has done everything it can to establish, document, maintain, and enhance water rights.
- Army headquarters should provide detailed, practical "how to" guidance on water rights for garrison commanders, installation managers, water managers, lawyers, and other personnel who have water responsibilities.
- The Assistant Secretary of the Army (Installations, Energy & Environment) (ASA(IE&E)), in collaboration with the Assistant Chief of Staff for Installation Management (ACSIM), should consider starting an Army working group on installation water rights and ensuring access to long-term water sources to identify and deal with installation water rights concerns and needs.
- Army installations should collaborate with state and local governments on water rights. Installation personnel should engage in relevant federal, state, and local water planning processes, including drought planning, to understand water development, allocation, and emergency plans.

We should note that since we first made these recommendations, the Army has been implementing the first three of them. In May 2014, the Secretary of the Army issued Directive 2014-08, "Water Rights Policy for Army Installations in the United States,"[2] and the Office of the Under Secretary of Defense issued a policy titled "Water Rights and Water Resources Management on Department of Defense Installations and Ranges in the United States and

---

[2] U.S. Department of the Army, "Water Rights Policy for Army Installations in the United States," Washington, D.C.: Secretary of the Army Directive 2014-8, May 12, 2014.

Territories."[3] Both provide policies and tasks that begin to implement the first two recommendations. One of those tasks is for ACSIM to provide implementation guidance on installation water rights. In addition, ASA(IE&E) and ACSIM have formed a working group on installation water rights and on ensuring access to long-term water sources.

Findings and Recommendations About Water Partnerships

Traditional installation water partnerships, such as Energy Savings Performance Contracts (ESPCs) and Utility Energy Service Contracts (UESCs), can be useful tools to acquire capital investments for installation water conservation and efficiency projects. While such partnership mechanisms tend to be used more for energy than water projects, water and energy projects can be bundled in an ESPC/UESC task so that short-term payback energy investments can be combined with longer-term water conservation investments payback.

Collaborations and partnerships play an important role in regional, state, and local water planning and management and are likely to increase. Such water collaborations and partnerships focus on managing and sharing water resources and supply; long-range water and drought planning; watershed management and planning; and developing and sharing water infrastructure. Such water collaborations and partnerships can be between two cities to share management of drinking and wastewater treatment systems or among states to help manage key regional water resources, such as the Great Lakes Partnerships. More regional collaborations and partnerships with multiple entities to manage water resources, such as state, regional, and local long-term water planning, are being developed and implemented.

State, regional, and local government and water utility partnerships are important to installations. They can help installations save money by lowering costs, maintain water rights, invest in shared infrastructure, and sustain access to water supply. Regional partnerships, involving multiple organizations, to help plan, manage, and/or share water resources are likely to grow and also are an important opportunity for the installations to increase long-term water security.

Because of the passage of the fiscal year 2013 National Defense Authorization Act (NDAA) Section 331, installations have expanded authority to partner with state and local governments. This authority gave new opportunity for installations to partner with nearby cities and towns to share water infrastructure and services. We already found some examples of such collaboration, including Fort Carson sharing costs of replacing a shared water pipeline with a state prison in Colorado and Fort Huachuca partnering with Huachuca City to process the city's wastewater. Local government collaborations are an emerging opportunity for helping to finance installation water infrastructure investments.

---

[3] U.S. Department of Defense, "Water Rights and Water Resources Management on Department of Defense Installations and Ranges in the United States and Territories," Washington, D.C.: Office of the Under Secretary of Defense, May 23, 2014.

Recommendations

- The Army should encourage installations' energy and water personnel to incorporate more water projects into their ESPCs and UESCs.
- Army headquarters and installation personnel should reach out more frequently to water utilities to encourage them to partner in installation UESC projects.
- Installation personnel should seek to take an active role in state and local governments' regional and local water planning and collaborative management by partnering with these organizations where they can.
- Army headquarters should ensure that installation strategic planning and guidance emphasizes the importance of collaborating and partnering with state and local governments regarding water issues, from water planning to mutually beneficial financing in shared infrastructure projects.
- ACSIM or Installation Management Command (IMCOM) should help installations conduct pilot experiments in partnering with a local government to increase investment in water infrastructure.

Other Findings and Recommendations

Water Utility Privatization

Privatization of an installation's water system limits its control and flexibility over water resources and can affect an installation's future water system investments, partnerships, and market opportunities. At some installations, the privatized water utility has limited incentives to invest in water conservation, especially when the utility sells water to the installation. Water privatization also can make it more difficult to partner with a local government in shared water infrastructure. Lastly, installations are likely to lose water management personnel and water expertise as installations rely more and more on the water utility contractors.

Recommendations

Before privatizing an Army installation's water systems, the Army should assess a number of different factors, including the importance of having long-term water expertise, water rights, and water-market opportunities at an installation. Such issues are more important for installations in arid areas.

- The Army should assess all the different water rights implications, the potential long-term water security pros and cons, and the implications for water conservation.
- When privatizing water utilities, any of the downsides just assessed should be minimized ahead of time as much as possible.
- The Army also should assess alternative approaches for funding the needed water infrastructure investments, including partnerships with local governments and other organizations.
- The Army should study the implementation of previous Army water utility privatization deals to develop lessons learned and incorporate them into future deals so those deals are

conducted more effectively and efficiently and can ensure installation long-term water rights, security, conservation and market opportunities.

Future Water Shortages and Successful Installation Water Conservation

Some proactive installations that have made significant investments in water conservation run the risk of being penalized by state and local government during times of drought or other water emergency.

Recommendations

- The Army should try to ensure that its installations' successful water-conservation efforts are considered during periods of regional and local water scarcity.[4]
- The Army should review state water plans and calculate their effects on Army installations. Army headquarters, regional, and installation personnel should be reviewing and working with state and local governments in those entities' regional and local water scarcity planning.

---

[4] Water scarcity refers to when human and environmental demands for water exceed the available water within a given region at a given time period.

Acknowledgments

We would like to thank Richard Kidd IV, Deputy Assistant Secretary of the Army Energy and Sustainability, for sponsoring this research. We also thank members of ASA(IE&E) staff who helped us in this study, including Marc Kodack and Bob Conley.

A special thanks goes to installation water and other personnel who hosted our visits and spent their valuable time with us at Forts Benning, Bliss, Carson, and Huachuca. We also thank other Army personnel we interviewed at other Army installations, at IMCOM, and at other organizations.

Finally, we thank RAND colleagues Nicholas Burger and Jerry Sollinger for their insightful reviews of this document and Julie Taylor for help with the water rights section. We also thank Christopher Weber from IDA for his thorough review of this document.

Any errors of fact or judgment that remain are solely those of the authors.

Abbreviations

AAFES	Army and Air Force Exchange Service
ACC	Arizona Corporation Commission
ACF	Apalachicola-Chattahoochee-Flint
ACSIM	Assistant Chief of Staff for Installation Management
ACUB	Army Compatible Use Buffering
ADWR	Arizona Department of Water Resources
af (or AF)	acre-feet
AMA	active management areas
ARRA	American Recovery and Reinvestment Act
ASHRAE	American Society of Heating, Refrigeration and Air-conditioning Engineers
AWWA	American Water Works Association
AWBA	Arizona Water Bank Authority
BA	biological assessment
C-BT	Colorado–Big Thompson
CAGWRD	Central Arizona Groundwater Replenishment District
CAP	Central Arizona Project
CAWCD	Central Arizona Water Conservation District
CBD	Center for Biological Diversity
CSP	central shortgrass prairie
CSU	Colorado Springs Utilities
CVP	California Valley Project
CWA	Clean Water Act
DOE	U.S. Department of Energy
DPTM	Directorate of Plans, Training and Mobilization
DPW	Department or Directorate of Public Works
DRID	Defense Reform Initiative Directive
DWR	Department of Water Resources
ECIP	Energy Conservation Investment Program
ECM	Energy Conservation Measure
ENRD	Environmental and Natural Resources Division
EO	Executive Order
EPWU	El Paso Water Utilities
ESA	Endangered Species Act
ESCO	Energy Service Company

ESPC	Energy Savings Performance Contract
FEMP	Federal Energy Management Program
FS	U.S. Forest Service
FY	fiscal year
GPCD	gallons per capita per day
IBWC	International Boundary Water Commission
ICC	International Code Council
IMCOM	Installation Management Command
INA	Irrigation Non-expansion Area
LADPW	Los Angeles Department of Public Works
LEED	Leadership in Energy and Environmental Design
M&V	measurement and verification
MAF	million acre-feet
MCA	Military Construction, Army
MDW	Metropolitan Water District
MGD	million gallons per day
MILCON	military construction
MWR	morale, welfare, and recreation
NAF	non-appropriated funds
NAS	Naval Air Station
NDAA	National Defense Authorization Act
O&M	operations and maintenance
PCMS	Piñon Canyon Maneuver Site
PPP	public-private partnership
PuP	public-to-public partnership
R&R	Repair and Restoration
RO	reverse osmosis
SCWD	Southeastern Colorado Water Conservancy District
SIR	savings-to-investment ratio
SPRNC	San Pedro Riparian National Conservation Area
SRM	Sustainment Restoration and Modernization
SRP	Salt River Project
SWP	State Water Project
UESC	Utility Energy Service Contract
UP	utility privatization
USACE	U.S. Army Corps of Engineers
USBR	U.S. Bureau of Reclamation
U.S. EPA	U.S. Environmental Protection Agency
USFWS	U.S. Fish and Wildlife Service

USGBC	U.S. Green Building Council
USGS	U.S. Geological Survey
WRMP	Water Resource Management Plan
WUI	water use intensity
WWTP	wastewater treatment plant

1. Introduction

Background

Water is a finite resource and critical to the survival and functioning of all aspects of the U.S. economy—communities, agriculture, industry, and government. As the U.S. population and the economy have grown, so too have the pressures on the increasingly stressed supply of water. In many parts of the country, lakes, rivers, and other surface water sources and groundwater sources are over-allocated, causing competition for the rights and access to water. Numerous entities—farmers, ranchers, cities, towns, oil and gas companies, other industries, environmental groups, federal landowners, including military installations, and others—compete for water. All have found it increasingly challenging to share water resources, especially during droughts. The long-term access to reliable sources of water is uncertain. Since water is a public good and is controlled by federal, state, and local governments, the ability to access water often is being determined by complex agreements among government agencies and increasingly in court battles. However, some market mechanisms, such as water auctions, also are starting to develop to help in the sharing of water between buyers and sellers. Similarly, public-to-public partnerships are being implemented to help government agencies better plan, manage, and share water resources and make investments in water infrastructure.

Because of the increasing shortages of water supplies (often due to ongoing droughts and other pressures on water supplies), cities and towns throughout the United States have begun water conservation programs. Similarly, the federal government and the Army have implemented policies and set goals for water conservation. Executive Order (EO) 13514 requires all federal agencies to set a baseline for potable water use at fiscal year 2007 and reduce potable water use intensity (WUI) by 2 percent per year based on the FY 2007 baseline through FY 2020, a total reduction of 26 percent. Army installations must meet this goal.[1] The Army also has implemented the Net Zero Water Installation Initiative, and in April 2011 it identified eight installations to participate in a pilot program to become net zero water installations by 2020.[2] A Net Zero Water Installation is defined as "an installation that reduces overall water use, regardless of the source, increases use of technology that uses water more efficiently; recycles and reuses water, shifting from the use of potable water to non-potable sources as much as possible; and minimizing inter-basin transfers of any type of water, potable or non-potable, so

---

[1] Even though this water-conservation goal is an Army-wide requirement, each installation is expected to strive toward meeting it at the individual installation-level. Namely, the Army has spread this goal equally to all installations rather than requiring more water conservation from some installations compared to others.

[2] Six of them are Net Zero Water pilots and two of them are integrated Net Zero pilots for water, waste, and energy.

that a Net Zero water installation recharges as much water back into the aquifer as it withdraws."[3]

To meet such water goals and manage their water resources, Army installation managers face a number of challenges. First, the costs of water at most installations are still very low, so it is challenging to find capital to invest in water conservation and infrastructure projects. For instance, conservation projects often do not have a good financial rate of return on the investment given the low cost of water. Similarly, it is extremely challenging to find capital for investments in new large-scale water and wastewater treatment plants and supporting infrastructure. Local governments and water utilities face similar challenges because aging water infrastructure needs to be replaced.

Second, the Army has limited funding sources for such water activities. Some limited Sustainment Restoration and Modernization (SRM) and Military Construction (MILCON) funds can be used to invest in some water efficiency projects and upgraded infrastructure. Similarly, some partnership mechanisms, such as Energy Savings Performance Contracts (ESPCs) and Utility Energy Service Contracts (UESCs), can be used to partner with industry and public utilities to implement installation water efficiency projects. However, these options have strict requirements and cannot be used to invest in large-scale water infrastructure, such as a new water treatment plant.

Third, given the demands for water and threats to water supply, concerns are increasing about water security concerns. Namely, installations need to ensure that they have sufficient water over the long term, even during a drought emergency, to perform their missions. Such long-term water security may be at risk at some installations.

Through partnership and evolving water market mechanisms, Army installations have an opportunity to leverage industry and local government activities to increase water conservation, improve water infrastructure, and enhance installation water security.

Purpose

The Assistant Secretary of the Army (Installations, Energy & Environment), Headquarters, U.S. Army, asked RAND Arroyo Center to assess existing market mechanisms and partnership opportunities that Army installations could use to possibly take advantage of industry, utility and other investments in installation water systems and to develop recommendations about how current policies and activities could be adjusted to facilitate the use of these mechanisms and opportunities. The project had three main tasks:

- Review the literature on market mechanisms and partnership opportunities for leveraging investments in installations' drinking and wastewater systems.

---

[3] U.S. Department of the Army, "Army Directive 2014-02 (Net Zero Installations Policy)," Washington, D.C.: Secretary of the Army, January 28, 2014.

- Conduct Army installation case studies of the drivers and barriers to using market mechanisms and partnership opportunities to help fund installations' water activities.
- Develop recommendations about how policies and activities could be adjusted so installations can more effectively take advantage of market mechanisms and partnership opportunities.

Methodology

This research was mostly conducted from May 2012 through September 2013.[4] Our assessment was conducted using three integrated methods: a review of relevant literature, conducting in-depth installation water case studies, and interviewing relevant experts.

Our literature review covered water market literature and activities, including the academic, business, financial, and economic literature regarding water market mechanisms. Documents about federal, state, and local government water rights, laws, activities, and policies also were reviewed. Similarly, OSD, Army, and other Service water policies and activities were reviewed. Installation documents, such as water management plans and Net Zero Water Installation roadmaps, ESPC and UESC documents, and U.S. Department of Energy (DOE) Federal Energy Management Program (FEMP) water-related documents and web sites were important resources. We also reviewed the water utility and management trade press.

We conducted in-depth case studies to understand installation water programs and state and local government water laws, policies, and market activities for four installations: Fort Carson, Fort Huachuca, Fort Benning, and Fort Bliss. These installations were chosen to highlight different water management issues. At each installation, RAND researchers spent one to two days interviewing water program, legal, contracting, Department or Directorate of Public Works (DPW), environmental, and other personnel, as well as some water utility contractors and partners. These installation visits helped us understand the installation's water program goals and activities, existing partnerships, water project challenges and successes (especially regarding project funding), and water rights and security concerns. Since the state and local water regulations and policies often determine water market opportunities, we also assessed state and local water rights, laws, and policies for the four cases study installations to understand water market opportunities in these states: Arizona, Colorado, Texas, Georgia, New Mexico, and Alabama. We also conducted phone interviews with installation personnel at several other installations and reviewed state policies and water market activities in several other states, such as California and Washington.

Lastly, RAND researchers also interviewed other relevant water, market and partnership experts at ASA(IE&E), Assistant Chief of Staff for Installation Management (ACSIM), Installation Management Command (IMCOM), communities, utilities, and industry. We attended

---

[4] We should note that some things have changed since this research was conducted, such as updates to federal policies and installation activities. Given the challenges to updating such a long document and the dynamic nature of the material, only the most relevant updates were made.

the American Water Works Association (AWAA) annual conference June 10–14, 2012, and interviewed public and private company water utility managers at this conference. Between the installation visits and the other interviews, we interviewed more than 50 people.

How to Read This Report

This report is not intended to be read cover to cover, although it can be. Rather, it is intended as a resource for those involved in installations' water management. Such readers will be familiar with many aspects of installation water management and will not need to read sections about topics with which they already are familiar. The intention is for such readers to read the sections that provide them with new information. For the most part, the sections are written so that they do not depend on the reader having read the preceding chapters. The installation case studies are an example. Below, we describe the general content of each chapter so that readers can identify the ones that best serve their interests. Because the chapters can be read independently, in some cases information is repeated from one chapter to another. Those reading the entire document can simply skip over the repeated sections.

Chapter Two. This chapter provides a brief history of trends in managing water in the United States and how they have changed over time. In short, it provides an overview of historical water policies and practices that have led to today's situation on water management. The report describes how water management historically has focused accessing water sources, such as the development of water transfer and storage infrastructures, and today focuses on efficient management of existing resources. Those include water conservation and using non-traditional sources of water, such as stormwater and reclaimed water. This chapter also discusses some of the main challenges that water managers currently face, such as aging water infrastructure and depleting groundwater aquifers. (New approaches to sharing or reallocating water using market approaches are discussed in Chapter Four.) Chapter Two also contains a section that briefly explains U.S. water rights, including those on military installations. This chapter functions as an introductory guide and reference to key water management trends, practices, and issues that set the context for understanding water partnership and market opportunities, so it covers a lot of material. We suggest that readers check the table of contents for this chapter and read only the areas with which they are unfamiliar.

Chapter Three. This chapter provides background information about Army installation water programs goals, funding sources, and water partnership opportunities. It contains information about Army installation water goals and the ways in which installations fund investment in water and wastewater systems, including some traditional partnerships activities, such as ESPCs and UESCs. It also provides an overview of different types of water partnerships (public-private, public-to-public, and regional) and describes some evolving opportunities for installations to participate in such partnerships to enhance long-term water security and to help finance installation water infrastructure investments.

Chapter Four. This chapter describes various water market mechanisms. Such mechanisms take a market-like approach to water but also take into account the unique characteristics that distinguish water from typical commodities. These mechanisms complement the water management policies and approaches discussed in Chapter Two. The chapter discusses basic leasing and selling, water auctions, water banks, block pricing, and water quality trading. It also describes the potential benefits and limitations of using market mechanisms as well as discusses several specific water market mechanisms. It ends by discussing the implications of water market mechanisms for Army installations.

Chapters Five and Six. These two chapters contain relatively extensive case studies of water issues for two states and Army installations in those states: Colorado and Fort Carson, and Arizona and Fort Huachuca respectively.[5] They can be read individually. These two case studies are provided to illustrate how resolving water issues can vary across installations. Generally, both chapters discuss the overall water situation, including water demand, supply, historical agreements, rights, and distribution. Each chapter presents examples that illustrate issues related to water markets as well as to water sources. They describe installations' water management programs, strategic goals, sources, and conservation activities.

Chapter Seven. The last chapter presents the findings and recommendations of the study.

---

[5] Since this document already was so long, we wrote in-depth case study chapters for only two of the four case studies. In addition, these other case studies are discussed more in an unpublished RAND document: Beth E. Lachman, Susan A. Resetar, Geoffrey McGovern, Katherine Pfrommer, and Jerry M. Sollinger, "A Guide to Army Installation Water Rights: How to Establish, Maintain, Document, and Enhance Your Water Rights," unpublished RAND Corporation research, March 14, 2015.

2. Background on U.S. Water Management Trends and Rights

This chapter provides an overview of the historical approach to water management, allocation, and infrastructure; the main current water management trends; pervasive challenges to these water activities and approaches; and U.S. water rights. The focus is on broader water management issues in the U.S. rather than on Army installation activities. The purpose of this chapter is to give installation personnel who may be unfamiliar with some of these topics a basic overview, because they are important to understand the context for trying to participate in water partnerships and markets. It serves as a quick reference for these topics.

Water resources are managed at the local, state, and federal levels. Moreover, unlike other resources, water is an essential resource that is not easily transferred without large investments in infrastructure. Nor do water-pricing approaches used today typically capture its full value. The historical development of water infrastructure, as well as water rights legal regimes, have left an indelible mark on current options for managing this resource. For all of these reasons, the story of understanding how water resources are and can be managed now and in the future is a complex one.

To make this story understandable, we have divided this chapter into four main parts:

- Historical Approaches to Water Management Focused on Accessing Water Sources
- Water Management Today Focuses on Efficient Management of Existing Resources
- Water Management Faces Key Challenges
- U.S. Water Rights.

First, this chapter describes the historical trends for managing water that focused on the development of water transfer and storage infrastructure for a developing nation. Approaches to water management in the United States have shifted significantly over time, as water availability, needs, and priorities have evolved. More than a century ago, westward expansion and the growth of new communities, agriculture, and mining spurred massive infrastructure projects for developing, storing, and moving water. We illustrate with some examples from California, Arizona, and Colorado, such as the developing of water infrastructure and storage for the cities of San Francisco and Los Angeles and construction of the Central Arizona Project to supply water to growing cities in Arizona. At the same time many complex intra- and interstate agreements, especially in the west, were negotiated for managing and sharing these water sources, and legal action frequently was needed to resolve increasingly complex conflicts over water rights. To explain this historical trend, we briefly describe the Colorado and Arkansas River Compacts and the failed attempt to develop an interstate river compact between Georgia, Alabama, and Florida for the allocation of water in the Apalachicola-Chattahoochee-Flint (ACF) River Basin and the resulting conflicts over this eastern river basin. Next, this section provides a brief discussion of some of the legal processes and claims and battles for water that started to develop in the U.S., such as river adjudication processes and lawsuits about water rights. Examples are presented for Texas and Arizona. Lastly, this section ends with a brief overview of how evolving federal policy regarding environmental concerns started to impact water policy, such as the passage and implementation of the Clean Water Act and Endangered Species Act.

Second, this chapter describes how today water managers in the U.S. are focused on the efficient management of existing water resources, including non-traditional approaches. Population, industry, environmental needs, and other demands continue to grow, but available water sources largely are already allocated and—in many cases—are over-allocated. Funding for large infrastructure projects has waned. Thus, the focus of water management has shifted over time: Communities and water management agencies seek to conserve and manage existing water resources more effectively and address unmet needs with non-traditional sources of water, such as taking advantage of effluent, stormwater runoff, and even desalinated water. Experimentation with market mechanisms also has been employed as a way of reallocating existing supplies to meet growing demands (discussed in Chapter Four). In this section, we briefly discuss how water conservation has grown as local, state, and federal governments have implemented conservation incentives and water utility usage charges; as local governments have adopted more water-wise building codes and as industry has developed more water-efficient appliances; and as our culture has shifted from more industry and consumer awareness about the benefits of conserving water. Then we describe some of the current and evolving approaches to use non-traditional water sources including better management and use of stormwater; increasing use of reclaimed water; and some development of desalination plants. National statistics are used and examples from cities in California, Florida, Texas, and Washington state are discussed to illustrate the range in implementation of such approaches. This section also provides a brief overview of state, regional, and local water planning, a key trend that is critically important to ensure the long-term availability for water for different users in U.S. communities as well as water market opportunities. Four states' examples are presented to illustrate the diversity in planning approaches: California, Colorado, Georgia and Texas. Lastly, this section discusses two current approaches, watershed management and sustainability, that government water managers are using to more strategically manage water resources.

Third, this chapter briefly discusses some of water managers' main challenges. Water management today must confront key challenges:

- Aging infrastructure
- Water quality concerns
- Depleting groundwater aquifers
- The pressures of population growth
- The effects of climate change on water availability
- Public demands for low-cost water (which does not fully account for its value).

We briefly explain how each of these issues presents a challenge to water managers and provide examples and statistics to illustrate the pressures they place on U.S. water supplies, effective water management, and goals for long-term water sustainability.

Finally, this chapter presents an overview of U.S. water rights, both regarding surface water and groundwater; how such water rights vary throughout the country; and how installations obtain such water rights.

As this chapter's overview just described, the complexities of water management require this chapter to be extensive. Readers may choose only to read topics with which they are unfamiliar. However, the context of this chapter is important to installation personnel, because the military operates within this larger system and because installations are affected by the laws, investments, and actions of others. Moreover, understanding U.S. water management trends and water rights sets the context for understanding U.S. water partnerships and market opportunities. In addition, installation personnel unfamiliar with the development of water resources would benefit from this material because:

- The major actors involved in water resources development still control large quantities of water in some areas of the country—water that military installations use today. The infrastructure and processes for storing and transporting water to users are long-lived capital investments that are still in use. And the legal and historical contexts provide installation personnel with an improved understanding of what water they are entitled to, how this entitlement can be maintained, and what legal agreements may cause the military to obtain less water.[1] This material will also give staff a sense of the political environment around water supply development and distribution, including other demands on water use and where potential risks to water supplies exist. Knowledge of the actors, existing infrastructure, and user community are needed to better understand the opportunities for and implementation of water investments, markets, and partnerships.

- State and local governments are responding to a tightening of water resources (full- or over-allocation of water) and changing demand patterns for water in ways that have, and will, affect the military. Naturally, installation personnel should understand how water supplies are being managed, since the military depends on these supplies in many cases and new requirements could extend to the military. In many cases, the military has adopted some of these approaches, but water resources, opportunities and practices vary among installations. As communities move forward with these practices, the military may learn from these experiences and apply these lessons to installations. Participatory processes should include the military's needs, as should potential changes to water allocation.

- Finally, water managers at all levels and types of government agency face the same key challenges and are seeking innovative ways to deal with them. Understanding how demands for water are changing and the challenges to water sources, availability,

---

[1] A more in-depth guide on maintaining installation water rights is in an unpublished RAND document: Beth E. Lachman, Susan A. Resetar, Geoffrey McGovern, Katherine Pfrommer, and Jerry M. Sollinger, "A Guide to Army Installation Water Rights: How to Establish, Maintain, Document, and Enhance Your Water Rights," March 14, 2015.

allocation, and conveyance will help personnel anticipate how their water supplies may be affected in the future so they can make the appropriate investments, partnerships, and outreach to ensure a stable water supply. As water managers in the community and the military address these challenges, opportunities may arise to share experiences and partner with others.

We should note that many of the examples discussed in this chapter are from the West because of the large investments, growth, and extreme water resource constraints in that region. In many areas of the western U.S. where water supplies tend to face more shortages because of a drier climate, more of some of the trends and approaches have been implemented. However, these issues are applicable to all areas of the country to varying degrees.

In addition, Army installations are experimenting with and using many of the management approaches and dealing with the challenges discussed in this chapter. But these activities are not the focus of this chapter, since it is focused on the broader national water management context. Exemplary in-depth installation examples are presented for Fort Carson and Fort Huachuca in Chapters Five and Six.

Historical Approaches to Water Management Focused on Accessing Water Sources

Starting in the late 1800s and into the middle of the 1900s or so, much water management focused on the development of water transfer and storage infrastructure, including the construction of dams, reservoirs, and conveyance canals, to meet the water demands of a growing country. In the 1900s, the development of intra- and interstate agreements for managing and sharing water sources was another important trend, as communities and industry competed for the same water resources. Legal claims for water rights and sources also became an important part of this history and part of the struggle over how best to manage and allocate limited water resources. Lastly, near the end of 1990s, federal water policy changes became more concerned about environmental issues regarding water and while giving less emphasis to large infrastructure projects, such as building dams. Next we provide an overview of this history, because understanding it helps set the context for understanding water management today and how water market mechanisms function.

Development of Water Transfer and Storage Infrastructures

Population growth and development significantly shaped water management from the late 19th century to the mid-20th century, particularly westward expansion and settlement. Farmers, miners, and prospectors traveled west in search of new opportunities in agriculture and livestock industries as well as mining. Businessmen, traders, and others followed to provide food, clothing, and other necessities while communities around them grew rapidly. Both the mining and agricultural opportunities that drove people westward, and the communities that supported those industries, significantly increased the need for water management both to provide water and to control flooding:

> In the era of western expansion, national economic and social policy favored development—principally mining and agriculture—and the establishment of communities by settlers, all of which demand water. Water was an instrument for realizing that policy.[2]

Mining, in particular, diverted water from streams to hillsides, and these early diversions set the stage for first-come, first-served prior appropriation water rights. The federal government sought to realize this policy of settlement and growth by providing the financial backing necessary to develop water. In 1902, Congress established the U.S. Bureau of Reclamation (USBR), which became the single largest funder of water-development projects in the West.[3] Thus began a period of both storage and conveyance infrastructure investments that sought to provide water to meet growing demands, at the same time fostering controversies that survive to this day. Water became available in a wide range of regions, and because water supplies were subsidized, additional uses, such as agricultural irrigation in previously arid regions, were developed. The following section describes the developments of some major conveyance and storage infrastructure projects in California, Arizona, and Colorado, mostly during the first half of the 20th century.

California's two major centers of development—San Francisco and Los Angeles—illustrate this history of supporting and encouraging growth through large water development projects. Just before the Gold Rush of 1849, San Francisco's population was approximately 1,000.[4] By 1852, a mere four years later, it had exploded to 35,000, and by 1900 it had grown to 340,000.[5] In response to the burgeoning population growth, San Francisco sought new water supplies to meet this growing demand. City officials had the idea that Hetch Hetchy Valley in Yosemite National Park would be an ideal place to build a dam and reservoir because of its steep rocky walls, narrow outlet, and large storage capacity. Hetch Hetchy Valley had the same "scenic grandeur" of Yosemite Valley, but because it was in a more remote area and lacked wagon road access it did not get many visitors. San Francisco officials started lobbying Congress and acquiring the private lands and water rights around Hetch Hetchy Valley and Lake Eleanor. The first congressional decision came in 1901 when Congress passed a Right of Way Act, which granted rights-of-way through government reservations for water conveyance infrastructure that had "domestic, public or other beneficial uses." Then, in 1913, Congress passed the Raker Act,

---

[2] David H. Getches, "The Metamorphosis of Western Water Policy: Have Federal Laws and Local Decisions Eclipsed the States' Role?" *Stanford Environmental Law Journal,* Vol. 20, No. 3, January 2001, p. 6.

[3] Getches, 2001, p. 11.

[4] Nancy E. Bobb and Greg A. Kolle, "Bridging the Bay," *Public Roads*, FHWA-HRT-12-006, Vol. 76, No. 2, September/October 2012.

[5] Campbell Gibson, "Population of the 100 Largest Cities and Other Urban Places in the United States: 1790 to 1990," Washington, D.C.: U.S. Census Bureau Population Division, Working Paper No. 27, June 1998.

which authorized the City of San Francisco to build reservoirs on federal lands. The act was a controversial congressional decision over the appropriate use of public lands and is credited with contributing to the development of the environmental movement. Indeed, more than 70 years later, a federal law passed in 1988 (Public Law 100-563), designed to prevent the expansion of Hetch Hetchy and Lake Eleanor in Yosemite National Park, prohibits the expansion of reservoirs in national parks without further congressional authorization.[6]

The first major dam of the system was the Lake Eleanor Dam, constructed between 1917 and 1918 with a capacity of 27,000 acre-feet (af). The much larger O'Shaughnessy Dam on the Tuolomne River in the Sierra Nevada, which flooded the Hetch Hetchy Valley and created the Hetch Hetchy reservoir, was constructed between 1915 and 1923 and is shown in Figure 2.1. The dam was completed in 1923 at a cost of $8.3 million ($111 million in today's inflation-adjusted dollars) and held 206,000 af of water. It was expanded in the late 1930s at an additional cost of $4.3 million ($70 million today) to accommodate a growing population. Water flows 160 miles through aqueducts and tunnels (including a 26-mile-long tunnel through the Coast Range) from the reservoir to the city of San Francisco, providing approximately 85 percent of the city's water needs. Today, a century later, Hetch Hetchy continues to provide water to San Francisco and nearby cities, though significant debates recently have occurred about removing the dam and restoring the valley.[7]

While Los Angeles today is many times larger than San Francisco, its beginnings were relatively modest. In 1850, Los Angeles' population was only about 1,000. It grew to approximately 100,000 by 1900, driven by a mix of agriculture, oil discoveries, new rail lines, and explicit promotion by developers.

As its population and industry increased, water needs in Los Angeles also grew:

> Between 1880 and 1900, the population of Los Angeles grew five-fold, from 50,393 people to 250,000, and, given the city's climate, links via the intercontinental railroads, and its position as a major West Coast port, prospects for the city's continued growth seemed promising. Yet one major obstacle stood in the way of Los Angeles's continued growth: the absence of sufficient water. By the turn of the century, there was growing concern among city boosters that more water had to be found if the city were to achieve prominence on the west coast.[8]

---

[6] Richard Amero, "Lessons From Hetch Hetchy," BalboaParkHistory.net, June 30, 2000.

[7] Water Education Foundation, "Layperson's Guide to California Water," 2008, p. 8.

[8] Gary Libecap, "Owens Valley and Western Water Reallocation: Getting the Record Straight and What it Means for Water Markets," *Texas Law Review*, Vol. 83, No. 7, June 2005, pp. 2055–2089.

**Figure 2.1. The Hetch Hetchy Valley Before (top) and After (bottom)
Construction of the O'Shaughnessy Dam**

SOURCES: "Hetch Hetchy Valley, Sierra Nevada Mountains, CA," Library of Congress Prints and Photographs Division, Washington, D.C., 1911.
"Hetch Hetchy Reservoir," City of Mountain View, undated.

Under the direction of its water chief William Mulholland, Los Angeles sought to meet these needs by, in part, aggressively claiming water and land rights from the distant Owens Valley in the Eastern Sierra. President Roosevelt granted rights to this water in 1906, and the city constructed the 233-mile Los Angeles Aqueduct from 1908 to 1913 to transport water from the Owens River to the San Fernando Valley. As one study noted:

> At the time, the Los Angeles Aqueduct was one of the nation's largest public works projects, second only to the Panama Canal. By 1920 Owens Valley provided a flow of 283 cubic feet per second of water, whereas the entire Los Angeles River basin, the water source for Los Angeles prior to the construction of the aqueduct, supplied a flow of just 68 cubic feet per second. The new water supported the city's growth from 250,000 people in 1900 to over 2.2 million by 1930.[9]

However, the residents of Owens Valley strongly resisted the export of "their" water, and the effects it would have on their community, environment, and economy. The resistance became violent in the 1920s, as residents attacked the infrastructure in what has been termed the

---

[9] Libecap, 2005, pp. 2055–2089.

California Water Wars. The anger of residents was not without merit because the effect on the Owens Valley was severe: by the late 1920s, the once navigable Owens Lake had dried up.[10]

> By 1935, Los Angeles had acquired 95 percent of the farm acreage and 88 percent of the town properties in the valley. The water transferred from Owens Valley, a marginal agricultural area, made possible the growth of Los Angeles, and Owens Valley remains the largest single source of water for the Los Angeles Basin.[11]

Over time, a second aqueduct was built and groundwater pumping to support this aqueduct led to a drop in the aquifer, as well as loss of springs. This caused groundwater-fed meadows to dry up, leading to desertification.[12]

Other states share similar histories of water management and conflict. In Arizona, one of the country's driest states, water development has been a feature for centuries. The Hohokam Indians built irrigation canals more than 1,000 years ago to develop agriculture in the arid lands. In the 1800s, ranchers and miners seeking copper, silver, and gold drove Arizona's development. The communities that grew around these endeavors needed water, and many of their needs were met with irrigation canals built on top of original Hohokam canals. Later, starting in 1917, Phoenix and other towns developed numerous dams along the Salt and Gila rivers and East Clear Creek, making up the Salt River Project (SRP).[13] The USBR constructed two of these dams and modified several others.[14] One of these, the Theodore Roosevelt Dam, was Reclamation's first major project and completed at a cost of $10 million ($232 million in 2012 dollars).[15] The SRP has been the site of significant water rights disputes. While the current adjudication process of the Salt, Verde, Gila, and San Pedro rivers began in the 1970s, adjudication first began in 1905 and led to the 1910 "Kent Decree" which determined rights in the Salt River.[16]

Another major infrastructure project, in Arizona, was developed to provide water for rapidly growing communities from the Colorado River. Arizona was a reluctant party to the Colorado River Compact out of great concern that California would take more than its fair share of water. However, at the same time, it sought to secure its own water sources and needed federal approval to build the Central Arizona Project to deliver Colorado River water to users across the state. The

---

[10] Los Angeles Aqueduct, "The Story of the Los Angeles Aqueduct," undated; National Aeronautics and Space Administration, "Owens Lake, California," September 12, 2011.

[11] Libecap, 2005, pp. 2055–2089.

[12] Named the "Owens Valley Syndrome" by some, the perceptions and the controversies surrounding the transfer from the Owens Valley to Los Angeles, the first large rural to urban water transfer, established a negative precedent in the eyes of farmers for future agricultural to urban transfers and is credited with contributing to certain limitations on water transfers in California. See Libecap, 2005, pp. 2055–2089.

[13] Gary Pitzer, Susanna Eden, and Joe Gelt, "Layperson's Guide to Arizona's Water," 2007, Tucson, Ariz.: Water Resources Research Center and Water Education Foundation, p. 10.

[14] U.S. Bureau of Reclamation, "Salt River Project," last updated August 19, 2011.

[15] National Park Service, "Withdrawal of National Historic Landmark Designation: Roosevelt Dam," undated.

[16] Pitzer, Eden, and Gelt, 2007, p. 10.

state legislature ratified the Colorado River Compact of 1922 in 1944 in part to win federal approval. However, final approval for the Central Arizona Project (CAP) did not come until 1968, nearly 15 years later, in part because of legal uncertainties related to water rights. The USBR began construction in 1973, but the first water deliveries were not made until 1985. At the time of its completion in 1993, the entire projection cost $4 billion ($6.3 billion in 2012 dollars), making it the most expensive water project undertaken by Reclamation and the most expensive water system in the United States.[17] Today, the system consists of a 336-mile canal from the Colorado River to Tucson, with numerous aqueducts, tunnels, pumps, and pipelines. It is part of the state's critical infrastructure: "More than 4 million people and 300,000 acres of irrigated farmland in Central Arizona depend on the CAP for delivery of over half their water supply."[18]

As was the case in Arizona and California, Colorado's mining and agriculture drove its development and water needs. Unlike its neighbors, however, Colorado has significant sources of water from the numerous rivers and streams with headwaters in the Rockies. (As such, the vast majority of water originating in Colorado leaves the state.) Much of Colorado's water management efforts bring water from the Western Slope of the Rockies to the Eastern Slope, the site of most of the state's agriculture and development. In the late 1890s and early 1900s, farmers and locals approached the USBR to find alternative, trans-mountain sources of water, given that local sources were insufficient. However, it was not until 1937 that the specific plan for the Colorado–Big Thompson project was pushed through the Colorado Legislature. The Colorado–Big Thompson project is the largest trans-mountain water diversion in the state, transporting water from the headwaters of the Colorado River in the Western Slope of the Rocky Mountains, under the Continental Divide, to the Eastern Slope to support agriculture. Construction began in 1938 and lasted for 20 years at a total cost of $162 million, though the first deliveries began in 1947. The massive project consists of 12 reservoirs, 35 miles of tunnels, and 95 miles of canals. Approximately 260,000 acre-feet of water are provided annually to supplement 700,000 irrigated acres and more than 750,000 people in the South Platte River Basin.[19]

In 1962, Congress approved the Fryingpan-Arkansas Project, another trans-mountain, trans-basin diversion in Colorado that brought water from the Western Slope of the Rocky Mountains to growing settlements on the Eastern Slope. A close second to the Colorado–Big Thompson project in scope, the Fryingpan-Arkansas Project required the construction of six storage dams, 17 diversion dams, and hundreds of miles of canals, conduits, and tunnels, among other structures. Water from this project originally provided water to the over-appropriated Arkansas River Valley, supplementing 280,600 irrigated acres, increasing agricultural productivity, and

[17] Jennifer E. Zuniga, "The Central Arizona Project Bureau of Reclamation," 2000, p. 3; Jim Turner, "Arizona: A Celebration of the Grand Canyon State," Layton, Utah, Gibbs Smith, 2011; Pitzer, Eden, and Gelt, 2007, p. 10.

[18] Pitzer, Eden, and Gelt, 2007, p. 11.

[19] Robert Autobee, "Colorado–Big Thompson Project," U.S. Bureau of Reclamation, 1996; Northern Colorado Water Conservancy District, "The Colorado Big Thompson Project: Historical, Logistical, and Political Aspects of This Pioneering Water Delivery System," undated.

providing water for municipal and industrial uses. Over time, the demand for municipal and industrial uses has grown, and approximately a quarter of the agricultural acreage has been transferred to the cities of Colorado Springs, Pueblo West, Aurora, and communities along the Fountain Creek watershed.[20] Both the Colorado–Big Thompson and Fryingpan-Arkansas projects were part of the USBR's extensive water development portfolio in Colorado.[21]

Development of Intra- and Interstate Infrastructure Projects and Compacts

Major interstate water infrastructure projects also were undertaken. As the above stories suggest, the race to develop water resources has not been without contention. Complex interstate compacts and international agreements, both of which require congressional approval, were developed simultaneously to resolve disputes between states and to help govern new water supplies. States enter into compact agreements to divide water rights to shared resources. Since 1950 there have been 26 compacts among states seeking to allocate shared water.[22]

(More recent efforts have focused on the ability of these compact agreements to adapt water allocation and management to changing demographics, environmental conditions, and political interests, particularly in light of the effects that climate change is predicted to have on droughts and water availability.[23])

The first, and perhaps the most well known of these agreements, are the Boulder Canyon Project and the Colorado River Compact that, respectively, developed and govern the Colorado River. The 1,450-mile Colorado River flows from the Rocky Mountains, across Colorado and Utah and along the borders of Arizona, California and Nevada, into Mexico to the Gulf of California. The Colorado River's tributaries make up the Colorado River Basin, which also covers the states of Wyoming and New Mexico. Figure 2.2 shows a map of the Colorado River Basin and the seven states that share it.

---

[20] U.S. Bureau of Reclamation, "Fryingpan-Arkansas Project," last updated April 4, 2013; Jedediah S. Rogers, "Fryingpan-Arkansas Project," U.S. Bureau of Reclamation, 2006.

[21] Rogers, 2006.

[22] In addition to compacts, water can be allocated among states in two other ways. First, Congress can pass a law to apportion water rights. This has been done only twice, in 1928 between Arizona and California and in 1990 between California, Nevada, and the Piute Tribe. Second, the Supreme Court can use federal common law to determine water allocation. Often this approach is seen as a last resort since decisions are piecemeal, and the process is lengthy and expensive. In fact, the Supreme Court has encouraged states to use compact agreements to resolve water allocation issues in lieu of the court system. See Bill Ganzel, "State to State Water Agreements," Ganzel Group, 2006; and Nathan C. Johnson, "Protecting Our Water Compacts: The Looming Threat of Unilateral Congressional Interaction," *Wisconsin Law Review*, Vol. 875, 2010, pp. 876–905.

[23] Johnson, 2010, pp. 876–905;Jason Robison, Katja Bratrschovsky, Jaime Latcham, Eliza Morris, Vanessa Palmer, and Arturo Villanueva, "Forging ahead in the Era of Limits: The Evolution of Interstate Water Policy in the Colorado River Basin: Colorado River Basin Background Paper," Water Federalism Project, Harvard University, Cambridge, Mass., April 19–21, 2012; Richard A. Wildman, Jr. and Noelani A. Forde, "Management of Water Shortage in the Colorado River Basin: Evaluating Current Policy and the Viability of Interstate Water Trading," *Journal of the American Water Resources Association (JAWRA)*, 2012, pp. 1–12.

Figure 2.2. Map of the Colorado River Basin

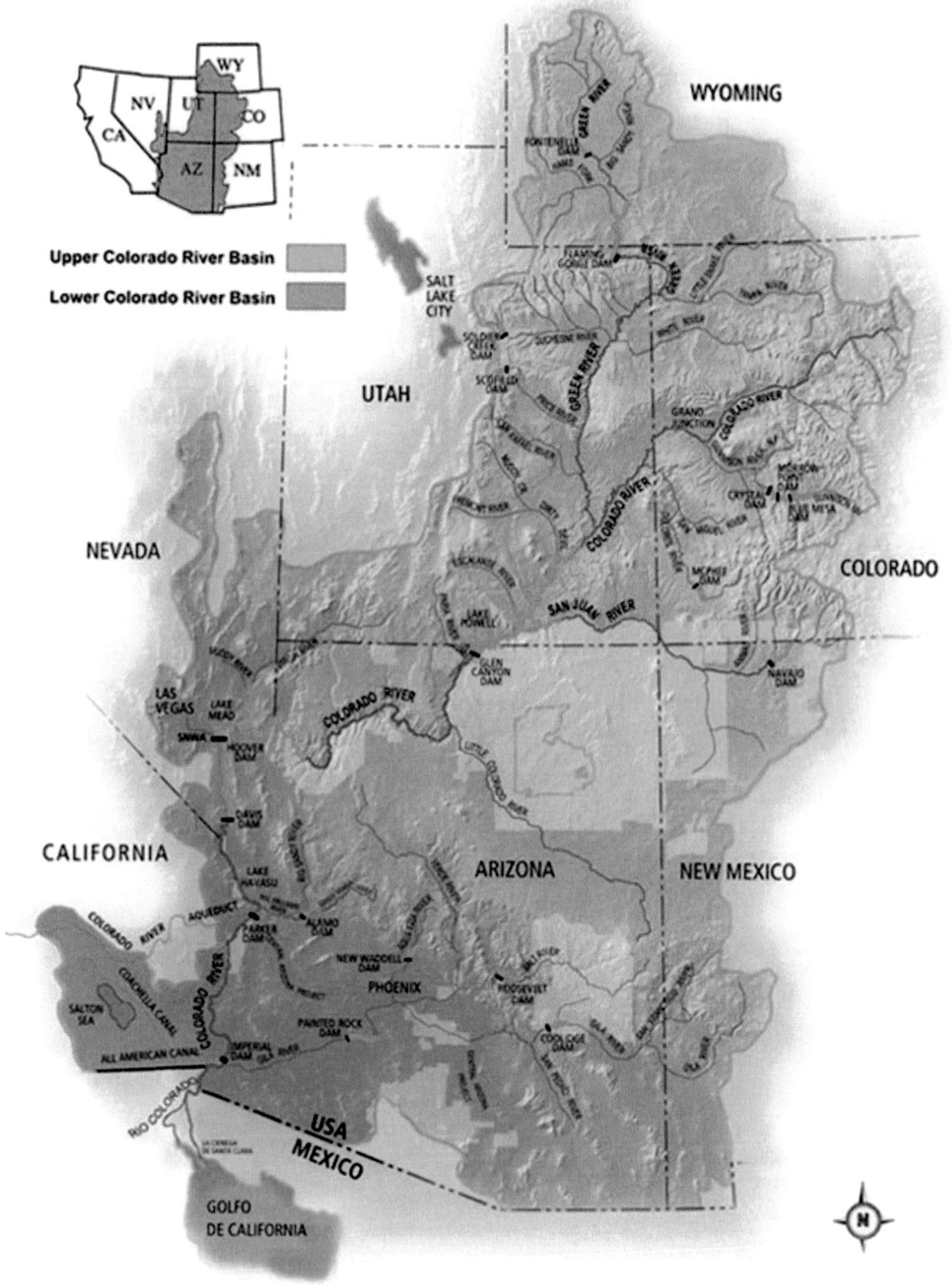

SOURCE: Glen Canyon Dam Adaptive Management Plan, undated.

From early history to recent times, the basin has been a major source of water for the people of the West. In the late 1800s and early 1900s, the basin enabled the region's rapid mining, agriculture, and economic development. States raced to develop the water but competing uses led to contention between users.

The Boulder Canyon Project, enabled by the Boulder Canyon Project Act of 1928 and undertaken by the USBR, involved massive efforts to develop the Colorado River. The $165 million authorization ($2.2 billion today when adjusted for inflation) principally funded the construction of the Hoover Dam (which impounds Lake Mead), the Imperial Dam, and the All-American Canal, which provides water to California's Imperial Valley, as well as other measures for flood control and power generation. The Hoover Dam also separates the states into "Upper Basin" states (Colorado, New Mexico, Utah, and Wyoming) and "Lower Basin" states (Arizona, California, and Nevada). On the surface, the project developed the Colorado River for the potential benefit of all states. But it was driven by California's interest in developing the southern portion of the state. One report notes:

> The technological and legal initiatives . . . all emanated from California and were designed to facilitate (1) the agricultural development of the Imperial Valley, and (2) the municipal growth of Greater Los Angeles. The irony of this circumstance is that California provides very little water to the flow of the Colorado; nonetheless, California occupies a geographical and topological relationship to the river that facilitated its ability to utilize the flow before any other states in the Southwest could develop projects of comparable scale or economic importance.
> . . .
>
> Because the flow is so limited (at least in terms of the amount of land that can conceivably benefit from it) and the possible uses so vast, it did not take long for citizens of various states to perceive other states as competitors. For many years, this competition would remain hypothetical, but things began to changes in the 1920s. ... With the completion of the Hoover Dam, utilization of the river's resources would pass from the realm of the possible to the realm of reality."[24]

The shared use of the Colorado River Basin among the seven states was controversial and required extensive negotiation. The upper basin states (Colorado, New Mexico, Utah, Wyoming) primarily were interested in preserving their access to water for future growth, while in general the lower basin states (Arizona, California, Nevada) were interested in ensuring access to additional water supplies to support growth. The "Law of the River" collectively describes numerous federal laws, court decisions, interstate compacts, and regulations designed to allocate resources and address disputes. Central to the "Law of the River" is the Colorado River Basin Compact, a 1922 agreement among the seven states to equitably divide water from the Colorado River; establish the relative importance of different uses of water; ensure development of the basin; and reduce existing and future controversies.[25] The compact requires the Upper Basin states to deliver 7.5 million acre feet (MAF) of water per year on a 10-year rolling average to the

---

[24] David P. Billington, Donald C. Jackson, and Martin V. Melosi, *The History of Large Federal Dams: Planning, Design, and Construction in the Era of Big Dams*, Denver, Colo.: U.S. Bureau of Reclamation, 2005, pp. 129–130.

[25] Reed Watson and Brandon Scarborough, "Colorado River Water Bank: Making Conservation Profitable," Bozeman, Mont.: Property and Environment Research Center, 2009; Pitzer, Eden, and Gelt, 2007; Colorado Division of Water Resources, "A Summary of Compacts and Litigation Governing Colorado's Use of Interstate Streams," 2006.

Lower Basin states (Arizona, California, and Nevada). If the 10-year rolling average falls below this level, the Lower Basin states may make a "call" on the compact, forcing the Upper Basin states to reduce water consumption so the Lower Basin states get that water.[26] In 1928, Congress divided the Lower Basin allocation, giving California 4.4 MAF, Arizona 2.8 MAF, and Nevada 0.3 MAF each year.[27]

The compact itself was not without controversy. Arizona's fears that California would claim senior rights to some of Arizona's water led the state to turn out its armed forces. One report describes this tension:

> When construction began on a second dam at Parker, also on the lower Colorado River, and the point from which urban California would take its allocation through the Colorado River Aqueduct, Arizona objected. This time, taking matters into its own hands, it sent the Arizona National Guard to stop construction on Parker Dam from the Arizona side of the river. Cooler heads prevailed and California got the dam and aqueduct it needed to bring Colorado River water to Los Angeles and inland southern California.[28]

As of 2014, a call on the river had never occurred. However, several factors have changed since the compact was signed, and much concern remains over the possibility that this may occur under the Colorado River Basin Compact. The implications of such a call are unclear. The original delivery requirements of 7.5 MAF were based on years where snowpack and rainfall was wetter than average data, and so concern arose about meeting release requirements given recent severe droughts and a return to the lower, more typical, precipitation levels (by as much as 20 percent by recent estimates).[29] In particular, Front Range cities, which are junior water rights holders relative to agriculture-based rights holders, would be required to curtail their water use to meet the minimum flow requirements to lower basin states. Therefore, they are vulnerable to the ambiguity surrounding the implications of a compact call.[30] Climate change, droughts, demographic changes, and other factors are large stressors on the Colorado River Compact and threaten the sustainability of the Colorado River Basin itself. As a consequence of these factors, the USBR undertook the Colorado River Basin Study to assess and manage risks in the coming decades.

The Arkansas River Compact illustrates the interstate competition for water and the difficulties in establishing and sustaining mutually acceptable agreements. As shown in Figure

---

[26] Watson and Scarborough, 2009, p. 5.

[27] Pitzer, Eden, and Gelt, 2007, p. 12.

[28] Pitzer, Eden, and Gelt, 2007, p. 12.

[29] Sharlene Leurig, "Water Ripples: Expanding Risks for U.S. Water Suppliers," *Ceres*, December 2012.

[30] Many of the assumptions regarding a compact call are untested and therefore are uncertain. For example, there is some debate as to the definition of "extraordinary drought," the exact determination of minimum flow requirements, the treatment of tributaries, etc. For more on this issue see Colorado River Governance Initiative, "Colorado River Law and Policy: Frequently Asked Questions (FAQs)," Boulder, Colo.: Western Water Policy Program, Natural Resources Law Center, University of Colorado Law School, version 1.0, March 2011.

2.3, the Arkansas River flows from the Collegiate Peaks in Colorado to Kansas, and on to Oklahoma and Arkansas where it joins the Mississippi River. As early as 1902, Colorado and Kansas have disputed and litigated over the apportionment of Arkansas River waters. The states negotiated the Arkansas River Compact in 1948, seeking to settle existing disputes and remove future disputes, and to divide the water equitably between the states.

Figure 2.3. Map of the Arkansas River Basin

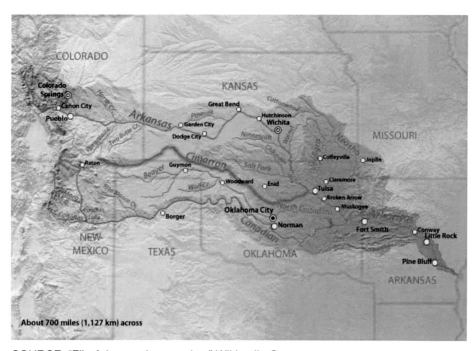

SOURCE: "File:Arkansasrivermap.jpg," Wikipedia Commons.

The Arkansas River Compact was intended to protect the status quo of water supply use in 1949 and to allocate remaining water supply between the two states. The Compact stipulates that future development and use in Colorado should not materially deplete flows to Kansas, but it does not specify a particular allocation. Instead, it allows each state to call for up to a maximum rate of water to be released from the John Martin Reservoir in Bent County, Colorado, regardless of whether the other state makes a similar call. The absence of a clear allocation led to a "race to the reservoir"—both states sought to use any stored water as quickly as possible before the other state could do the same.

In the late 1970s, the states sought an alternative, more efficient way of allocating water from the river. In 1980, the states negotiated an operating plan that allocated 60 percent of the water stored in the John Martin Reservoir to Colorado, with 40 percent allocated to Kansas. This allocation also did not prevent disputes. The Kansas Department of Agriculture documents one such dispute:

> After the compact . . . Colorado allowed high-capacity irrigation wells to be
> developed in the Arkansas River Valley. The well pumping reduced river flow

and materially depleted water that would have been available to Kansas. Kansas filed *Kansas v. Colorado,* No. 105, Original, in 1985 to enforce the terms of the Arkansas River Compact . . . In 1995 . . . the court found that Colorado's post-compact well pumping violated the compact. . . . In April 2005, Colorado paid Kansas more than $34 million in damages for Colorado's compact violations from 1950 through 1999 and more than $1 million in legal costs in June 2006.[31]

Water developments, and contentious claims for water, are not limited to the West. The 430-mile Chattahoochee River originates in northeastern Georgia, forms part of the boundary between Georgia and Alabama, and then joins the Apalachicola River flowing into the Gulf of Mexico at the Florida Panhandle. The river initially served as a major navigation route and a source of mechanical power for textile, saw, and grain mills. Initial development efforts of the 19th century sought to improve the river's navigability. In the early to mid-20th century, efforts shifted to providing hydropower for growing communities. The first large-scale hydroelectric dams were built between 1899 and 1924. Additional hydropower and dam projects followed and were seen as important measures of flood control as well as power generation. In 1953, Congress authorized the Apalachicola-Chattahoochee-Flint Project to construct dams for flood control, power generation, and navigation.[32] These investments aided the rapid growth of cities in the area, the largest of which is Atlanta.

The same growth led to disputes over water. For the past 20 years, the "Tri-State Water War" between Georgia, Alabama, and Florida has centered on allocation of water in the Apalachicola-Chattahoochee-Flint (ACF) River Basin. The three states failed to derive mutually agreeable allocations after several years of negotiations within an interstate compact and, as a result, have resorted to litigation to resolve their disputes.

The historical underpinnings of this dispute date arguably to the mid-century, when the Army Corps of Engineers built the Buford Dam on the Chattahoochee to create Lake Lanier for the purposes of flood control, power generation, and navigation. Although the lake was sought as a potentially significant source of water, the USACE did not assign it a monetary value and did not want to charge Atlanta for the water supply benefits because it said they "were all incidental to the purposes of hydropower and flood control and would 'not cost the Federal Government 1 cent to supply.'"[33] The allocation of water in Lake Lanier affects downstream users in Georgia, Alabama, and Florida. Alabama relies on the river system for potable water, irrigation, hydroelectric and thermoelectric power, and recreation; Florida, too, uses water from the system for irrigation, industrial uses, municipal water supply, and thermoelectric power. Florida in particular relies on in-stream flows to support oyster beds in the Apalachicola Bay that provide

---

[31] Kansas Department of Agriculture, "Kansas-Colorado Arkansas River Compact Fact Sheet," August 2009, p. 2.

[32] Lynn Willoughby, "Chattahoochee River," *New Georgia Encyclopedia*, July 18, 2003; Lynn Willoughby, *Flowing Through Time: A History of the Lower Chattahoochee River*, Tuscaloosa, Ala.: University of Alabama Press, 2012.

[33] Cited by Megan Baroni, "Lessons from the 'Tri-State' Water War," *American Bar Association*, Vol. 35, No. 2, Winter, 2012.

90 percent of Florida's oysters. All downstream users rely on sufficient in-stream flows to assimilate pollution from Atlanta and other uses along the river.[34]

During the last 25 years or so, the Atlanta region experienced tremendous growth, with the population increasing from 3.1 million in 1990, to 4.3 million in 2000, and 5.5 million in 2010.[35] As Atlanta required more water from the reservoir at Lake Lanier for drinking water, drought occurred, further straining water supplies available to the city as well as to downstream users. In 1958, Congress passed the Water Supply Act, which in part required the Corps to obtain approval for providing water and requiring that entities seeking water supplies to pay for the storage. Over time, however, the Corps approved water allocations without seeking congressional approval, including releasing water for Atlanta on an interim basis. As this went on, neighboring states and power companies sued the Corps, and in turn, Georgia sued the Corps in 2001 to secure additional releases. The case made its way through the District and Circuit courts, and in 2012 the U.S. Supreme Court refused to hear the case, upholding the Circuit Court of Appeals' ruling that the Corps has the authority to allocate water. The Corps is now in the process of updating its Water Control Manual, which governs dam operations for the Apalachicola-Chattahoochee-Flint (ACF) river basin, to balance the somewhat competing interests of water supply, water quality, flood control, navigation, hydropower, recreation, fish, and wildlife.[36] However, the ruling does not resolve Atlanta's request for added supplies. Atlanta and other cities continue to face uncertainty about the future of their water, which in turn affects their potential for growth and that of downstream users as well.[37]

In sum, the early and middle part of the 20th century saw massive efforts to develop water transfer and storage infrastructure around the country, and inter- and intra-state compacts were a key part of this development. This national trend can be seen in Figure 2.4, which shows the reservoir capacity in the United States from 1900 to 1995. Reservoir capacity grew from approximately 100 MAF in 1920 to approximately 700 MAF just 50 years later—a consequence of the rapid construction of dams designed to meet growing water needs and to facilitate development by managing droughts and floods, particularly in the West. Most of the existing storage and associated transfer infrastructure was financed by the federal government and was to be repaid or cost-shared by local entities later. Active federal agencies included the USBR, U.S. Army Corps of Engineers (USACE), and the Tennessee Valley Authority.[38]

---

[34] Janet C. Neuman, "Have We Got a Deal for You: Can the East Borrow from the Western Water Marketing Experience?" *Georgia State University Law Review*, Winter 2004.

[35] Atlanta Regional Commission, "Atlanta Region Plan 2040: Chapter 2: Trends and Forces Impacting the Future," 2011.

[36] Chattahoochee Riverkeeper, "Charting a New Course for Georgia's Water Security," July 2010.

[37] Cited by Baroni, 2012.

[38] Benedykt Dziegielewski and Jack Kiefer, "U.S. Water Demand, Supply and Allocation: Trends and Outlook," Institute for Water Resources, U.S. Army Corps of Engineers, Report 2007-R-3, December 2008.

Figure 2.4. Reservoir Storage for the Continental United States

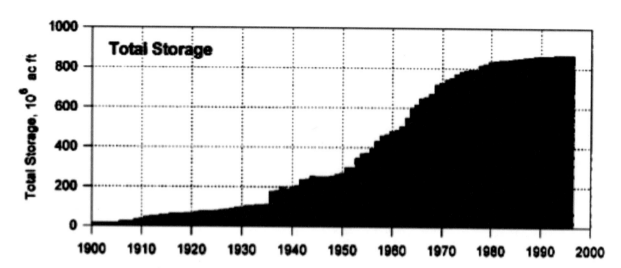

SOURCE: William L. Graf, "Dam Nation: A Geographic Census of American Dams and Their Large-Scale Hydrologic Impacts," *Water Resources Research*, Vol. 35, No. 4, 1999, p. 1309.

Through the USBR and the Army Corps of Engineers, the federal government had a significant role in changing the landscape of water in the West:

> The United States Bureau of Reclamation constructed 355 of the largest reservoirs plus 16,000 miles of canals ... Altogether, Reclamation reservoirs store over 119 million acre-feet of water and Corps reservoirs in the West store 103 million acre-feet.[39]

Additionally, development was not limited to major project construction. Significant infrastructure was needed to distribute the water supplies to homes, businesses, industries, and other users. One study estimates that there are more than 1 million miles of underground pipes to deliver drinking water, involving trillions of dollars of investment, as shown in Figure 2.5.[40]

---

[39] Getches, 2001, p. 14.

[40] American Water Works Association, "Buried No Longer: Confronting America's Water Infrastructure Challenge," 2012, p. 3.

Figure 2.5. Historic Investment in U.S. Water Mains, 1850–2000

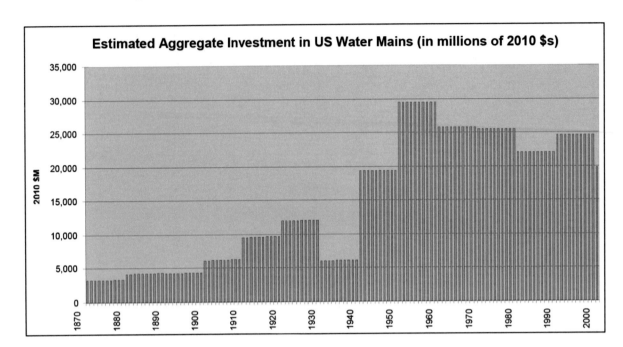

SOURCE: American Water Works Association, 2012, p. 6.

Legal Claims for Water Sources and Rights

Rapid growth, development of new water, new agreements, and uncertain water laws also resulted in a complex patchwork of water rights and significant water conflicts within states. Many claims for water rights and sources ended up facing legal challenges. Texas's history provides a compelling illustration of the conflicts over water that led to such challenges. Texas was governed first by Spanish law as a Spanish territory and then by Mexican law from 1821 to 1836 as part of Mexico. When it gained independence as a sovereign nation in 1840, Texas implemented British common law. It joined the United States in 1845, seceded in 1861, and then returned in 1870.[41]

This political, legal, and historical evolution shaped the state's water rights and development. In Texas, two competing systems of water rights evolved simultaneously. Its judicial system upheld a system of riparian rights consistent with common law established in 1840. However, under pressure to develop new sources of water, the Texas Legislature pursued a program of appropriative water rights through the Irrigation Acts of 1889, 1895, and 1913. The incompatibility of these two systems came to a head in the 1950s, when Texas suffered through a seven-year uninterrupted drought, considered the worst in 700 years. One report summarized the situation as follows:

---

[41] Glenn Jarvis, "Historical Development of Texas Surface Water Law: Background of the Appropriation and Permitting System," McAllen, Tex., April 2008.

24

From 1949–51, Texas rainfall dropped by 40 percent and West Texas experienced severe declines. By 1952, more than half the state received 20 to 30 inches less rainfall than normal. In 1952, Texas' monthly rainfall average fell to just 0.03 inches—the lowest level since the Weather Bureau began collecting records in 1888.[42]

This period saw years of litigation between individual parties seeking to establish rights to this increasingly scarce resource. In a case referred to as the "Valley Water Case," the state filed an injunction on the Lower Rio Grande, where this drought began, to start an adjudication process and clarify water rights. The 1969 lawsuit, *State v. Hidalgo County Water Control and Improvement District No. 18*, involved 42 special water districts, more than 2,500 individuals, and more than 90 lawyers. It cost an estimated $10 million.[43]

Other states have been drawn into similarly massive legal battles over water. During the 1970s and 1980s many lawsuits (called general stream adjudications) were filed as states grappled with competing interests for water, due in part to rapid development. At this time, a majority of the western states have initiated general stream adjudications in an effort to better manage water resources and resolve disputes over water rights, especially in cases where rights are over-appropriated or when droughts occur. The state court system typically holds adjudications for multiple purposes:

- To validate and quantify the rights of all water users within a river basin or watershed
- To develop water right decrees that can be the basis for better water supply management
- To create a centralized, current record of water uses.[44]

They are comprehensive in that they seek to resolve all water rights issues within the area of concern.[45]

Because they are comprehensive, multidisciplinary and technical by nature, and involve many parties with sometimes competing interests, river basin adjudications are complex, contentious, and time-consuming. These lawsuits are

> among the largest civil proceedings ever to be litigated in state and federal courts. For instance, 27,000 persons have filed more than 77,000–88,000 claims to water rights in the Arizona general stream adjudication. In Idaho, more than 110,000 persons have filed 150,000 claims for water rights in the Snake River system. In Montana, approximately 80,000 persons have filed more than 200,000 water

---

[42] Ric Jensen, "Why Droughts Plague Texas," *Texas Water Resources*, Vol. 22, No. 2, Summer 1996, p. 4.

[43] Jarvis, 2008; Doug Caroom and Paul Elliott, "Water Rights Adjudication—Texas Style," *Texas Bar Journal*, November 1981; Otis Templer, "Water Law," Texas State Historical Association, undated.

[44] Clarifying water rights facilitates market transfers of water rights, which is an emerging tool for reallocating water resources to meet changing demands for water change.

[45] Because of the McCarran Amendment, these processes typically include federal agencies' and Indian tribes' claims to water rights.

rights claims in the statewide adjudication (the largest adjudication in the country).[46]

Arizona has used general stream adjudication to resolve all water rights along two river systems in which water rights and claims are especially numerous and complex. The first is the adjudication process of the Gila River Basin, which stretches east to west across southern Arizona. The process began as petitions in 1974 and 1979 to resolve water rights claims in the Salt, Verde, Gila, and San Pedro rivers. This basin drains most of central and southern Arizona, and today includes 85 percent of Arizona's population (including the Tucson and Phoenix metropolitan areas and Ft. Huachuca) and most of its agriculture and industry.

In 1979, the legislature amended adjudication statutes to transfer these claims to the Maricopa County Superior Court, and in 1981, the Arizona Supreme Court consolidated the adjudications into a single proceeding. In late 2012, the adjudication concerned 83,500 claims by more than 24,000 claimants. The parties include the United States (to include Fort Huachuca), the state of Arizona, Native American tribes, municipalities, public and private utilities, agricultural irrigation districts, industrial corporations, and individual farms, ranches, and other private water users. The adjudication process has not yet resulted in a final decree, but it has resulted in significant revisions to Arizona's water code and multiple Arizona Supreme Court decisions.[47]

The adjudication process for the Little Colorado River, a tributary of the Colorado River in northern Arizona, began similarly. Phelps Dodge Corporation filed a petition in 1978 with the State Land Department to determine the water rights of the Little Colorado River. Under the same 1979 legislative amendments, the Little Colorado River Adjudication was transferred to the Apache County Superior Court.[48] By late in 2012, this adjudication process concerned the claims of 3,100 water users in northeastern Arizona.

By fall of 2012,

> Arizona, California, Idaho, Montana, Wyoming, and Washington are undertaking comprehensive, basin-wide adjudications of water rights. Utah, Colorado, New Mexico, Oregon, and Oklahoma are presently dealing with water rights on a more piecemeal basis, either because they have finished general adjudications, or because general adjudications are not necessary at this time. North Dakota has not attempted an adjudication, Texas has completed one, and South Dakota and Alaska abandoned their attempts.[49]

---

[46] National Judicial College, "Dividing the Waters," undated. Andrea K. Gerlak and John E. Thorson, "General Stream Adjudications Today: An Introduction," Universities Council on Water Resources, *Journal of Contemporary Water Research & Education*, No. 133, May 2006, pp. 1–4.

[47] Joseph M. Feller, "The Adjudication that Ate Arizona Water Law," *Arizona Law Review*, Vol. 49, 2007, pp. 405–440.

[48] Maricopa County Superior Court, "Arizona's General Stream Adjudications," Maricopa County, Ariz., undated.

[49] National Judicial College, "Dividing the Waters," undated.

Evolving Federal Water Policy: Environmental Concerns

The middle of the 20th century saw a growing concern among the public about the environmental effects of past water development projects, as well as broader concern about environment and ecosystem protection. In the 1950s and 1960s, environmentalists won major battles to halt the construction of three major dams on the Colorado River. However, it was not until the late 1970s that policy began to shift and herald today's era of increasing environmental protection. In this decade, Congress passed the Clean Air Act of 1970, the Clean Water Act of 1972, the Endangered Species Act of 1973, the Federal Land Policy and Management Act of 1976, and the National Forest Management Act of 1976.[50] Concerns about water quality started becoming an important part of water management, which is a key challenge for water management today and is discussed further below.

Jimmy Carter's presidency in 1977 spearheaded the change in federal water policy, which sought greater financial and environmental accountability. As one of the first steps in his term, Carter eliminated numerous water projects deemed to have the same major environmental problems seen in the past. The backlash to this move was severe, garnering vocal criticism from politicians and developers across the West. Nevertheless, future administrations continued these policies, which are now widely hailed as bold and forward thinking. One study notes:

> The hit list [of projects] was, at the time, considered a political blunder of epic proportions, but history should rightly mark it as signaling the end of the era of federal dam-building in the United States and perhaps a visionary policy.[51]

Since the 1970s, as the USBR's support for large projects has waned, the greatest federal influence on water projects has come from the Clean Water Act (CWA) and the Endangered Species Act (ESA). Later in this chapter, we discuss some CWA impacts. In Chapter Six, we provide a detailed description of how the ESA has impacted water use and conservation activities at Fort Huachuca. In fact, the ESA is a main driver of water conservation activities at Fort Huachuca.

Water Management Today Focuses on Efficient Management of Existing Resources

Traditional water demands such as population growth and economic development continue to grow, particularly in the West. Simultaneously, other demands for water use are receiving greater attention and new demands are developing. For instance, increasing concern over ecosystem

---

[50] Public Law 91-604, Clean Air Act Amendments of 1970, December 31, 1970; Public Law 92-500, Clean Water Act, October 18, 1972; Public Law 94-579, Federal Land Policy and Management Act, October 21, 1976; Public Law 93-205, Endangered Species Act, December 28, 1973; Public Law 94-588, National Forest Management Act, October 22, 1976; Getches, 2001, pp. 17–18.

[51] Getches, 2001, pp. 16–17.

health has led many to push for greater allocation of water for environmental purposes. Another stressor within several regions is that growth in water-intensive industries and new methods of extracting unconventional natural gas consume significant amounts of water, and concerns have arisen that wastewater from these processes could contaminate water sources if not properly managed.[52]

However, nearly all of the traditional water sources are already allocated or over-allocated, and past approaches to water development rarely are viable options for addressing today's needs. Moreover, water supply variability, especially frequent and prolonged droughts in parts of the country, has exposed weaknesses in past approaches as states, communities, and regional entities look for alternatives.

Today, water management is geared toward extending the supply of water, improving our understanding of water use, and reallocating it as new purposes are growing. Key elements of water management efforts include such practices such as water conservation; use of alternative sources of water; state regional, and local water planning; watershed management; and environmental sustainability. These efforts are being undertaken at the federal, regional, state, and local levels. We next briefly discuss each of these five areas. In this discussion, we present some illustrative examples to show the different types of approaches being implemented. There also is some experimentation with market mechanisms, many of which seek to facilitate transfers of the finite supply of water among users (discussed in Chapter Four).

Water Conservation

Water conservation programs are becoming increasingly important to address water sustainability. Three main trends have contributed to increased water conservation. First, local governments (as well as federal and state agencies) and utilities recognize the need to reduce water usage to help save limited water supplies, and have started implementing policies to encourage water conservation. Second, stricter building codes regarding water efficiency and the increased use of water-efficient appliances in society have helped reduce per capita water use. Third, overall industry and consumer awareness has increased regarding the need to conserve water and the benefits of doing so, also helping Americans to reduce per capita water use. In Arizona, for instance, a recent survey of 15 municipalities found that most had reduced their per

---

[52] While not large relative to total water withdrawals in aggregate, new demand for water for these purposes can stress local resources. For natural resource extraction such as hydraulic fracturing, water, typically drawn from local surface and groundwater sources, makes up 99 percent of the fluid used to fracture wells for natural gas extraction from coal beds and shale. The amount of water needed for a single well ranges from 50,000 gallons to 350,000 gallons if in a coal bed and 2 million to 4 million gallons if in a shale formation. Between 15 percent and 80 percent of the fluid is recovered and may be recycled in future operations. There are concerns that unrecovered fluid, which may contain a variety of potentially hazardous materials, may infiltrate and contaminate water sources. See U.S. Environmental Protection Agency, "Study of the Potential Impacts of Hydraulic Fracturing on Drinking Water Resources: Progress Report," December 2012, p. 80.

capita water use by more than 10 percent between 2003 and 2008. Three had reduced use by 15 percent and one by 77 percent.[53]

To increase water conservation, communities have employed a mix of options. They include policy changes, such as building codes and standards, market-based demand management such as the use of block rates, improvements in technology, and public education and awareness. As one example, Phoenix has used a number of market incentives and policies to achieve a decline in water use from 217 gallons per capita per day (GPCD) to 184 GPCD, a decrease of 15.3 percent. It uses a block-pricing structure in which a certain block of water is allocated for a fixed fee. Above that allocation, users are charged at a higher rate, which varies depending on whether the higher use occurs during "high," "medium," or "low" months. The city offers rebates for low-income homes to retrofit high-efficiency fixtures and check for leaks. The city applies a Water Resource Acquisition Fee for each new connection added to its water system, but offers credits to developments that exceed current conservation standards. The city has a number of ordinances related to conservation, including limits on irrigation, unrepaired leaks, and site design and desert preservation. Phoenix also uses a number of education and outreach programs to promote a culture of water conservation. The city also implements supply-site measures, such as using automatic leak-detection technology in its distribution system, and uses effluent (which can be thought of as a non-traditional water source as well as a conservation measure).[54]

As another example, Colorado Springs Utilities estimates that system-wide water usage averaged 186 GPCD from 1990 to 2006, but that water conservation measures implemented from 2002 to 2005 subsequently reduced that amount to 172 GPCD. Through these earlier conservation measures, it estimates savings 4.7 percent of its 2007 water demand relative to 1999 levels. It further estimates that the programs outlined in its 2008–2012 water conservation strategic plan will save an additional 7.6 percent of 2017 demand. Its measures include:

- Residential block rates and commercial seasonal rates for water use
- Efficient toilet, washer, and other appliance rebates and incentives
- Ordinances, policies, and audits on landscaping and irrigation
- Education campaigns.[55]

Because so much water is used to irrigate grass in drier areas, some western cities such as Las Vegas and Los Angeles have incentive programs for residents and other customers to remove their lawns and replace them with less water-intensive landscaping. Las Vegas Valley Water District started a turf-removal rebate program in 2003 and in the 10 years since has paid nearly $200 million to remove 165.6 million square feet of grass from homeowners' and

---

[53] Western Resource Advocates, "Arizona Water Meter: A Comparison of Water Conservation Programs in 15 Arizona Communities," October 2010.

[54] Western Resource Advocates, 2010, pp. 81–85.

[55] Colorado Springs Utilities, "2008–2012 Water Conservation Plan," January 30, 2008. We discuss these CSU activities further in Chapter Five.

businesses' properties, saving 9.2 billion gallons of water. Grass lawns are now even banned in the front yards of new developments in Las Vegas. Los Angeles started a program in 2009 to pay people for replacing their lawns with plants that require less water. By 2013 the city had paid $1.4 million for the removal of over one million square feet of grass in people's yards. The program was saving about 47 million gallons of water each year. Given the program's success, in July 2013, the city increased the rebate incentive from $1.50 per square foot of grass removed to $2 per square foot.[56] Because of the California drought in November 2014 the rebate rates were increased by the Los Angeles Department of Water and Power (LADWP) to $3.75 per square foot.[57]

Building standards and codes also are evolving to conserve more water.[58] For example, the American Society of Heating, Refrigeration, and Air-Conditioning Engineers (ASHRAE) Standard 90.1 is widely referenced in local codes as the minimum in what building requirements should include. First issued in 1975, this standard is updated approximately every three years to be increasingly efficient. In response to the growing importance of sustainability and exemplary building design, ASHRAE Standard 189.1P, the "Standard for the Design of High-Performance, Green Buildings except Low-Rise Residential Buildings," has been developed for buildings whose owners wish to exceed the minimum requirements of Standard 90.1. It is known as "The Green Standard" and sets standards for energy efficiency, recycling, water conservation and efficiency, and other green building components. In response to the requirement that federal government facilities uphold the standards of "high performance," The Army has adopted ASHRAE 189 for all new buildings.[59]

Other standards also address water conservation and efficiency. The International Code Council (ICC), which developed the International Building Code that is widely adopted in the United States, released the International Green Construction Code. Like the ASHRAE 189 standard, this code addresses many aspects of green design. For water efficiency, its provisions include standards for fixtures and appliances and for rainwater storage and greywater systems.[60]

---

[56] Ian Lovett, "Arid Southwest Cities' Plea: Lose the Lawn," *New York Times*, August 11, 2013.

[57] For more information, see Los Angeles Department of Water and Power, "Mayor Garcetti Announces Increased LADWP Rebate for Residential Turf Removal to Highest Level in Southern California—$3.75," November 3, 2014.

[58] Codes and standards are related but different concepts. State and municipal building codes establish mandatory minimum performance requirements for buildings in a jurisdiction. These are typically based on standards developed by industry and government experts and address almost all aspects of buildings, including the floor plan, energy and water consumption, structural integrity, materials use, and the performance of subsystems such as plumbing and appliances. Some of the codes and standards are for residential buildings while others are for commercial buildings.

[59] U.S. Department of the Army, "Sustainable Design and Development Policy Update," Washington, D.C.: Office of the Assistant Secretary of the Army, Installations, Energy and Environment (ASA(IE&E)), October 27, 2010.

[60] Paula Melton, "International Green Construction Code Passes," *Environmental Building News*, November 8, 2011.

A 2002 U.S. Environmental Protection Agency (U.S. EPA) study that surveyed conservation efforts around the country observed:

> The incidence of water conservation and water reuse programs has increased dramatically in the last 10 years. Once associated only with the arid West, these programs have spread geographically to almost all parts of the United States. In many cities, the scope of water conservation programs has expanded to include not only residential customers, but commercial, institutional, and industrial customers as well.[61]

The Energy Policy Act of 2005 set federal standards for more efficient plumbing fixtures and appliances that use water, and provided incentives for the purchase of efficient appliances, including water heaters.[62] As another example, U.S. EPA's WaterSense program is a campaign and labeling program to identify products that are water efficient, increasing public awareness and demand for these products and encouraging manufacturers of water-consuming products to be innovative.[63]

The importance of water conservation can be seen in Figure 2.6, which compares aggregate trends in population growth and water withdrawals. The purple line tracks the total population of the United States, which has grown steadily from 1950 to 2005, suggesting a growing need for water. However, water withdrawals, indicated by the vertical bars, peaked in 1980 and have declined since then. This indicates that per-capita use has declined and speaks to the importance of conservation and efficiency in recent water management efforts.[64] (These are aggregate level data. Localized shortages may occur when regional growth outpaces available locally available water.)

---

[61] U.S. Environmental Protection Agency, (U.S. EPA), "Cases in Water Conservation: How Efficiency Programs Help Water Utilities Save Water and Avoid Costs," 2002.

[62] Public Law 109-58, "Energy Policy Act," August 8, 2005.

[63] U.S. Environmental Protection Agency, "Water Sense: An EPA Partnership Program," undated.

[64] A great deal of information is synthesized in this graphic. Additional USGS data show that water withdrawals by end use have remained relatively constant despite the overall increase in population (and those depending on public water supplies) and irrigated acreage. Withdrawals for thermoelectric power, irrigation, livestock, public supply, and other industrial have remained constant, suggesting that conservation and efficiency improvements have indeed supported population and economic growth. An exception is the industrial sector, where withdrawals have affected by two factors over time: pollution control standards, which have encouraged conservation and greater efficiencies, and a decline in manufacturing. Therefore, not all of the overall reduction of water use intensity per capita can be attributed solely to conservation and efficiency for this sector. See U.S. Geological Survey, "Trends in Water Use in the United States, 1950 to 2005," October 31, 2012; U.S. Geological Survey, "Estimated Use of Water in the United States in 2000," undated.

Figure 2.6. Trends in Water Use in the United States, 1950–2005

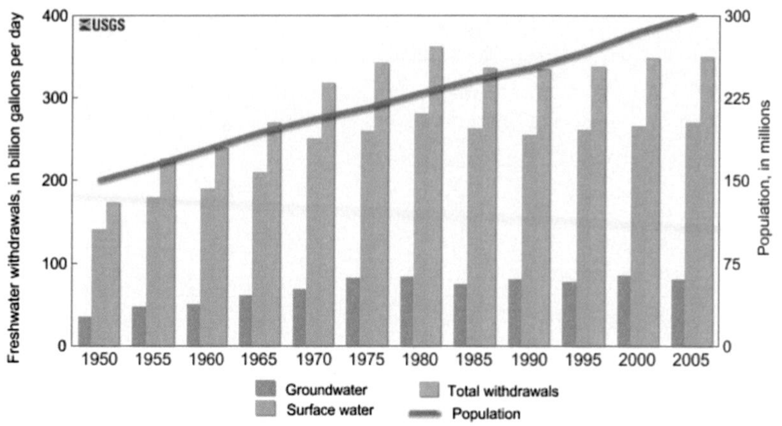

SOURCE: U.S. Geological Survey, "Trends in Water Use in the United States, 1950 to 2005," October 31, 2012.

We should note that other federal policies and requirements are in place for implementing water conservation at federal facilities, including military installations. We discuss these in Chapters Three, Five, and Six. Many Army installations have active water conservation programs, as we illustrate with Forts Carson and Huachuca in Chapters Five and Six, respectively.

Non-Traditional Water Sources

Water managers and communities also are emphasizing non-traditional sources of water to meet their needs. Such non-traditional approaches include the better management and use stormwater, increased use of recycled water (also referred to as water reclamation),[65] and desalinating water. We discuss each of these three areas below. To extend water supplies, water managers may use a combination of such approaches, as well as using water transfers and water markets (discussed in Chapter Four). One 2011 study of 11 water agencies in Southern California found that all 11 intend to diversify their supply portfolio with additional local supplies that include such measures.[66] The Los Angeles Department of Public Works (LADPW) illustrates this trend. As shown in Figure 2.7, LADPW's 2012 water management plan relied heavily on water from the Los Angeles Aqueduct and Metropolitan Water District (MWD), with groundwater, recycled water, and conservation accounting for only 16 percent. In contrast, in 2035, LADPW plans to

---

[65] "Water recycling (or water reclamation) involves treating municipal wastewater to remove sediments and impurities for reuse." See Water Education Foundation, "Layperson's Guide to Water Recycling," Sacramento, Calif., 2004, p. 2.

[66] Caitrin Phillips, "Imported vs. Local Water Supplies: The Planning Decisions Facing Southern California Water Agencies," paper for the Natural Resources Defense Council, August 3, 2011, p. 1.

increase these alternative water options substantially to 42 percent, from a mix of stormwater capture, water transfers, recycled water, and conservation.[67]

Figure 2.7. Composition of the Los Angeles Department of Public Works' Future Water Management Strategy

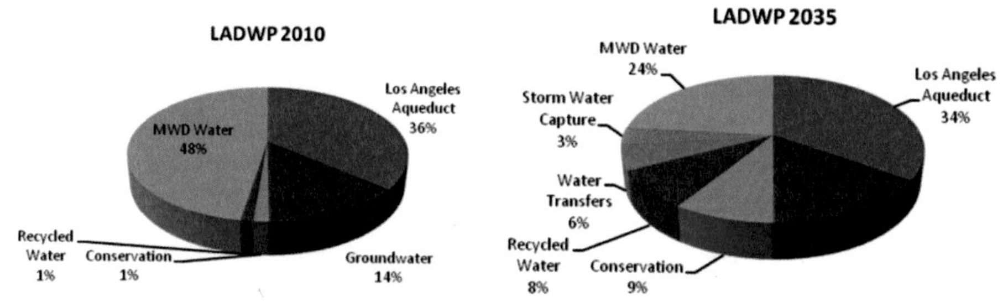

SOURCE: Barry Nelson, "Southern California's New Wave of Local Water Supplies," February 3, 2012.

Better Management and Use of Stormwater

Water managers across the country also are more effectively managing and using stormwater as another water source for selected water uses, such as toilet flushing and irrigation. Before we explain this trend, we provide some background about stormwater. Stormwater is rainwater and melted snow and ice that runs off roads, driveways, lawns, buildings and other sites. On fields, woods, and other natural or undeveloped areas, such water runoff is absorbed into the ground, is filtered, and then flows into streams and rivers or replenishes underground aquifers. In developed areas, with more impervious surfaces such as roads, buildings and sidewalks, the water is prevented from soaking into the ground. Instead, the water runs into storm drains, drainage ditches, and sewer systems. In many developed areas, such runoff can cause problems with downstream and building flooding, erosion, habitat destruction, combined sewer overflows, and pollution in streams, rivers, and coastal waters. The pollution is caused as the stormwater runs over yards, streets, and other developed surfaces. It picks up oils, sediments, lawn treatment chemicals, nutrients, trash, and other pollutants carrying them throughout the watershed.[68] The result is that government water managers, including those at Army installations, have taken new approaches to managing water runoff to help prevent these problems, as well as to use stormwater as a supplemental water source.

Federal, state, and local regulations and policies have been the main drivers behind improved stormwater management. The Clean Water Act (CWA), passed in 1972, established a regulatory

---

[67] Barry Nelson, "Southern California's New Wave of Local Water Supplies," Switchboard (Natural Resources Defense Council staff blog), February 3, 2012.

[68] For more on these problems, see U.S. Environmental Protection Agency, "Stormwater Management," undated; Fairfax County, Va., Department of Public Works and Environmental Services, "What is Stormwater Management?" 2013.

control process called the National Pollutant Discharge Elimination System (NPDES). The NPDES program requires permits to develop and implement a stormwater management program that includes control measures. The CWA sought to clean up the nation's polluted surface waters with the goal of making them "swimmable and fishable." At the time, many of the nation's surface waters were highly impaired due to the continued discharge of "non-point" source pollutants carried by stormwater runoff, which is still a problem today and is discussed further below. Amendments to the Clean Water Act established specific requirements that regulated stormwater discharges from Municipal Separate Storm Sewer Systems (MS4s). Because of the CWA, in 1994, U.S. EPA also created the Combined Sewer Overflow (CSO)[69] Control Policy to provide guidance on how communities with combined sewer systems can meet Clean Water Act goals in as flexible and cost-effective a manner as possible.

Traditional stormwater management design had been focused on developing piped networks and channels to transport stormwater runoff into a stream, river, large stormwater basin, or to a combined sewer system that flows into a wastewater treatment plant. Such practices sometimes contributed to some of the problems mentioned above. Consequently, today, more communities are focused on innovative approaches to stormwater management such as low impact development (LID) and wet weather green infrastructure. "LID aims to restore natural watershed functions through small-scale treatment at the source of runoff."[70] LID practices include rain gardens, green rooftops, rain barrels, bioretention facilities and permeable pavements. "Wet weather green infrastructure encompasses approaches and technologies to infiltrate, evapotranspire, capture, and reuse stormwater to maintain or restore natural hydrologies."[71] By implementing such principles and practices, water can be managed to reduce the impact of stormwater runoff while promoting the natural movement of water within an ecosystem or watershed. Applied on a broad scale, such practices also can also help maintain or restore a watershed's hydrologic and ecological functions. Basically, these new innovative stormwater management practices provide a range of benefits, including a means of reducing traditional water demands for toilet flushing and irrigation; reducing pollution from urban runoff, and thus improving water quality; replenishing aquifers; and managing flooding of drainage systems.

---

[69] "Combined sewer systems are sewers that are designed to collect rainwater runoff, domestic sewage, and industrial wastewater in the same pipe. Most of the time, combined sewer systems transport all of their wastewater to a sewage treatment plant, where it is treated and then discharged to a water body. During periods of heavy rainfall or snowmelt, however, the wastewater volume in a combined sewer system can exceed the capacity of the sewer system or treatment plant. For this reason, combined sewer systems are designed to overflow occasionally and discharge excess wastewater directly to nearby streams, rivers, or other water bodies. These overflows, called combined sewer overflows (CSOs), contain untreated human and industrial waste, toxic materials, and debris in addition to stormwater. They are a major water pollution concern for the approximately 772 cities in the U.S. that have combined sewer systems." See U.S. Environmental Protection Agency, "Combined Sewer Overflows," February 16, 2012.

[70] For more information, see U.S. Environmental Protection Agency, "Stormwater Management," undated.

[71] For more information, see U.S. Environmental Protection Agency, "Stormwater Management," undated.

Many communities are implementing these practices and experiencing such benefits. They include Chicago, Illinois; Alachua County, Florida; Philadelphia, Pennsylvania; Arlington, Virginia; Emeryville, California; Seattle, Washington; and Lenexa, Kansas.[72] For example, to address stormwater runoff problems such as pollution and flooding, Chicago has been "promoting landscape-based, green infrastructure approaches that infiltrate, evapotranspire or harvest rainwater before it enters the sewer system."[73] In the past, Chicago has offered incentives for building green roofs.[74] Chicago Department of Transportation (CDOT) has had a Green Alley Program in which it implements permeable paving materials in city alleys to reduce flooding and increase infiltration of runoff. Through a Sustainable Streetscape Program, CDOT has also implemented rain gardens, a permeable plaza, permeable asphalt parking lanes, vegetated swales, and other green infrastructure projects. Chicago also has a Stormwater Management Ordnance that requires new developments and redevelopments of a certain size to detain at least half an inch of rain on site or reduce the imperviousness of the site by 15 percent. Green infrastructure benefits have included reducing stormwater runoff by 50 percent, reducing energy use significantly, improving water and air quality, and enhancing the pedestrian environment. Green infrastructure features, such as green roofs and increased tree plantings, have helped save energy by more cost-effectively addressing the high summer city heat which is exacerbated by the urban heat island effect because 58 percent of the city consists of impervious cover.[75]

Seattle Public Utilities has taken steps to realize the benefits of stormwater capture. It undertook its first major stormwater capture project in 2001 over a two-acre area. By summer 2012 it had retrofitted more than 900 acres with stormwater capture capabilities and was expanding its strategies to include vegetation, permeable pavement, and cisterns.[76] In one neighborhood, Seattle's Belltown neighborhood on Vine Street, residents and property owners started a small, local effort twenty years ago called Growing Vine Street "to turn their stretch of a former industrial neighborhood into an urban watershed."[77] This project now contains a large cistern, native plants, stormwater basins, and other LID techniques, and is considered a long-running success story for green infrastructure and LID implementation. This example helps illustrate the benefits of such practices:

> Returning some of nature's hydrology to the cityscape can make an enormous difference—or could—as more individuals, businesses, and neighborhoods

---

[72] For more information, see U.S. Environmental Protection Agency, "Green Infrastructure Case Studies," EPA-841-F-10-004, August 2010.

[73] U.S. Environmental Protection Agency, 2010, p. 37.

[74] Around 2008, such incentives were eliminated from the city budget because of its financial problems.

[75] U.S. Environmental Protection Agency, 2010, pp. 37–39.

[76] Nancy Ahern, "Green and "Living" Buildings in Seattle: Seattle Public Utilities' Perspective," American Water Works Association Annual Conference & Exposition, June 12, 2012.

[77] Cynthia Barnet, "Water Works: Communities Imagine Ways of Making Every Drop Count," *Orion Magazine*, July/August 2013.

remake their bit of the terra firma. Washington State University scientists have found that streets with rain gardens clean up 90 percent or more of the pollutants flowing through on their way to the sound. Green roofs reduce runoff between 50 and 85 percent and can drop a building's energy costs by nearly a third. Cisterns like the one on Vine Street solve two problems, reducing runoff and capturing water for outdoor irrigation—which in summer can account for half a city's freshwater demand.[78]

The Rincon Heights neighborhood in central Tucson provides an interesting neighborhood example. This community partnered with a local non-profit to install rainwater capture facilities along two main roads. Landscaped basins and gravel-filled trenches collect stormwater to passively irrigate native trees and shrubs. These efforts seek to improve water quality, reduce flooding, and use stormwater effectively.[79]

Army installations also are doing more stormwater management projects. For example, because of its sustainability program, LID is the "preferred method of stormwater management" at Fort Bragg.[80] This post has implemented a range of LID projects, including stormwater detention basins with native plants, rain barrels, and rain gardens. In fact, Fort Bragg has installed a couple hundred rain barrels on different buildings to help use rainwater from roofs for irrigation (see Figure 2.8 for a photo of a rain barrel at Fort Bragg).

---

[78] Barnet, 2013.

[79] Watershed Management Group, "WMG Receives American Planning Association Award for Green Projects," undated.

[80] Fort Bragg, "Sustainable Fort Bragg: Water," undated.

Figure 2.8. Rain Barrel at Fort Bragg

SOURCE: Photo by Beth Lachman.

In Chapter Five, we discuss what Fort Carson is trying to do with stormwater, and in Chapter Six we discuss Fort Huachuca's stormwater related activities. Chapter Six also describes some of the erosion problems that Fort Huachuca faces because of stormwater runoff.

The use of stormwater runoff as a water supply also is called rainwater or water harvesting, and is defined as runoff captured from rooftops or other hard surfaces that can be used for beneficial use after minimal or no treatment.[81] More and more communities are looking to stormwater as an alternative water source for selected applications. In fact, the Southern California Water Committee released a report in early 2012 that examines how the region can utilize stormwater capture as a key element of long-term water sustainability. It notes:

> Finding new sources of drinking water must include local supplies developed within the coastal plain of Southern California, including the increased use of stormwater. This is a departure from historical practices which relied primarily on augmenting supplies with imported water. …[P]olicy and planning

---

[81] The use of stormwater runoff may not always be practicable in areas where return flows are critical for maintaining downstream water rights. In Colorado, for example, rainwater harvesting is legal in only narrow circumstances because of water rights laws.

frameworks are shifting toward more integrated management of these water resources.[82]

In summary, where feasible, stormwater is being used increasingly as a non-traditional water supply source, especially for irrigation applications.

Increasing Use of Reclaimed Water

The use of reclaimed water, i.e., treated wastewater, also has received significant attention. Also called water recycling, the use of reclaimed water offers opportunities to recharge aquifers and displace demand for traditional potable water uses (such as agriculture, landscape irrigation, and toilet flushing). It also reduces wastewater discharges into oceans and rivers. Many communities are starting to see reclaimed water as an important and stable water supply.

> With water management shifting away from the construction of new dams, reservoirs, and conveyance canals, the viability of water recycling as a water source has increased over the years. New water treatment technologies are now considered a realistic way of maintaining a water supply to meet demand.[83]

Before discussing the different uses of recycled water, it is important to clarify the terminology regarding water reuse. Table 2.1 presents some basic definitions of different water reuse terms.

---

[82] Southern California Water Committee, "Stormwater Capture: Opportunities to Increase Water Supplies in Southern California," January 2012.

[83] Water Education Foundation, "Layperson's Guide to Water Recycling," Sacramento, Calif., 2004, p. 2.

Table 2.1. Water Reuse Terminology

Term	Definition
Blackwater	Water captured from toilets and urinals along with kitchen waste.
Direct potable reuse	The introduction of highly treated reclaimed water either directly into the potable water supply distribution system downstream of a water treatment plant or into the raw water supply immediately upstream of a water treatment plant.
Greywater*	Water captured from sinks, baths, showers, and residential laundries that can be treated and reused. It does not include water from kitchen sinks or dishwashers.
Indirect potable reuse	The planned incorporation of reclaimed water into a raw water supply, such as in potable water storage reservoirs or groundwater aquifer, resulting in mixing and assimilation, thus providing an environmental buffer.
Rainwater harvesting	Runoff captured from rooftops or other hard surfaces that can then be used for beneficial use after no or minimal treatment.
Reclaimed water	Municipal wastewater that has gone through various treatment processes to meet specific water quality criteria with the intent of being used in a beneficial manner such as irrigation. The term *recycled water* often is used synonymously with reclaimed water.
Wastewater	Used water discharged from homes, businesses, and industry.
Water reuse	The use of treated wastewater for a beneficial use, such as irrigation or industrial cooling.

SOURCE: Richard J. Scholze, "Water Reuse and Wastewater Recycling at U.S. Army Installations: Policy Implications," U.S. Army Corps of Engineers, ERDC/CERL SR-11-7, June 2011, p. 3.
* Some organizations do accept a definition of "greywater" that does include kitchen and dishwasher wastewater along with wastewater from soiled-diaper washing. This greywater has higher levels of risk.

Using reclaimed water is not a new idea, but it has met with some strong resistance in the past. For instance, to help cope with the 1987–1992 drought in California's San Gabriel Valley, the Upper San Gabriel Municipal Water District sought to use reclaimed water to recharge the San Gabriel Aquifer. The aquifer provided drinking water to the region, and opponents to the project nicknamed it "toilet-to-tap." A lawsuit from a nearby brewing company and negative public perception resulted in a change in the project's location and reduction in scope.[84] Numerous other localities reclaimed water projects have faced opposition, and some of these efforts are thought to involve little public input and were perceived to have a "good enough" approach to quality.[85]

Scientific uncertainties about the health effects of extremely small levels of contaminants, such as antibiotics, in reclaimed water have driven some of this concern. Namely, emerging

---

[84] Water Education Foundation, 2004.

[85] California Assembly Committee on Water, Parks and Wildlife, "SB 918 Senate Bill - Bill Analysis," June 1, 2010.

contaminants that may remain in wastewater, such as pharmaceuticals, have raised concern about even appropriately treated reclaimed water being used for human consumption.[86]

As water has become an increasingly stressed resource and treatment technology has improved, these perceptions have changed. Reclaimed water is increasingly seen as a valuable resource. The American Water Works Association notes:

> In the past several years, an emphasis has been placed on water reuse and its benefits. As technology has changed, reuse has become more socially acceptable and affordable. Therefore, more utilities are looking to reclaimed water as a source to satisfy demands, particularly demands that do not require potable-quality water.[87]

Florida, for example, uses reclaimed water in a variety of ways. As shown in Figure 2.9, these include irrigating public access areas like golf courses and parks, agriculture irrigation, industrial use, groundwater recharge, and wetlands.[88] Groundwater recharge is particularly valuable in Florida and other coastal areas to reduce seawater intrusion into the state's aquifers, a consequence of over-extraction of groundwater. This increases the quality of the aquifer and helps maintain an often-essential source of drinking water.[89]

Figure 2.9. Florida's Use of Reclaimed Water, by Flow

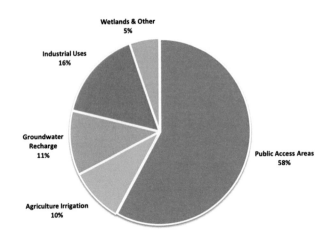

SOURCE: Florida Department of Environmental Protection Water Reuse Program, "2011 Reuse Inventory," May 2012, p. 6.

The largest reclaimed water facility in the country is in Orange County, Calif. This four-year-old facility recharges 20 percent of the region's aquifer with reclaimed water, approximately 70 million gallons per day. This water goes through a treatment process that involves fewer

---

[86] Water Education Foundation, 2004, p. 15.

[87] American Water Works Association, "Water Reuse Rates and Charges: Survey Results," 2008, p. 4.

[88] Florida Department of Environmental Protection Water Reuse Program, "2011 Reuse Inventory," May 2012, p. 6.

[89] Southwest Florida Water Management District, "Reclaimed Water Guide," 1999.

chemicals and more filtering than past approaches and produces water that is of approximately the same quality as distilled water.[90]

More and more communities, especially in dry Western areas, are using reclaimed water for irrigation and other purposes. In 2003 in California, about 580,00 acre-feet of recycled water from municipal wastewater was used in agricultural and landscape irrigation (48 and 20 percent respectively), for groundwater recharge (14 percent) and other uses.[91] Some communities have even developed extensive "purple pipe" systems to supply the reclaimed water resources to community customers. For example, in El Paso, Texas, El Paso Water Utilities (EPWU) has more that 51 miles of purple pipe that carried 2.2 billion gallons of reclaimed water in 2012. EPWU supplies over 5.83 million gallons per day of reclaimed water to city parks, golf courses, school grounds, apartment landscapes, construction operations, and industrial sites. EPWU reclaimed water also is used for the operation of some of their treatment plants (in-plant use) and to recharge the Hueco Bolson aquifer through injection wells and infiltration basins.[92] EPWU's Fred Hervey Water Reclamation Plant supplies El Paso Electric Company annually with about 900 million gallons of reclaimed water for its cooling towers.[93]

Recycled water already is used on many military installations for a variety of purposes, including vehicle washing, irrigation, cooling tower makeup, and aquifer recharge. For example, Fort Houston in San Antonio, Tex., draws upon the city's reuse pipeline for irrigation and cooling tower needs. Fort Carson in Colorado uses recycled water for its central vehicle wash facility and reclaimed water to irrigate the golf course. Fort Huachuca, Arizona, uses reclaimed water to irrigate the golf course and recharge the local aquifer.[94] Chapters Five and Six provide more details about how Fort Carson and Fort Huachuca take advantage of their wastewater.

As shown in Figure 2.10, the WateReuse Association finds that while 90 percent of water reuse occurs in just four states—Arizona, California, Florida, and Texas—it is growing in others.

---

[90] Felicity Barringer, "As 'Yuck Factor' Subsides, Treated Wastewater Flows from Taps," *New York Times*, February 9, 2012.

[91] Water Education Foundation, 2004, p. 4.

[92] El Paso Water Utilities, "Reclaimed Water," 2007.

[93] El Paso Water Utilities, "Wastewater Treatment: Fred Hervey Water Reclamation Plant," 2007.

[94] For more Army examples, see Scholze, 2011.

**Figure 2.10. States That Account for 90 Percent of Water Reuse (dark purple)
and States Where It Is Growing (light purple)**

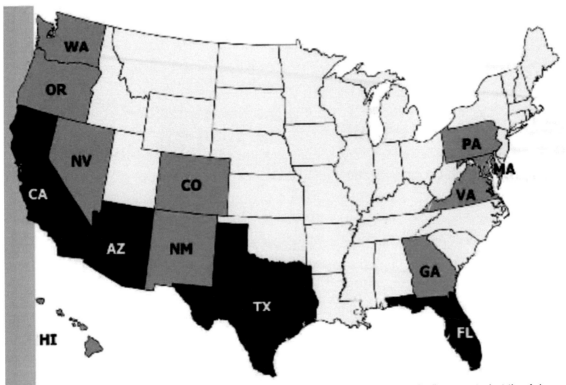

SOURCE: Guy Carpenter, "Reclaimed Water Trends Nationally and Internationally," presented at the Arizona Governor's Blue Ribbon Panel on Water Sustainability, February 5, 2010.

In addition to social and health-related issues, entities considering water recycling need to consider its second-order effects, since recycled water alters patterns of water use. These changes can affect the ecosystem, as well as downstream users, and they must be assessed, especially where reuse flows are a large proportion of the affected hydrologic system. For example, in some waterways wastewater treatment plant effluent contributes to water body's continuous flow, so a decrease in this effluent could reduce overall flow and cause more variable flows. On the flip side, reuse could raise aquifer levels and increase stream flows where the two are hydrologically connected. In terms of quantity and discharge water constituents (chemical, physical, and biological), these alterations to the hydrological cycle must be reviewed and assessed. Related is the issue of water rights and the effects an upstream user may have on the availability of water for downstream rights holders. State law may determine whether or not the use of recycled water is acceptable within the water rights regime. For example, Utah municipalities or other governmental agencies may recycle sewage effluent as long as it is put "to a beneficial use consistent with, and without enlargement of, those water rights." Those interested in reuse also must provide information to the state engineer on:

- The water rights to be reused

- The diversion and depletion limits of the water rights as originally approved
- The quantity and location of proposed reuse including any unused effluent
- An assessment of the total amount of water removed from the local hydrologic system, both for the original water uses and those anticipated with reuse.[95]

States are evolving regulations related to water reuse as this practice develops. The U.S. EPA has issued guidelines for water reuse to assist parties considering water reuse.[96] Finally, in addition to viewing wastewater as a steady water source, some researchers and managers also are starting to see it as a source for recovering energy and other materials, such as fertilizer. One such vision holds that:

> Waste treatment systems will become resource recovery centers, enabling 1) low-cost production of water that offsets demands for high quality water; 2) energy production that offsets fossil fuel demands; and 3) nutrient recovery that offsets fertilizer demands. Distributed, networked bioreactors will foster development of local markets for wastewater recovery, with low transport costs and the ability to match product quality and quantity to regional needs.[97]

The challenge is to develop cost-effective technology for recovering the clean water, energy, and materials from wastewater. The importance of wastewater/effluent and recycling water, such as how it is being used, viewed, and planned to be used at Fort Carson and Fort Huachuca also is discussed in Chapters Five and Six.

Desalination

Desalination is another non-traditional approach that some communities are using to meet water needs. A desalination facility removes salt and other dissolved minerals from water and allows municipalities to use alternative water sources, expanding available water supply. "Desalination plants have treated brackish groundwater, irrigation runoff, seawater and domestic wastewater for decades in California."[98] Although desalination provides an alternative water source, energy costs, the need to properly manage the brine discharge, and other environmental effects (notably on intake water) must be balanced with the need for high-quality water. In this section, we illustrate some different examples of desalination plant implementation in Texas and California, including one on an Army post.

Desalination plants have high energy costs, and the resulting brine byproduct must be disposed of properly. The higher the salt content in the water, the more energy is needed, so most desalination plants have been smaller plants that process less-salty water sources. Plants that treat

---

[95] Utah Department of Natural Resources, Water Resources Division, "Water Reuse in Utah," April 2005.

[96] U.S. Environmental Protection Agency, "Guidelines for Water Reuse," EPA/600/R-12/618, September 2012; National Research Council, "Water Reuse: Potential for Expanding the Nation's Water Supply Through Reuse of Municipal Wastewater," Washington, D.C.: National Academy Press, 2012.

[97] Craig Criddle, "Wastewater as a Valuable Resource," Woods Institute for the Environment Solution Brief, Stanford University, May 2010.

[98] Water Education Foundation, 2004.

brackish groundwater are cheaper than ones that treat seawater, which can be ten times saltier or more. Texas actually has 44 desalination plants, none using seawater, which were built for public water supplies, mostly in rural areas. Most of these plants are small, providing less than three million gallons per day.[99] The state's largest plant is located on Fort Bliss in El Paso, Texas, but it is owned and operated by EPWU. The $91 million Kay Bailey Hutchison Desalination Plant (see photo in Figure 2.11), a partnership between Fort Bliss and EPWU, was funded by different federal sources and EPWU. The total plant capacity is 27.5 million gallons per day (MGD) of potable water; although it averages about 3 MGD of reverse osmosis (RO) water; it can make up to a total capacity of 15 MGD RO water, which is blended with Fort Bliss well water to produce the potable water.[100] Even this plant has high energy costs associated with forcing water through a membrane to remove the salts. Production of this desalinated water from brackish well water in El Paso costs 2.1 times more than local nearly potable groundwater, and 70 percent more than surface water.[101]

[99] Kate Galbraith, "Texas' Water Woes Spark Interest in Desalination," *Texas Tribune*, June 10, 2012.

[100] For more information, see El Paso Water Utilities, "Desalination: Setting the Stage for the Future," 2007.

[101] Galbraith, 2012.

Figure 2.11. Kay Bailey Hutchison Desalination Plant

SOURCE: Photo by Beth Lachman.

In California, some desalination facilities remove seawater. In fact, one desalination plant saved the small town of Sand City, which has 330 residents. New requirements from the state were going to cut the town's groundwater pumping limit by half, which meant that the town "faced a tough choice—find a new source of water, or cease to exist." In response, the town built a desalination plant that produces just 300 acre-feet of water, or around 600,000 gallons when running at full capacity. This was enough to meet the town's needs after its groundwater pumping was curtailed.[102] Larger desalination facilities in California, which have larger energy costs and environmental concerns, face more challenges. An example is the recently completed $1 billion Carlsbad, California, desalination plant. Opened in December 2015, this plant was projected to produce up to 56,000 acre-feet a year (50 million gallons per day), which would sell for roughly $2,000 per acre-foot. Environmentalists have opposed the plant because it would intake 300 million gallons a day of water from the sensitive Agua Hedionda lagoon, and because

---

[102] Jaymi Heimbuch, "Desalination Plant Helps Save a California Coastal Community," The Learning Channel, 2013.

the plant uses an "open ocean intake" technology that they claim hurts fisheries and habitat. Environmentalists contend the "majority of organisms pulled in are killed."[103]

Figure 2.12 shows the growth in capacity of desalination plants that were installed or contracted from 1952 to 2005.[104] This volume is small—only 0.4 percent of total U.S. water use. Particularly in California, numerous recent proposals have been made to increase this capacity. The issues of high energy-use requirements, high cost, and negative environmental impacts mean that efforts have been limited so far. But as options for developing new water supply are fewer and more costly, the need and investments in desalination plants will rise. Increasing periods of drought also will lead to more examination and interest in desalination plants, as is being experienced in California with its recent drought.[105]

Figure 2.12. Desalination Capacity in the United States, 1952–2005

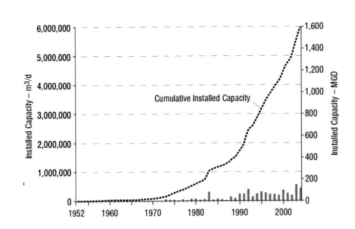

SOURCE: Heather Cooley, Peter H. Gleick, and Gary Wolff, "Desalination, with a Grain of Salt: A California Perspective," Pacific Institute, June 2006, p. 21.

State, Regional, and Local Water Planning

Many state, regional, and local water managers and planners have been developing state, local, and regional plans for managing and ensuring long-term water resources. Special drought plans also are being developed. Regional, state, and local water plans are being written to help different jurisdictions within a state, region, or local area better coordinate, collaborate, and

---

[103] San Diego Coastkeeper, "Desalination in San Diego," undated. For more information on this plant, see Bettina Boxall, "Seawater Desalination Plant Might be Just a Drop in the Bucket," *Los Angeles Times*, February 17, 2013; and Felicity Barringer, "In California, What Price Water?" *New York Times*, February 28, 2013; and Matt Rascon and R. Stickney, "Carlsbad Desalination Plant Opens," San Diego 7 NBC News, December 14, 2015.

[104] Actual production is less than this, given that the database from which this data was gathered includes plants that were contracted but never built, plants that were built but never operated, and plants that were once operated but are now closed. See Heather Cooley, Peter H. Gleick, and Gary Wolff, "Desalination, with a Grain of Salt: A California Perspective," Pacific Institute, June 2006, p. 21.

[105] Amanda Little, "Can Desalination Counter the Drought?" *The New Yorker*, July 22, 2015.

manage water resources over the long term. These managers increasingly are realizing that improved water management for economic, environmental, and domestic purposes requires long-term planning. Such water management plans also have been developed in response to many of the issues affecting water supply and use. They include over-appropriation of water resources, competition and conflicts over water resources, insufficient or aging infrastructure, floods and droughts, impaired water quality, and habitat and species loss. While there is variability in the focus and implementation of these plans, comprehensive and effective planning processes will better assess water supply and uses (both in terms of quantity and quality) and better identify where potential conflicts might arise to improve the long-term management of this scarce resource. Ideally, these plans also will comprehensively assess surface water and groundwater, incorporate stakeholder input, and lead to actions to implement the plans.

Collaboration and partnership are major parts of developing and implementing state, regional, and local water management plans. First, agencies at all levels of government (local, state, and federal) have specific data, specialized expertise, and varying perspectives on water management that they bring to the process. Second, to develop a comprehensive and integrated plan it is necessary to have participation from a wide range of groups that engage in water management and use. Third, stakeholder groups offer different points of view on the range of water uses, values, and needs.

State, regional and local plans usually are integrated and coordinated, though different states take different approaches. State plans generally are implemented at the local level, and states offer incentives, loans or grants to assist with projects. In some cases, implementation of these plans is a major concern in places where funding or enforcement provisions are weak. Summarized in Table 2.2, we highlight state water-planning efforts in Georgia, Texas, Colorado, and California to illustrate the different ways in which states address water management and planning. We include more detail in these discussions than in some other parts of this chapter because these plans are an important opportunity for Army installations to help with their long-term water security, which is discussed in other parts of this document. We should also note that in the next section of this document, there is a brief discussion of watershed approaches, including Washington state's watershed planning and watershed management within the Chesapeake Bay.

Table 2.2. A Comparison of Select State Planning Processes

State	Statewide Plan	Regions	Implementation
California	Prepared by the Department of Water Resources with stakeholder input. Scenario-based assessment of water supplies, uses, and dependent natural resources. Updated every 5 years. State Drought Plan 2008 as directed by the governor.	Developed at the state level using 10 hydrological regions. Integrated Regional Water Management Plans since 2002 covering all aspects of water supply and use. Approximately 50 regional plans.	Strategy only—by law, the California Water Plan cannot mandate actions nor authorize spending. Local and regional analysis required to implement projects in support of the plan. state-level and regional management actions. Partnerships at the regional level with state funding available for projects.
Colorado	New process in 2013 led by the Colorado Water Conservation Board. Drought Management Plan led by the Colorado Water Conservation Board.		Plan will guide state water projects, studies, funding, financial assistance, project permitting and reviews. Interagency Coordinating Group directs state-level response. Water providers are encouraged to implement their own individual drought plans.
Georgia	State plan provides data, methodology, and structure. State plan does not integrate regional plans.	11 regions along political boundaries. Addresses water supply and water use within the region.	State permitting and grants. Regional-level implementation of other actions.
Texas	State plan integrates regional plans from a statewide perspective to guide policy. Focus is on droughts and floods.	16 regions (originally developed using hydrological, water infrastructure, political, and socioeconomic characteristics).	State funding of water projects, allocation of water rights permits, and policy recommendations to the state legislature.

In response to droughts, population growth, and conflicts with neighboring states over water, Georgia has begun efforts in the last 10 to 15 years to improve statewide water management. Legislation has been issued to improve water supply and planning and to encourage water conservation. In 2004 the Georgia Comprehensive State-wide Water Management Planning Act created a water planning process that, among other things, has integrated consideration of uses that affect both water quantity and quality.[106] The essential purpose of the plan is for "Georgia to manage water resources in a sustainable manner to support the state's economy, to protect public health and natural systems, and to enhance the quality of life for all citizens."[107] In January 2008 the statewide water plan was formally adopted. In addition to the state plan, Georgia has 11 regions, based on political boundaries, which develop water plans. Ten regional plans were developed and completed between 2009 and November 2011 subject to statewide review and comment.[108]

The process statewide is coordinated through the Water Council, which consists of seven state agency heads, including heads of the Environmental Protection Division (EPD), Soil and Water Conservation Commission, Department of Natural Resources, and Department of Agriculture. Citizen members and four elected state officials are involved as well. The State Water Plan establishes the framework and guidelines for the regional plans, though it does not integrate those plans. This framework directs the EPD to work cooperatively with the regions to develop water development and conservation plans through developing and providing water resource assessments for watersheds and aquifers for use by the regions, evaluating regional plans, providing opportunities for public review and comment, and, in terms of implementation, making permitting decisions consistent with the plans and ensuring that regional plans are implemented in a consistent and equitable manner.

At the regional level, the planning councils, which are authorized by the legislation and whose members are appointed by the governor, lieutenant governor, and speaker of the Georgia House of Representatives, were required to develop specific actions that will ensure water supply (both groundwater and surface water) and will satisfy demand (quantity and quality) over a 40-year horizon. Each regional planning council consisted of approximately 25 residents of the planning region representing various water users: agriculture, forestry, industry, commerce, local

---

[106] The Georgia General Assembly in 2010 passed a second law, The Water Stewardship Act. The act contains several measures to improve water conservation. These include a year-round restriction on select daytime watering; metering requirements for multi-tenant buildings; high-efficiency plumbing codes for new construction; and high-efficiency industrial cooling system requirements for new industrial buildings (the last three measures went into effect on July 1, 2012). In addition to these acts, in 2011 the state committed more monies toward building reservoirs and other supply enhancement activities by creating its Water Supply Program. See Chattahoochee Riverkeeper, "Georgia's Response to the Tri-State Ruling."

[107] Official Code of Georgia Annotated (O.C.G.A.), § 12-5-522.

[108] The North Georgia Water Planning District was established under an earlier statute in 2001 for a total of 11 planning regions.

governments, water utilities, regional development centers, tourism, recreation, and the environment. These councils considered the water uses and sources within each region to develop a long-term regional plan that incorporates the issues within the regional boundaries (and coordinates with others as needed). The regions are responsible for: developing water demand forecasts and determining the assimilative capacity of water bodies within the boundaries, identifying opportunities to improve water management within the region, incorporating public input, and guiding water users in implementing the regional plan. Those ultimately responsible for implementing the plan are the water users within each region, those developing water infrastructure and permit-holders or local governments and other entities that are eligible for state grants and loans.[109]

To illustrate the process, we briefly discuss one regional plan for the middle Chattahoochee River area where Fort Benning is located. A 30-member council adopted the Middle Chattahoochee Regional Plan in November 2011. The vast majority of water in the Middle Chattahoochee Region comes from surface water sources, such as the Chattahoochee River, while less than 10 percent comes from groundwater sources.[110] The Middle Chattahoochee is comprised of 11 counties and 34 municipalities; the water users (in descending order) are energy, municipalities, agriculture, and industry. The Middle Chattahoochee Water Planning Council says it would like to see improved management and operations of the USACE's impoundments upstream to ensure minimum stream flows within the vicinity of Columbus, Georgia, as well as more coordination on dealing with potential interstate compacts. The Council is advocating for changes to the USACE Master Control Manual and improved modeling of the Appalachicola-Chattahoochee-Flint River Basin flows. The plan itself includes recommendations for demand management, wastewater returns management, enhanced water supply,[111] and improved water quality.[112]

Overall water planning in Georgia is a collaborative effort involving regional, state, and federal agencies. At least eight different state agencies are involved. Federal agencies include the U.S. EPA, USACE, USGS, U.S. Department of Agriculture Natural Resources Conservation Service (USDA-NRCS), the U.S. Fish and Wildlife Service, the U.S. National Park Service, National Oceanic and Atmospheric Administration (NOAA), and the Federal Emergency Management Agency (FEMA). University of Georgia technical experts also assist in the

---

[109] U.S. Army Corps of Engineers, "Building Strong Collaborative Relationships for a Sustainable Water Resources Future: State of Georgia Summary of State Water Planning," Washington, D.C., December 2009.

[110] The major water resources within this region are the Chattahoochee River, West Point Lake, Walter F. George Lake, and four aquifers (Cretaceous, Clayton, Claiborne, and Floridan).

[111] Water supply is enhanced through improved surface water storage, ponds for irrigation, regional coordination, drought contingency planning, and potential use of aquifer storage, etc.

[112] Georgia Department of Natural Resources, "Middle Chattahoochee Regional Plan," September 2011; Fort Benning, "Joint Land Use Study," 2008; Georgia Environmental Protection Division, "Georgia's Water Future in Focus: Highlights of Regional Water Planning, 2009–2011."

planning process. Public participation is required at the state level as well as within the individual regions. The Water Council has held a series of public and town hall meetings to solicit public comment on the State Plan while the regional councils have solicited local input from the citizenry and other groups.

Texas Water Planning

Texas has a long history of water planning. The first statewide water management plan was adopted in 1961 after the state's most severe drought period in the 1950s, and subsequent plans were adopted in 1968, 1984, 1990, 1992, and 1997. After a drought in 1996, the state passed Senate Bill 1 to improve the development and management of the water resources in the state. The bill created a framework for water planning that is still used. Plans were developed from the local and regional levels, where stakeholders were required to develop a consensus-based plan for meeting water needs during droughts. These regional plans are incorporated by the Texas Water Development Board (TWDB) into a comprehensive statewide plan. The Texas Water Development Board is the state agency responsible for long-range water-related planning, as well as wastewater project financing. The Commission on Environmental Quality (CEQ), the state agency that allocates surface water rights, issues withdrawal permits, and is responsible for water quality programs, also helps in the planning process.

The state's most recent water plan was released in 2012. As stated in the plan, the purpose of water planning in Texas is "to ensure that our state's cities, rural communities, farms, ranches, businesses, and industries will have enough water to meet their needs during a repeat of this great drought."[113] The Texas Water Code indicates this is to be performed through the

> orderly development, management, and conservation of water resources ... to ensure that sufficient water will be available at a reasonable cost to ensure public health, safety, and welfare; further economic development, and protect the agricultural and natural resources of the entire state.[114]

Water planning is performed regionally in the 16 regions established within the state in an open process. Texas law requires the regional planning groups to have at a minimum the interests of agriculture, industry, business, environment, counties, the public, municipalities, water districts, river authorities, water utilities, and power generation represented within its membership. These planning groups assess water availability (both surface and groundwater) and demand projections during drought periods to identify classes of water users that are potentially at risk of receiving too little water. In addition, they recommend strategies and associated costs to address these potential shortages; identify risks and uncertainties involved in planning; and evaluate water management strategies' effects on the state's water, agriculture, and natural resources. Once adopted by the regional planning group, the regional plan is submitted to the

---

[113] Texas Water Development Board, "Water for Texas: 2012 State Water Plan," January 2012.

[114] Texas Water Code, Chapter 16.051.

Texas Water Development Board (TWDB) for approval and integration within the state water plan. The state water plan, in turn, provides the basis for water policy and recommendations to the Texas Legislature. As prescribed by the legislation, TWDB can provide financial assistance for water supply projects only if the project is consistent with the needs identified in the regional water plans and the state water plan. (This also applies to the Texas Commission on Environmental Quality's granting of water-right permits.)[115] Each step of the process is open to the public and provides numerous opportunities for public input. The plans are updated every five years.

In addition to incorporating multiple stakeholder group perspectives and public comments, the planning process is collaborative within the regions and among various state and federal agencies. The agencies provide critical expertise and resources, represent their particular interests, and implement and enforce the plan.

In the most recent plan, surface water availability is expected to increase by 6 percent; groundwater is expected to decrease by 30 percent; and reuse is expected to increase by 27 percent in 50 years (from 2010). Demand is expected to exceed supply during drought by 3.6 million af in 2010 growing to 8.2 million af per year in 2060 with shortfalls in irrigation and municipal users at 45 percent and 41 percent of total needs respectively. Total capital costs were estimate at $231 billion over the 50 years for water management, flood control, water treatment and distribution, and wastewater treatment. A majority of the financing is expected to come from local sources, but some state financing will be necessary. Four planning groups were able to identify strategies to meet all of the needs for water identified in their regions; while 12 planning groups were unable to identify economically feasible strategies to meet all water-supply needs for each water user group in their planning areas (while all categories of water users had unmet needs, irrigation accounted for the majority of unmet needs). The consequences of doing nothing accounted for lost business opportunities (from tens of billions of dollars to more than $100 billion), lost jobs, and lost state tax revenues; less population growth, and a projection that more than 50 percent of the state's population in 2060 would only receive just over half of their water needs.[116]

In addition to state financing of water management activities and infrastructure needs identified in the regional plans, TWDB developed several recommendations for improving water management for drought conditions. These recommendations, based on the regional plans, include:

- Providing for three additional reservoirs and nine river segments of ecological value protections in the state water code.
- Acquiring reservoir sites to support water supply development that would be available beyond the 50-year planning horizon.

---

[115] Texas Water Development Board, 2012.

[116] Texas Water Development Board, 2012.

- Eliminating unreasonable restrictions on the voluntary inter-basin transfer of surface water through legislation.
- Limiting TWDB to technical review in the petition process on reasonable future conditions.
- Requiring all retail public utilities to conduct water loss audits on an annual basis instead of every five years.
- Developing long-term affordable and sustainable methods for financing assistance for state water plan projects.[117]

Colorado Water Planning

Colorado does not have a statewide water management plan but instead has relied in the past on a series of complementary planning efforts for water supply, drought, and water quality. However, an executive order issued by the Governor in 2013 directed the Colorado Water Conservation Board (CWCB) to initiate a statewide water plan after water demands exceeded supplies and because of on-going droughts problems. In addition, threats to Colorado's water rights from nine interstate compacts and two equitable apportionment agreements suggest the state could be more attentive toward protecting these rights. Moreover, the state recognizes the need for a more integrated approach to water-supply planning that incorporates quantity with quality. Some water rights transfers from irrigated lands have caused undue economic, social, and environmental losses. The governor noted in the Executive Order that discussions held by the Interbasin Compact Committee and Water Basin Roundtables over intra- and inter-basin challenges have created momentum for planning activities.[118] As a result, Colorado's Basin Roundtables,[119] the Interbasin Compact Committee,[120] and Colorado Water Conservation Board are collaborating on water supply planning to identify and develop measures to meet future water demands while achieving "healthy watersheds and environment, robust recreation and tourism economies, vibrant and sustainable cities, and viable and productive agriculture."[121] Among

---

[117] Texas Water Development Board, 2012.

[118] Colorado Governor's Office, "Directing the Colorado Water Conservation Board to Commence Work on the Colorado Water Plan," Executive Order D2013-005, May 14, 2013.

[119] Nine basin roundtables were established by the Colorado Water for the 21st Century Act enacted in 2005. The roundtables (for the Denver area and the eight major river systems in Colorado—Arkansas, Colorado, Gunnison, Metro, North Platte, Rio Grande, South Platte, Southwest, and Yampa/White) create water needs assessments for their respective river basins and develop locally-based solutions collaboratively that address stakeholders' water issues in each region.

[120] The IBCC, which also was established by the Colorado Water for the 21st Century Act, is chartered to encourage dialogue on water issues among various stakeholder groups and to increase participation of these stakeholders in a locally-driven decision making process. It also assists Basin Roundtables and assesses Colorado's ability to utilize and meet its interstate compact agreements. It is comprised of 27 members: two each from the nine Basin Roundtables, six appointed by the governor with expertise in either environmental, recreational, governmental, industrial or agricultural issues; two others as appointed by the Senate Agriculture Committee chair and the House Agricultural Committee chair. The Director of Compact Negotiations chairs the IBCC.

[121] Colorado Water Conservation Board, "Colorado Water Plan Interim Website," undated.

other things, the CWCB has been directed to align state water projects, studies, funding, water project financial assistance, water project permitting and reviews (expediting those that include conservation, innovation, and collaboration and possibly efficient infrastructure, healthy watersheds, and smart water conservation), and overall efforts with the Colorado Water Plan. The CWCB also is conducting an inventory of water rights held by state agencies. A final plan is expected by the end of 2015.[122]

In addition to the appointed members of the Interbasin Compact Committee and the nine Basin Roundtables, many state agencies will have pertinent input to the water plan. The agencies that must cooperate on the plan include, but are not limited to, the Colorado Water Resource and Power Development Authority, Department of Agriculture, the Energy Office, and the Department of Public Health and Environment (including the Water Quality Control Division).

The Colorado water plan will build on previous planning efforts, such as water supply planning and drought planning performed by the Colorado Water Conservation Board. planning by the Colorado Department of Public Health and Environment Water Quality Control Division, and work by the nine Basin Roundtables and the Interbasin Compact Committee. Briefly, the water supply plan of 2010 called for:

- Cooperation and dialogue among classes of intra- and inter-basin water users.
- Identifying funding and investing in projects and approaches that extend water supplies including during periods of drought.
- Performing risk analyses to ensure that Colorado maximizes its compact and decree entitlements.
- Investing in projects that serve multiple objectives, serving agricultural water supply needs.
- Supporting demand management and conservation.

The Basin Roundtables, on the other hand, have focused on water needs assessments for both consumptive and non-consumptive uses, identifying unappropriated waters within the basin of interest and proposing projects for meeting user needs through an inclusive process that incorporates input from local governments, water providers, and other stakeholders.[123] Another Colorado water planning effort, the Colorado Drought Management Plan, is discussed next.

Colorado Drought Management Plan

Colorado has recognized the importance of planning for drought since the late 1970s, when it initiated its first statewide plan to "provide an effective and systematic means for the State of Colorado to deal with emergency drought problems which may occur over the short or long term."[124] The plan does not create a new government agency to deal with drought, but provides a

---

[122] Colorado Governor's Office, 2013.

[123] Becky Mitchell, "Colorado Water Plan," American Water Resources Association—Colorado Section, August 2013.

[124] National Drought Mitigation Center, "State Drought Planning," 2008.

forum and process for coordinating all aspects of drought planning and response among relevant public and private entities, as well as communicating drought information to the public and local governments. Therefore, this process is a collaborative effort that encompasses local, state, and federal agencies along with water utilities, academia, and the private sector. The most recent plan went through a comprehensive revision in 2010 and was released for public comment.[125] In 2012 the plan was activated statewide by the governor.[126]

The Colorado Water Conservation and Control Board oversees the process, which incorporates many public agencies and the private sector. Other disaster management agencies and the Governor's Office also provide approvals. At this time, the planning process is based on federal requirements in the Disaster Mitigation Act of 2000 for Colorado to be eligible for federal disaster aid. Participants and contributors to the plan have come from academia, federal and state agencies, and interest groups. Outreach also is being made to a larger group of public and private audiences.

The planning process includes four elements: monitoring, assessment, mitigation, and response. Different task forces are responsible for each element, drawing on expertise within various agencies at the state and federal levels. Monitoring is performed by the Water Availability Task Force, which is comprised of water supply specialists from state, local, and federal governments, as well as experts in climatology and weather forecasting. Several Impact Task Forces perform assessments that focus on the effects to the economy, citizens, and environment. Although nine Impact Task Forces initially were formed, they were consolidated into seven: municipal water, wildfire protection, agricultural industry, tourism, wildlife, water availability, and energy.[127]

The chairs of the Water Availability Task Force and each Impact Task Force incorporate these effects into an overall assessment. The group also provides recommendations. Drought response is led by an Interagency Coordinating Group, which is comprised of agencies that might lead a specific response to drought, to ensure the response is coordinated and comprehensive. The Interagency Coordinating Group and governor can recommend activating specific Impact Task Forces during droughts.

Local governments or utilities are not legally required to develop drought management plans,[128] but the state provides resources and tools to water providers to develop local plans. Their plans may be submitted to the Colorado Water Conservation Board (CWCB) Office of Water Conservation and Drought Planning for review and approval. This office also provides

---

[125] Associated Press, "Colorado Releases Draft Drought Mitigation Plan," *Denver Post*, July 23, 2013.

[126] Colorado Department of Natural Resources, "The Colorado Drought Mitigation and Response Plan," prepared pursuant to Disaster Mitigation Act 2000 & Section 409, PL 93, September 2010; Colorado Water Conservation Board, "Drought," undated.

[127] Colorado Department of Natural Resources, 2010; Colorado Water Conservation Board, undated.

[128] However, large water utility providers are required to implement water conservation plans.

guidelines, a planning toolbox, and drought mitigation planning grants for water providers or state and local governmental entities. Despite these efforts and research showing that additional drought preparedness planning is needed at the local level, in 2013 only 27 percent of Colorado municipal water providers had a drought-response plan and only 37 percent had an individual assigned to drought planning. (Colorado Springs Utilities, which services Fort Carson, had only a conservation plan, much like most of the state's small- to medium-sized water providers.) Large water providers typically did have drought plans, although they varied in their comprehensiveness and were not necessarily approved by the state.[129]

California Water Planning

California has been involved in water planning for more than 100 years. Water planning has become even more important in recent years because of the on-going drought.[130] A scenario-based plan, the California Water Plan, presents the status of its water supply and demands to include agricultural, urban, and environmental demands abiding by the requirements set forth in the California Water Code. Also presented in the plan are alternative regional and statewide strategies to reduce water demand, increase supply, reduce flood risk, improve water quality, and enhance environmental conditions. The California Department of Water Resources (DWR) develops the plan using a collaborative framework for incorporating input from elected officials, agencies, tribes, water and resource managers, businesses, academia, stakeholders, and the public. The 2013 update, which was nearing the public review period, was coordinated through a Water Plan State Agency Steering Committee, which is comprised of representatives from state agencies. This differs from previous years in which plans were coordinated solely through the DWR and had little formal input from other state agencies. Regional forums develop Integrated Regional Water Management (IRWM) Plans and provide opportunities for local interests to provide input to the plan and may include discussions of Integrated Regional Water Management activities, financing strategies and opportunities, flood control programs and policies, and regional priorities for the short term.

Integrated Regional Water Management (IRWM) Planning is a collaborative effort to manage water resources within a region in a more comprehensive manner. This was initiated at the state level by Senate Bill 1672 (which contains the Integrated Regional Water Management Act). It was passed in 2002 to encourage and assist agencies within a geographic region to develop integrated approaches to water management that incorporate the physical, environmental, societal, economic, legal, and jurisdictional aspects within an open and collaborative stakeholder process to improve the quality, quantity, and reliability of local water supplies. Various individuals, agencies, and organizations involved with water supply and

---

[129] Colorado Water Conservation Board, undated.

[130] This section was written in 2013 and does not include the more recent planning because of the recent California drought.

quality, stormwater runoff, flood control, watershed and natural resources, along with land use; environmental issues; business; technology; and academia are involved with the process. The state provides grant funding for developing and implementing integrated water management plans.[131] IRWM planning is examined further in Chapter Three in the regional partnerships discussion.

Watershed Management

Another current water management trend is managing water resources based on natural watersheds. According to U.S. EPA, "A watershed is the area of land where all of the water that is under it or drains off of it goes into the same place."[132] From an ecological point of view, it makes more sense to manage water resources based on a watershed than on an artificial boundary. Indeed, many federal, state, and local government agencies started implementing policies and activities for managing water systems based on their natural watersheds rather than by city, county, or other jurisdictional boundaries. Washington state has been at the forefront of using watershed management approaches. In 1993, its Department of Ecology began a new watershed management framework that designated 23 water-quality management areas and established a process for issuing permits, assessing water quality, and facilitating decision-making in each such area.[133]

Washington has taken particular action to manage Puget Sound, one of the state's most critical but at-risk natural resources, and the watershed that drains into it. In 1996, the state Legislature passed the Puget Sound Water Quality Protection Act, which created the Puget Sound Partnership to "coordinate and lead the effort to restore and protect Puget Sound."[134] The partnership involves a wide range of stakeholders, including state and local governments, Native American tribes, academic institutions, non-profit and citizen groups, water and sewer utilities. It is charged with developing an "Action Agenda" that will prioritize cleanup and improvement projects, coordinate federal, state, local, tribal, and private resources, and ensure that all entities are working cooperatively.[135]

In a broader move, the Washington Legislature passed the Watershed Planning Act in 1998 to manage watersheds. The Act states:

> The legislature finds that the local development of watershed plans for managing water resources and for protecting existing water rights is vital to both state and

---

[131] California Department of Water Resources, "Integrated Regional Water Management Grants," undated; Greater Monterey County Integrated Regional Water Management Program, website, undated.

[132] U.S. Environmental Protection Agency, "What Is a Watershed?" March 6, 2012.

[133] Washington State Department of Ecology, "An Overview of Washington State's Watershed Approach to Water Quality Management," last updated April 2008.

[134] Washington State Legislature Revised Code of Washington, Chapter 90.71, "Puget Sound Water Quality Protection," 1996.

[135] Puget Sound Partnership, "About the Partnership," undated.

local interests. . . . The development of such plans serves the state's vital interests by ensuring that the state's water resources are used wisely, by protecting existing water rights, by protecting in-stream flows for fish and by providing for the economic well-being of the state's citizenry and communities.[136]

As one key feature, the legislation puts watershed planning under local control and provides a process by which members of a watershed community can assess watersheds' status and determine how to manage them in ways that balance competing demands. This process is at work in several areas.[137]

On the East Coast the Chesapeake Bay watershed, shown in Figure 2.13, drains into the largest estuary in the country and also has received significant attention in recent decades. Major studies in the 1980s found that, since the 1950s, significant areas in the bay have suffered from hypoxia, or a severe lack of oxygen, which has resulted in "dead zones" lacking marine life. Moreover, studies found that hypoxia had increased from the 1950s to 2000 because of increased pollution from surface water that drains from the watershed into the bay.[138] The decline in water quality and loss of marine life has significant water resource, economic, and other implications for the region.

Significant efforts to improve the Chesapeake Bay began in the 1980s. Maryland and Virginia established the Chesapeake Bay Commission in 1980 to cooperatively manage the bay.[139] In 1983, the commission joined with U.S. EPA, Washington, D.C., and the states of Maryland, Pennsylvania, and Virginia to sign the Chesapeake Bay Agreement, which created the Chesapeake Bay Program and outlined goals and strategies for managing the bay. This agreement was further revised in 1987 and 2000 to establish specific objectives for the bay's water quality, regional growth, public engagement, and governance, and it outlined actions to achieve those goals.[140] Today, the program's partners include other neighboring states, federal agencies, academic institutions, non-governmental organizations, and local agencies and utilities. The program's projects include monitoring the bay and its tributaries, assessing resource lands, and reducing pollution in the bay to meet U.S. EPA's pollution targets.[141] Individual states also have taken action to, among other things, develop permitting programs for discharging nutrients into the bay.[142] As a next major step, the program is developing Watershed Implementation Plans

---

[136] Washington State Legislature Revised Code of Washington, Chapter 90.82, "Watershed Planning," undated.

[137] Washington State Department of Ecology, "The Watershed Planning Act," undated.

[138] James D. Hagy, Walter R. Boynton, Carolyn W. Keefe, and Kathryn V. Wood, "Hypoxia in Chesapeake Bay, 1950–2001: Long-term Change in Relation to Nutrient Loading and River Flow," *Estuaries*, Vol. 27, No. 4, August 2004, pp. 634–658.

[139] Chesapeake Bay Commission, "History of the Commission," undated.

[140] Chesapeake Bay Program, "1987 Chesapeake Bay Agreement," undated.

[141] Chesapeake Bay Program, "Programs & Projects," undated.

[142] National Pollutant Discharge Elimination System, "Chesapeake Bay Watershed, Virginia: Watershed-Based General Permit for Nutrient Discharges and Nutrient Trading," undated.

for each of the seven bay watershed jurisdictions that document how water quality standards will be achieved and maintained.[143]

Figure 2.13. Map of the Chesapeake Bay Watershed

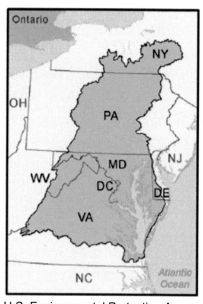

SOURCE: U.S. Environmental Protection Agency, undated.

In Arizona, the San Pedro Watershed is a third key example of the growing importance of watershed management. It is one of the last undammed rivers in the Southwest and is a critical habitat for more than 100 species of birds, 84 species of mammals, and 41 species of reptiles and amphibians. However, development and over-extraction of groundwater has threatened the watershed.[144] A number of organizations are working to restore the watershed's health. The Upper San Pedro Partnership is a coalition of nearly two dozen organizations, including Fort Huachuca, that works to identify, prioritize, and implement comprehensive policies and projects to meet the areas water needs while protecting the San Pedro Riparian National Conservation Area (SPRNC).

Watershed management, sustainability, and ecosystem restoration debates continue around the country and are growing concerns for water managers. As noted earlier, debate continues as to whether San Francisco should drain the Hetch Hetchy reservoir to restore the Hetch Hetchy Valley to its original state. The initiative to restore the valley was put to a vote in local elections in 2012 but was defeated amid arguments that the supply cannot be replaced by other sources and that restoration would be prohibitively expensive.[145]

---

[143] Chesapeake Bay Program, "Watershed Implementation Plans," undated.

[144] Nature Conservancy, "Arizona San Pedro River," last updated January 8, 2013.

[145] John Wildermuth and John Coté, "Hetch Hetchy Measure Swamped by Voters," *San Francisco Chronicle*, November 7, 2012.

A 2012 survey of 35 water utilities found that 53 percent strongly agreed and that an additional 43 percent agreed with the assertion that being "a leading steward of the environment and conservation" was an appropriate role for water utilities. This was second only to the role of "providing safe, affordable, and reliable tap water," to which 96 percent strongly agreed and 4 percent disagreed.[146]

Sustainability

Emerging water resources management also employs a more comprehensive framework—sustainability—for assessing water resources. Sustainability is a growing priority that may be incorporated into planning activities. It explicitly incorporates efforts to restore and maintain aquatic, riparian, and ecosystems' overall health, including water quality. These uses present another competing need of water resources, but they also may help ensure long-term water availability.

Sustainability and sustainable development are terms and concepts that emerged from global development efforts beginning in the late 1980s. Although these terms have varying definitions, the United Nations World Commission on Environment and Development (known as the Brundtland Commission) established the most commonly used one in its 1987 report, *Our Common Future*.[147] It defined sustainable development as "development that meets the needs of the present without compromising the ability of future generations to meet their own needs." Practical definitions usually recognize sustainability as a process that tries to balance and address environmental, social, and economic issues over the long term.

U.S. EPA defines sustainability as follows:

> Sustainability is based on a simple principle: Everything that we need for our survival and well-being depends, either directly or indirectly, on our natural environment. Sustainability creates and maintains the conditions under which humans and nature can exist in productive harmony, that permit fulfilling the social, economic and other requirements of present and future generations. Sustainability is important to making sure that we have and will continue to have, the water, materials, and resources to protect human health and our environment.[148]

Sustainability, in the water context, often focuses on long-term water planning and management for a region that accounts for all the long-term potential needs—human, economic, and environmental needs such as having enough water drinking, irrigation, and industrial uses while ensuring aquatic and riparian ecosystem health. These concepts may be implemented through regional, state, and local water management plans, which were discussed above. Some

---

[146] Janet Clements and Robert Raucher, "Expanding Utility Services Beyond Water Supply to Improve Customer Satisfaction and Utility Effectiveness," presentation at the AWWA 2012 Annual Conference, June 13, 2012.

[147] United Nations World Commission on Environment and Development, *Our Common Future*, 1987.

[148] U.S. Environmental Protection Agency, "What is Sustainability?" undated.

of the plans incorporate sustainability concepts by managing water resources by watersheds, as was just discussed.

Sustainability also can be viewed as a water-management challenge, since water managers still are developing and experimenting with this complex approach. However, we chose to place it here under current management approaches because of the focus on sustainability as an ongoing evolutionary implementation approach rather than as a challenge to be addressed, which is a natural transition to our next topic. Next, we discuss some key challenges that water managers face today.

Water Management Faces Key Challenges

Today water management is facing a number of key challenges, including aging infrastructure, water quality concerns, depleting groundwater aquifers, pressures of population growth, climate change effects on water availability, and public demands for low-cost water. We discuss each of these next.

Aging Water Infrastructure

Our U.S. water infrastructure developed mostly in the 20th century is now aging and needs rehabilitation and replacement. Figure 2.14 shows the proportion of infrastructure for each region over the years, illustrating that the bulk of the U.S. water system was built in the early to mid 20th century.[149]

---

[149] American Water Works Association, "Buried No Longer: Confronting America's Water Infrastructure Challenge," 2012, p. 19.

Figure 2.14. Proportion of the U.S. Drinking Water System Built, by Decade and Region

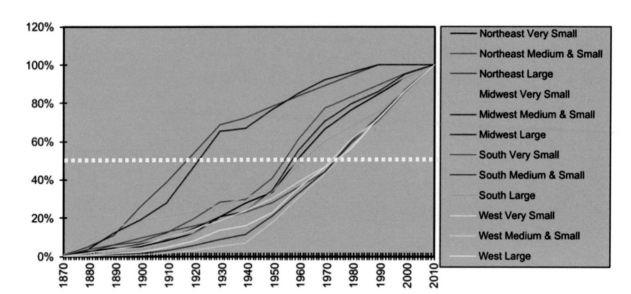

SOURCE: American Water Works Association, "Buried No Longer: Confronting America's Water Infrastructure Challenge," 2012, p. 19.

A 2009 U.S. EPA study found that the nation's nearly 75,000 public water systems will require approximately $335 billion in investment between 2007 and 2027 to install, upgrade, and replace drinking water infrastructure.[150] In a broader and more recent assessment, the American Water Works Association (AWWA) reports that restoring and expanding existing water supplies will cost $1 trillion over the next 25 years. The AWWA coined the term the "Replacement Era" to describe a new phase in water management "in which our nation would need to begin rebuilding the water and wastewater systems bequeathed to us by earlier generations."[151] The report further notes that the needs are not balanced geographically, finding that "the South and West will face the steepest investment challenges, with total needs accounting for considerably more than half the national total." It attributes these needs to these regions' rapidly growing populations.[152] These investments will need to be financed through fees and increase in water rates, which may prove socially and politically challenging given Americans' expectations for historically low water rates.[153]

---

[150] U.S. Environmental Protection Agency, "EPA's 2007 Drinking Water Infrastructure Needs Survey and Assessment Fact Sheet," February 2009.

[151] American Water Works Association, 2012, p. 3.

[152] American Water Works Association, 2012, p. 11.

[153] American Water Works Association, 2012.

Water Quality Concerns

Changes in land use and loss of natural habitat—combined with increasing pollution from industry, transportation, agriculture, and other sectors—have resulted in declining water quality in both surface and groundwater. Concerns about water quality are evident around the country, fueled by changes in infrastructure and use, the Clean Water Act (CWA), urban runoff and pollution from power plants, industry, and agriculture. Poor water quality can increase the cost of treatment or limit the potential uses of the water. We illustrate the range of concerns with several different examples from across the United States.

Historical records of salinity for the Arkansas River suggest that total dissolved solids and sulfate contents of low river flows doubled between 1906 and 1973. This increase in salinity is attributed to the increase in water consumption as well as the accumulation of salts due to increasing diversion of water to irrigation. This has delayed the movement of salts through the river waters.[154]

In the Great Lakes, mercury is a growing concern. The lakes are the largest system of freshwater lakes in the world and contain nine-tenths of the country's fresh surface water. One hundred and forty-four coal-fired power plants around the lakes emit many tons of mercury into the air, much of which finds its way into the Great Lakes through the soil and streams. This contributed to extensive advisories against consuming fish from the lakes, which has implications for public health, local economies, and wildlife that rely on fish for food. Mercury advisories have been growing across the country over the last 20 years.[155]

In New York state, the Hudson River is severely contaminated with polychlorinated biphenyls, or PCBs. They are the byproducts of manufacturing electrical devices and are "probable human carcinogens and are linked to other adverse health effects, such as low birth weight; thyroid disease; and learning, memory, and immune system disorders."[156] Like mercury, PCBs enter into human systems through consumption of contaminated fish, and affect other wildlife. New York state has issued fish advisories since 1973 due to high levels of PCBs in the Hudson River. The river is one of many federal Superfund sites, designated as natural hazardous waste sites that are priorities for cleanup.[157]

Metals and elements such as arsenic, lithium, and manganese are a particular concern in private drinking water wells, which often are untested but provide water to approximately 60 million Americans. These contaminants can result in a host of health problems. Concerns can

---

[154] Donald O. Whittemore, "Water Quality of the Arkansas River in Southwest Kansas: A Report to the Kansas Water Office," Kansas Geological Survey, 2000.

[155] The increase in advisories is partially due to greater monitoring. See Vicki Stamper, Cindy Copeland, and Megan Williams, "Poisoning the Great Lakes: Mercury Emissions from Coal-Fired Power Plants in the Great Lakes Region," National Resources Defense Council, 2012; and U.S. Environmental Protection Agency, "2010 Biennial National Listing of Fish Advisories," EPA-820-F-11-009, September 2011.

[156] U.S. Environmental Protection Agency, "Hudson River Cleanup," undated.

[157] U.S. Environmental Protection Agency, "Hudson River Cleanup," undated.

for almost a year and a half, until late September 2015, when the city issued a warning about the water. This use of contaminated water resulted in Flint residents having significantly elevated levels of lead in their blood, skins rashes, and other health problems. University researchers and a local pediatrician helped provide evidence for the seriousness of the problem. Bottled water and water filters have been distributed by different government agencies and volunteers. Michigan National Guard troops also were mobilized to help distribute thousands of bottles of water and filters. In October 2015, the U.S. EPA established the Flint Safe Drinking Water Task Force to provide technical assistance. On January 16, 2016, President Obama declared a federal emergency to provide federal aid.[160]

This water contamination could have been prevented for about $80–$100 per day by adding chemicals to prevent the corrosion, but this was not done.[161] "Disagreements and miscommunication between state and local officials about what federal law requires of so-called corrosion control measures further delayed fixing the problem."[162] Technical solutions are being implemented to fix the problems. As the investigation about what happened, determination of the health impacts, and other repercussions of this Flint water crisis continue, it illustrates the significance and complexities of water quality concerns.

The growing importance of water quality is also demonstrated by a shift among states away from developing water toward addressing issues of quality. For example, nearly all the elements of the "Law of the River" governing the Colorado River Basin from 1922 to the middle of the century dealt with water development and appropriation. Later augmentations concerned water quality, in particular reducing salinity. These include Minute 242 of the U.S.-Mexico International Boundary and Water Commission of 1973 and the Colorado River Basin Salinity Control Act of 1974.[163] Ensuring acceptable water quality is another factor in extending water supply to meet demand.

Depleting Groundwater Aquifers

Groundwater is a major source of water for drinking as well as for agriculture, mining, and energy extraction. In fact, about half the United States' population (and nearly all of the rural population) relies on groundwater for domestic drinking water use. Groundwater also is critical

---

[160] U.S. Environmental Protection Agency, "Flint Drinking Water Response," last accessed February 13, 2016; Sanburn, 2016; and Abby Goodnough, Monica Davey, and Mitch Smith, "When the Water Turned Brown," *New York Times*, January 23, 2016.

[161] Sanburn, 2016.

[162] Goodnough, Davey, and Smith, 2016.

[163] U.S. Bureau of Reclamation, "The Law of the River," March 2008.

to agriculture, providing more than 50 billion gallons per day, or 60 percent of the water used for irrigation nationwide.[164]

Groundwater is of particular concern as withdrawal rates often have exceeded recharge rates in many areas resulting not only in groundwater depletion but also infiltration of salts and minerals, increased pumping and treatment costs, land subsidence, sinkholes, saltwater intrusion near the coast, and less available surface water in rivers and lakes.

Land subsidence takes place when the land-surface elevation lowers because of changes occurring underground. A major cause is the pumping of water, though oil and gas pumping can also contribute to land subsidence. For example, California has a long history going back to the early 20th century of land-subsidence problems from extracting groundwater. In the San Joaquin Valley near the City of Mendota, the land subsided nearly 30 feet between 1925 and 1977 because of "intensive groundwater pumping." Other parts in this valley, such as farms near Tulare and Wasco, sank more than 12 feet, and areas near Kettleman City sank more than 25 feet. During the 1987–1893 drought, increased pumping demands on the aquifer reduced the water table in the Central Valley as much as 100 feet, causing portions of the San Joaquin Valley to sink by 8 additional feet. In fact, half of all U.S. land subsidence has occurred in California, mostly from groundwater pumping.[165] Land subsidence is expensive:

> Throughout California subsidence has damaged buildings, aqueducts, well casings, bridges and highways. As the land shifts, flooding problems are aggravated. Subtle changes in land gradient can adversely impact sewer lines and storm drainage. In all, subsidence has resulted in millions of dollars of damage.[166]

Florida has significant problems with sinkholes[167] because of the over-pumping of groundwater. For example, in 2010 farmers in Florida who were "trying to protect their crops during a freeze pumped so much water that the aquifer dropped 60 feet in just days" and caused 140 sinkholes to open in the region.[168]

Florida and California, as well as many other populated coastal states, have ongoing problems with saltwater intrusion[169] from groundwater pumping. In Southern California,

---

[164] U.S. Geological Survey, "Groundwater Depletion," USGS Water Science School, undated; Bridget Scanlon et al., "Groundwater Depletion and Sustainability of Irrigation in the U.S. High Plains and Central Valley," Proceedings of the National Academy of Sciences, May 29, 2012.

[165] Water Education Foundation, 2011, pp. 12–13.

[166] Water Education Foundation, 2011, p. 13.

[167] A sinkhole is when a hole in the ground occurs because of some form of collapse of the surface layer.

[168] Craig Pittman, "Florida's Aquifer Models Full of Holes, Allowing More Water Permits and Pollution," *Tampa Bay Times*, January 27, 2013.

[169] Saltwater intrusion is when ocean water or other saltwater moves into fresh groundwater, causing the groundwater to be contaminated by salt. Most coastal aquifers experience some naturally occurring seawater intrusion because seawater is denser than fresh water. However, human activities, especially intense groundwater

seawater intrusion began as early as the 1920s in Ventura, Los Angeles, and Orange counties. In fact, problems in California are not just along that region's coast:

> The Salinas and Pajaro river valleys of Central California's coast have struggled with a serious saltwater intrusion problem. Fifty years of overdraft in the Salinas Valley have allowed the seawater to move 5 miles inland, contaminating coastal groundwater aquifers and threatening the water supply for a $3 billion agricultural economy and the public water source for communities around Salinas.[170]

New Jersey is another state where coastal areas face saltwater intrusion problems from groundwater extraction. For instance, in Cape May County between 1960 and 1990, groundwater withdrawals caused the aquifer chloride concentration to increase from less than 50 mg/L to more than 200 mg/L in Atlantic City and from less than 50 mg/L to more than 500 mg/L at Cape May Point, Cape May City and the Wildwood communities.[171]

Addressing such saltwater intrusion problems is expensive. For instance, in some parts of Florida, "the aquifer—which most Floridians rely on for drinking water—has dropped by 60 feet. Some coastal communities are now getting saltwater in their wells, which costs millions to treat."[172]

In many parts of the country, well owners have to drill water wells deeper and deeper as aquifers decline from so much use in addition to increasing demand, competing water uses and drought. During drought, water users start using more groundwater as surface water sources run dry. For example, EPWU provides water to the city and county of El Paso, Texas, from surface waters taken from the Rio Grande River system and groundwater from the Hueco Bolson and Mesilla Bolson aquifers. EPWU's goal is to use 50 percent surface water but because of an ongoing drought for a couple years it had to use more groundwater. This groundwater use lowers the aquifer at a faster rate. In some parts of the country, wells are even starting to run dry. This occurred in January 2012 in Spicewood Beach, Texas, a community of about 1,100 about 30 miles northwest of Austin. As a result, the Lower Colorado River Authority (LCRA) had to truck in water via tanker tanks that was taken from fire hydrants about 10 miles away.[173]

The water levels are dropping in aquifers around the country as a result of increased use of groundwater for agricultural irrigation and energy needs, facilitated by the development of more powerful, high throughput pumps. In addition, lower precipitation in some areas has contributed

---

pumping, can make it much worse. See Ted Johnson, "Battling Seawater Intrusion in the Central and West Coast Basins," Water Replenishment District of Southern California Technical Bulletin, Vol. 13, Fall 2007.

[170] Water Education Foundation, 2011, p.14.

[171] Pierre Lacombe and Glen Carleton, "Hydrogeologic Framework, Availability of Water Supplies, and Saltwater Intrusion, Cape May County, New Jersey," U.S. Geological Survey, Water-Resources Investigations Report 01-4246, 2002.

[172] Greg Allen, "Now Endangered, Florida's Silver Springs Once Lured Tourists," NPR, April 13, 2013.

[173] Terrence Henry, "When Wells Run Dry: Spicewood Beach, Texas is Out of Water," National Public Radio, January 31, 2012.

to persistent droughts that limit the availability of surface water and increase the reliance on groundwater sources. Groundwater depletion is difficult to assess, because it cannot be measured directly and because the hydrology of aquifers can be complex. However, one USGS estimate suggests the United States has lost the equivalent of two Lake Eries' worth of groundwater, with the greatest declines occurring in the Southern Great Plains (Ogallala), the Mississippi River Delta (Mississippi River alluvial), and the California's Central Valley (Central Valley). Two aquifers, the Ogallala and the Central Valley, contribute approximately half of this decline. One study has estimated that the southern Ogallala, which provides water to parts of Kansas and Texas, could run out of water by 2045. The study found that about 4 percent of the land area above the aquifer (in parts of Kansas and Texas) is responsible for about one-third of its water losses.[174] Figure 2.17 shows where aquifer depletion is occurring throughout the country for major aquifers.

Figure 2.17. Cumulative Groundwater Depletion in the United States for Select Aquifers (1900–2008)

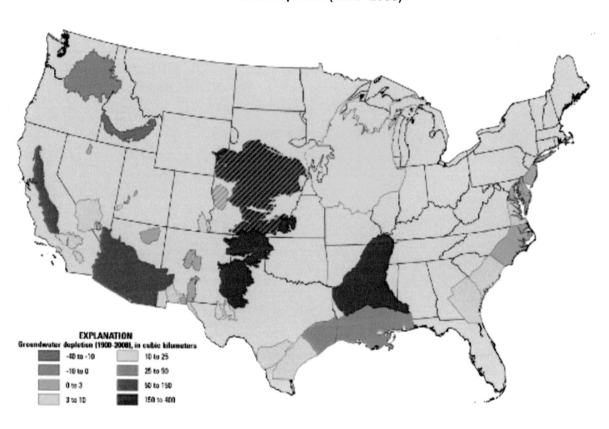

SOURCE: L. F. Konikow, "Groundwater Depletion in the United States (1900–2008)," U.S. Geological Survey Scientific Investigations Report 2013–5079, 2013.

---

[174] Waterwired, "Groundwater Depletion in the United States 1900–2008"; Scanlon et al., 2012; Pete Spotts, "Southern Great Plains Could Run Out of Groundwater in 30 Years, Study Finds," *Christian Science Monitor*, May 30, 2012.

Groundwater depletion problems also have affected some military installations. At California's Fort Irwin, land subsidence of up to 12 feet has occurred in some areas because of groundwater withdrawals, and cracks in the airfield, located on a dry lake bed, have begun to affect training operations.[175] Similarly, land sinking and fracturing problems from groundwater extractions have occurred at nearby Edwards Air Force Base, where a 12-foot deep, 4-foot wide, 2,000-foot-long fissure took a runway out of service. Despite Air Force and local government efforts to address groundwater overdraft issues, continued population growth in the region has prompted concerns that the base will lack sufficient water by the year 2035.[176] Fort Huachuca water managers have put a priority on conserving water and helping to recharge the local aquifer because of the impacts that potentially declining groundwater could have on the nearby Upper San Pedro River basin, which is discussed further in Chapter Six.

Regional, state and local government water managers respond to these groundwater challenges in a range of ways. They include expanded water conservation, groundwater recharging projects, and restricting groundwater use. Although we already have discussed water conservation activities, we briefly explain and illustrate the other two responses. In addition, another management approach to address groundwater depletion problems, groundwater banking, such as the Central Kansas Water Bank Association, is explained in Chapter Four.

Groundwater recharge is the process of replenishing groundwater with waters from the surface. It is being conducted all across the United States not only to help ensure enough groundwater supplies, but also to help prevent problems of groundwater contamination, land subsidence, sinkholes, and saltwater intrusion. As discussed above, better stormwater management and reclaimed water are used to help replenish groundwater aquifers. We already have provided examples of reclaimed water being used for groundwater recharge along the California coast, in Florida and in Texas. In Chapter Six, we discuss groundwater recharge efforts at Fort Huachuca in Arizona. Here we elaborate on the massive groundwater recharge efforts in Southern California.

In the 1950s and 1960s along the Southern California coast, water managers started recharging groundwater with water from the Colorado River and from Northern California. The recharge water from injection wells helps to create a hydraulic barrier blocking the fresh water in the aquifer from incoming seawater. The water managers created "a pressure ridge" or "wall" with a line of injection wells in three separate projects called the West Coast Basin Barrier Project, the Dominguez Gap Barrier Project, and the Alamitos Gap Barrier Project. By 2007 these three projects consisted of 290 injection wells and more than twice as many observation wells. By 2008, the injection wells were estimated to inject over 30,000 af of water. In

[175] Elisabeth M. Jenicek, Rebecca A. Carroll, Laura E. Curvey, MeLena S. Hessel, Ryan M. Holmes, and Elizabeth Pearson, "Water Sustainability Assessment for Ten Army Installations," U.S. Army Corps of Engineers, Engineer Research and Development Center, TR-11-5, March 2011.

[176] Water Education Foundation, 2011, p.13.

2007/2008, the cost of the injection well water was about $14 million and maintenance costs were between $4 million and $5 million.[177] By 1995, some recycled water was being injected, and by 2011 the recharge barrier water was 100 percent recycled water.[178]

Some state and local governments also have taken action to restrict groundwater use. In 1980, for example, Arizona passed the Groundwater Management Act[179] to help control severe overdraft and to combat rapid depletion of groundwater quantity and quality through groundwater permitting, developing and implementing water management plans, controls on new agricultural irrigation, and water conservation activities.[180] However, groundwater laws in different states are quite diverse (discussed later in this chapter), providing some water managers with more leverage to manage groundwater than others. Texas groundwater is governed by the rule of capture. This grants landowners the right to capture the water beneath their property, which essentially allows unlimited groundwater pumping. Because no judicially created "reasonable use" restriction exists, water managers in Texas have a harder time than other states in trying to control or restrict groundwater use.

As aquifers are overdrawn, surface water supplies become more unreliable and different interests compete over the limited water supplies, resulting in greater numbers of groundwater-depletion problems ending up in court. For example, in a long-standing dispute between Kansas and Colorado over the Arkansas River, Kansas filed suit in 1985 claiming that Colorado wells had overdrawn groundwater that feeds the river, causing flows to decrease in violation of the Arkansas River Compact. In 2013, Texas entered a suit against New Mexico claiming its wells were overdrawing from aquifers that feed the Rio Grande River violating the states' Rio Grande Compact (to which Colorado is also a party).

Dealing with aquifer depletion is, and will continue to be, a challenge for several reasons, including:

- Increasing demands for water and aquifer use
- The complexities in understanding the hydrology and science of groundwater systems
- Different groundwater laws
- Competing water interests
- Multiple landowners with water rights.

Water managers across the country are "straddling the divide between private ownership of land, and by extension the water under it, and a growing need to manage what often turns out to be a shared resource."[181]

---

[177] Tad Johnson, "Battling Seawater Intrusion in the Central and West Coast Basins," Water Replenishment District of Southern California, Technical Bulletin, Vol. 13, Fall 2007.

[178] Water Education Foundation, 2011, p. 14.

[179] This act is explained in more detail in Chapter Six.

[180] Arizona Department of Water Resources, "Overview of the Arizona Groundwater Management Code," undated.

[181] Spotts, 2012.

Water Supply Stressors: Population Growth and Climate Change Effects

All of these challenges are exacerbated by a still-growing population, particularly in the West, and by increasingly tight water resources due to competition and overuse. One study suggests that the United States' population may increase from 296 million in 2005 to 438 million in 2050.[182] As shown in Figure 2.18, much of this growth is expected to be in the West (percent labels), which also currently has higher levels of per-capita water use than other parts of the nation (color coding). These patterns and trends may further strain already scarce resources and require potentially high cost solutions to meet growing demands with limited supply.

Figure 2.18. Current Per Capita Domestic Daily Water Use and Projected Population Change by 2030

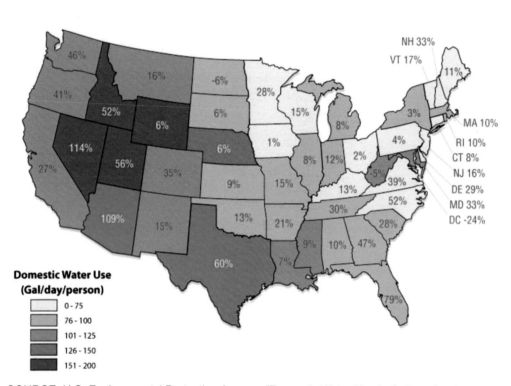

SOURCE: U.S. Environmental Protection Agency, "Domestic Water Use in Gallons Per Day Per Person and Projected Percent Population Change by 2030," undated.

These water challenges driven by population growth may be further exacerbated by the change and uncertainty resulting from global climate change, which could further strain water resources and infrastructure. Changing temperatures, precipitation, wind patterns, and other global processes will affect both water supply and demand. Figure 2.19 summarizes the effect higher temperatures may have on the water cycle for the interior West and the coastal areas and the East. It suggests that while temperatures may increase across the entire nation, it will affect

---

[182] Jeffrey Passel and D'Vera Cohn, "U.S. Population Projections: 2005–2050," Pew Research Center, February 11, 2008.

areas of the country differently. Overall precipitation may decrease in the West, while it may increase in much of the East. In the West a decrease in precipitation, and proportionately less snowfall relative to rainfall, will change the patterns of reservoir refill, meaning that less water will be available in the summer months, when demand is greatest. Higher temperatures also will result in increased demand for water for domestic use, agricultural use, energy production and other uses. In contrast, the eastern United States may experience increased precipitation in the form of more frequent and more intense rain events. Runoff from these rain events, flooding, or sea level rise likely will reduce water quality as runoff carries increasing amounts of pollutants into watersheds, overloads stormwater management systems, and damages infrastructure. These changes will have cascading and feedback effects for many sectors, such as on electricity supply, if hydropower output changes.

Figure 2.19. Predicted Effects of Climate Change on the Water Cycle in Regions of the United States

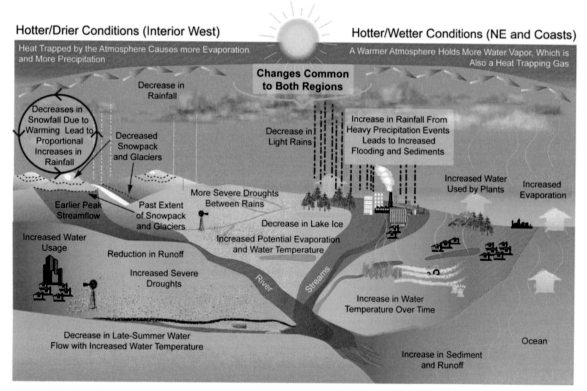

SOURCE: Karl, T.R. J.M. Melillo, and T.C. Peterson (eds.), "Global Climate Change Impacts in the United States," New York: United States Global Change Research Program, Cambridge University Press, 2009.

These shifting patterns will challenge water managers to simultaneously meet the needs of growing communities, sensitive ecosystems, and agricultural and industrial users under greater variability in water supply. These changes are daunting and deeply uncertain, compounding the challenges of long-term water management. For example, studies of the stresses on the Colorado River Basin exemplify the complex decision-making with regard to climate change uncertainties.

Simulations indicate that while the entire basin will become warmer, the Upper Basin (where the River derives most of its source water) could experience precipitation changes of anywhere from a decline of 15 percent to an increase of 11 percent over the next 50 years—although overall water shortages are expected. In order to explore robust strategies for responding to the changes brought on by climate change, an analysis of thousands of individual supply and demand scenarios, several reservoir operations scenarios, and several possible response portfolios[183] was performed to evaluate the ability of each one to meet key performance measures for meeting water needs. Given the level of uncertainty and varying preferences, no singular approach is appropriate. But the simulations and analysis found that several water management responses are effective under most circumstances, and therefore are high-priority options—those that result in municipal and industrial conservation, agricultural conservation and transfers, and desalination of ocean water, agricultural drain-water, or groundwater. The urgency of implementing each of these responses depends on the specific scenario. As the precise effects of climate change become apparent, additional refinement of these response portfolios is advised. While this is just one approach among many that water managers may use, it illustrates the complexity water managers face when responding to climate change.[184] Water managers across the country are starting to plan for the uncertainty and potential strains that future population growth and climate change effects may place on water resources, but the effects are highly uncertain.[185]

Water Pricing Challenges

As we have discussed, water is an essential commodity, which requires large investments in infrastructure to convey and treat it. Yet another concern facing water managers is that the public is accustomed to paying little for it. As U.S. EPA noted: "To the extent that water is underpriced, it will be used in quantities that exceed the economically efficient amount, and in applications in which its marginal value is less than its true opportunity cost."[186] In other words, underpriced water will be put to uses where the returns are not as great as the value of the water. Since water is essential to life and is a shared resource, most, if not all, states treat water differently from other commodities.[187]

---

[183] Four portfolios were developed that emphasized different preferences, such as technical feasibility and long term reliability, or environmental goals such as lower energy intensity and in-stream flows and excluded risky legal or policy changes.

[184] This is an extremely simplified view of the study. For more detail see David G. Groves, Jordan R. Fischbach, Evan Bloom, Debra Knopman, Ryan Keefe, *Adapting to a Changing Colorado River: Making Future Water Deliveries More Reliable Through Robust Management Strategies*, Santa Monica, Calif.: RAND Corporation, 2013.

[185] U.S. Environmental Protection Agency, "Climate Impacts on Water Resources," undated.

[186] U.S. Environmental Protection Agency, "The Importance of Water to the U.S. Economy Part 1: Background Report," September 2012.

[187] Water differs from standard commodities in other ways as well, and many parameters affect its value (quantity, quality, supply reliability, timing, location, etc.).

Water utilities offer one example of how water pricing contributes to some of the key challenges facing water managers and the complexities they encounter in trying to address them. Ostensibly, water utilities base their rates on their cost to provide the water. The issues of equity, affordability (by the poor), monopoly power, and access also must be considered. Public institutions, such as city councils, local water boards, or state regulatory agencies typically regulate water prices that ensure the price is reasonable and based on the cost to provide the water. It is a decentralized and frequently highly politicized process. Utilities often will attribute costs to specific customer classes such as residential, commercial, and industrial, to ensure that the resulting rates are equitable and to reduce cross-subsidization.[188]

Water utility rates traditionally have failed to adequately incorporate the full cost of providing water, such as the cost of obtaining water supplies in the longer term given projected usage, as well as environmental and other non-market valued consequences and long term infrastructure replacement. Additionally, the federal government has subsidized much water infrastructure (especially in the West), and this subsidy is not reflected in water prices. These views and practices make it difficult to implement water rates that incorporate the costs of replacing aging infrastructure, or that respond to short- or long-term water scarcity.[189]

Water utilities have several options for developing prices, or rate structures, for water usage. To cover costs fully and achieve any secondary objectives (such as conservation), the specific structure selected must take into account such features as the area's particular climate, population density, and likely demand. Rate structures may include variable components (those that rise with water usage) or fixed components (fees that are charged regardless of water usage) or a combination of the two. Common water utility rate structures are flat rates (a fixed charge for supplying water, regardless of usage); uniform rates (where charges are based on usage but the rate stays the same for all levels of use); tiered, or block rates (where charges are based on usage

---

[188] Leurig, 2012; Sheila M. Olmstead, and Robert N. Stavins, "Comparing Price and Non-price Approaches to Urban Water Conservation," Working Paper 14147, National Bureau Of Economic Research, Cambridge, Mass., June 2008; Sheila M. Olmstead, "Managing Water Demand: Price vs. Non-Price Conservation Programs," Pioneer Institute, July 2007; Janice A. Beecher, "Primer on Water Pricing," Michigan State University, Institute of Public Utilities Regulatory Research and Education, November 1, 2011; Brett Walton, "The Price of Water 2012: 18 Percent Rise Since 2010, 7 Percent over Last Year in 30 Major U.S. Cities," *Circle of Blue*, May 10, 2012; Bill Zieburtz and Rick Giardina (Eds.), *Principles of Water Rates, Fees and Charges* (Sixth Edition), Denver, Colo.: American Water Works Association, 2012.

[189] Olmstead and Stavin, 2008, Olmstead, 2007, Beecher, 2011, Walton, 2012. We should also note that the term "water scarcity" may have different meanings in different contexts since it does not have a standard definition; nor is there a consensus on exactly how it should be measured. In this report we frequently refer to "water scarcity" in the broad physical sense, where human and environmental demands for water exceed the available water for a given region and time period. Water scarcity may lead to declining aquifer levels, environmental degradation, and unequal water usage. As used in this report, water scarcity is a relative concept since it is determined by water availability and water use. Both of these are influenced by a large number of factors that are both nature- and human-driven. Several contributors to water scarcity include, but are not limited to population growth, land use and development, water quality, water resource management, and climate change.

and the rate varies in steps with the amount used); and seasonal rates (where the rate varies by time of year).[190]

Recall that the water rates that utilities charge are expected to recover costs. Water and wastewater utilities are capital-intensive, even compared to other utilities, so they have a high proportion of fixed costs that don't vary with water use. Fixed costs for utilities include water supply infrastructure, treatment facilities, and transmission and distribution infrastructure. The variable components of utilities' costs include labor, supplies, energy, chemicals, and any water that may be purchased. The ratio of fixed to variable costs for utilities has been noted at about a 4-5:1 ratio. However, in contrast to *cost* structures, *rate* structures frequently are the opposite, and have a small fixed portion (around 20 percent of revenues derived from fixed fees) and larger variable portion (around 80 percent of total revenues).

This dissonance between cost and rate structures can cause problems. Both Newport News, Virginia, and Austin Water in Texas, for example, generate around 90 percent of their revenue from variable rates. Austin Water's revenues dropped by 13 percent in 2010 when an unusually high level of precipitation led to a drop in water usage. In the Great Lakes region, the loss of industrial customers created excess capacity and a loss of revenues. Rate structures with fixed charges are advantageous in that they improve revenue stability for utilities since revenues are less dependent on water usage. The disadvantages, however, are that rates are regressive (they impose a greater burden on low-income households) and no price signal is available to customers regarding their water usage. In contrast, rate structures with a larger variable portion will send stronger signals to customers regarding water usage, but they also reduce revenue stability for the utility. Demand, or usage, may drop for any number of reasons, such as economic recession, declining population, conservation incentives and technologies, or drought. As a result, the ability of utilities to recover all of their costs through rate structures with a large variable proportion is diminished, because as usage drops, total revenues fall by more than the variable costs of supplying water. This adds levels of instability, uncertainty, and risk that costs will not be recovered.[191] For the case of utilities that may have undervalued water to begin with, the challenge is even greater.

In sum, establishing rate structures that generate revenues sufficient to cover costs in an equitable and financially responsible manner is no small task. It involves balancing multiple objectives, which may include cost recovery, promoting conservation, and encouraging industrial development. It also involves rigorous analysis of customer demand, historical and future supply needs in light of existing infrastructure, demographics, political sensitivities, and climate.[192]

---

[190] Equinox Center, "Water Pricing Primer," October 2009.

[191] Utilities that have historically underpriced their water either because they didn't have good information on real costs, have deferred investments, or have relied on tax subsidies. See Beecher, 2011.

[192] Beecher, 2011; P. Brandt and B. Ramaley, "Balancing Fixed Costs and Revenues: Newport News Waterworks' Shifting Revenues Recovery to Better Match Fixed vs. Variable Costs," American Water Works Association Conference Proceedings, June 13, 2012; and Walton, 2012.

In addition to revenue uncertainties, utilities' expenses are rising. As mentioned, droughts, the economic downturn, and stagnant growth in the customer base in recent years all have led to revenue shortfalls within the water utility industry. Concerns over water security have also risen with recognition that water supplies have been over-allocated (such as in the Colorado River Basin by 20 percent), litigation over interstate allocations persist (such as Apalachicola-Chattahoochee-Flint River Basin tensions between downstream users and the city of Atlanta), and groundwater aquifers are at risk of depletion or contamination (such as with Central California, High Plains and coastal aquifers in Massachusetts and Florida). As a result, ratings of water utility bonds, once considered a safe investment, have begun to drop. For example, Colorado Springs Utilities, which supplies water to Fort Carson, may face a possible downgrade by Moody's given the slow economic recovery and its expensive (nearly $1.5 billion) infrastructure investment to build the Southern Delivery System, which will pipe water from the Arkansas River to the city. Downgrades occur most often when rates have not been increased enough to cover maintenance costs or debt servicing as in the case of Altoona, Pennsylvania, and Bowling Green, Kentucky, where years of deferred maintenance have led to massive water loss rates and subsequent downgrades to their credit valuation. The consequences of perceived risk for revenue shortfall, as well as concerns over water security by investors, are that utilities have higher expenses since the cost of borrowing to invest in infrastructure is greater. Water rates are likely to rise (and meet political resistance) as utilities struggle with increased cost of funds, aging infrastructure that will require new investment, low federal funding, and reduced demand.[193]

As discussed above, utility rate structures may not raise enough revenue to cover long-term costs, such as replacing aging infrastructure and replenishing water that is being depleted in aquifers and elsewhere. As these pressures continue, rates eventually may rise. In many cases, Army installations purchase water from utilities. As utilities address these challenges and if rates increase sufficiently, more water-related investments on Army installations eventually will become viable on financial merits alone. Water managers also must recognize that since water prices do not always reflect the true costs of providing this resource compelling reasons may exist (such as potential scarcity or poor supply reliability, critical infrastructure failures, or environmental needs) that would warrant investment nevertheless. Also, any internal pricing schemes would be more economically efficient should they incorporate the full cost of providing water to more completely reflect the opportunity cost of its use.[194]

---

[193] Leurig, 2012.

[194] There are other reasons why current pricing schemes are not efficient in an economic sense having to do with average costs versus marginal costs, and the value of water in different uses. A full discussion of how economic theory treats water is beyond the scope of this report. Interested readers are referred to W.M. Hanemann, "The Economic Conception of Water," in *Water Crisis: Myth or Reality?* edited by P.P. Rogers, M.R. Llamas, L. Martinez-Cortina, London: Taylor & Francis, 2005; and U.S. Environmental Protection Agency, "The Importance of Water to the U.S. Economy Part 1: Background Report," September 2012.

U.S. Water Rights

In addition to knowing water management trends, it also is important to understand water rights. A water owner cannot participate in a water market, such as a water auction, unless he or she understands his or her legal rights to the water, which can be a complex process. This section provides an overview of U.S. water rights, including military installations' water rights.

Water is treated as both a public and private good, causing its access and allocation to be determined by a complex web of laws aimed at protecting the rights of individual water users, while at the same time ensuring maintenance of an adequate supply to serve public needs. Under most circumstances, determining water rights is the province of states, but in particular locales local laws may determine water rights as well. In addition, when deemed necessary, federal authorities can preempt state and local laws to bring water resources under federal control. Maintaining water's status as both a public and private good allows government's greater powers to intervene to protect public welfare, especially if it believes the activities of private water accessors are harming the public good. However, the power to intervene comes at a price: Complex water rights regimes, public oversight, and government regulation distort water markets from the full effects of supply and demand, often reducing the efficiency of water allocation and pricing.

Another factor complicating water rights is that there are two main types of different water resources: surface water and groundwater. Surface water, which is found in lakes, streams, and rives, is renewed by rain and snowmelt and also can be fed by groundwater. In contrast, groundwater, which is found in underground aquifers that collect water percolating down from the surface, is considered an exhaustible resource. Even though water seeps into aquifers, once an aquifer is tapped, the rate of usage typically outpaces the rate of replenishment. Hence groundwater, on the whole, is more susceptible to depletion and exhaustion than surface water, although droughts or excessive use can exhaust both types of sources.

The following section details the types of water rights regimes found in the United States for surface water and groundwater. It is followed by a discussion of military installations' water rights.

Surface Water Rights

Surface water rights differ by state, with a contrast between states to the east and those slightly west of the Mississippi River. In the eastern United States, riparian rights—meaning rights that are based upon land ownership—determine the right to surface water use. In the western United States, water access is based on prior appropriation rights, which allocate rights based on the place in history when a user first put the water to a continuous beneficial use. The divide is a historical artifact from conditions found in both parts of the country during their periods of settlement. In addition, as the map in Figure 2.20 shows, a few states with hybrid systems apply both legal approaches, depending on the circumstances.[195] We next explain these different approaches.

Figure 2.20. Surface Water Right Doctrines, by State

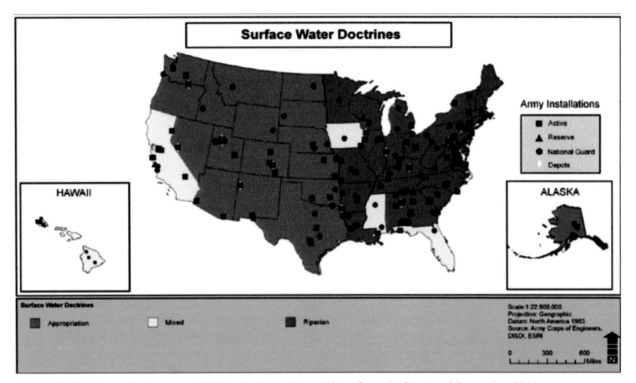

SOURCE: U.S. Army Environmental Policy Institute, "Army Water Security Strategy," December 2011.

Riparian Rights in the East

According to riparian doctrine, if water is on or flows through a parcel of land, the property owner can appropriate it as long as it is used for a "reasonable" purpose (a concept clarified below). Because rivers and streams provided an abundant water supply during the settlement of the eastern United States and government involvement in resource allocation was something that settlers wanted to avoid, the designation of water rights became a natural extension of the establishment of property rights. Due to water's abundance, the amount or type of water usage on one parcel of property did not tend to impact those with access rights upstream or downstream. The only usage rules were that each riparian owner had the right to access water as long as they did not significantly change the rate of flow or quality of the water.[196] Water use was not a requirement for maintaining the right; the right was tied to the property, rather than the

---

[195] U.S. Army Environmental Policy Institute, "Army Water Security Strategy," December 2011, p. A-25.

[196] Southwestern Water Conservation District, "The Water Information Program: Providing Water Information to the Communities of Southwest Colorado," undated.

usage.[197] Over time, as human actions were able to drastically alter the quality and variability of the water supply, riparian rules were modified. "Reasonable use" was redefined as not causing unreasonable harm to other riparian users, regardless of the flow.[198]

Another change to the riparian system evolved as populations and water demands increased. Tying access rights strictly to those users with adjacent properties severely restricted water allocation and could not meet the needs of the expanding population. Additionally, environmental and ecological concerns began to support the need for a new water management regime. As a result, most riparian states began to establish greater control over water allocation and established permit systems, a system known as "regulated riparianism." Under regulated riparianism, permit holders are subject to the same "reasonable use" standards as those whose water rights are based on land ownership.[199] Moreover, as part of the official management of the water resource, many states demand that permit applicants—and even accessors with riparian rights based on land ownership—file water usage plans for the state to review, regardless of whether their usage falls under the accepted notions of "reasonable use." Regulated riparian permits also place time limits on water access for permit holders and typically allow the state to demand usage reductions in times of drought or if it otherwise is deemed in the public interest.[200]

Prior Appropriation Rights in the West

Based on the "first in time, first in right" principle, the prior appropriation doctrine allows those who first appropriate a water source to use a specified amount for a beneficial use wherever they choose.[201] The doctrine of prior appropriation evolved during the 19th century with the growth of mining and agricultural opportunities in the West that often were located far from water sources that were considered sufficient for supporting these enterprises and the many settlers seeking to exploit them.[202] Originally, "beneficial use" referred to diverting water for industrial and agricultural purposes, but as societal needs changed the definition expanded to later include other purposes, including power generation, ecological preservation, and municipal use.

Under prior appropriation, water can be diverted far from the source. The first entity to access water holds in perpetuity the right to appropriate the same amount of water originally accessed, for the same "beneficial purpose." Other claims can be made on the remaining water as

---

[197] National Agricultural Law Center, "Water Law: An Overview," Fayetteville, Ark.: University of Arkansas School of Law, undated.

[198] Southwestern Water Conservation District, "The Water Information Program: Providing Water Information to the Communities of Southwest Colorado," undated.

[199] Southwestern Water Conservation District, undated.

[200] National Agricultural Law Center, undated.

[201] Gary D. Libecap, "Water Woes: Using Markets to Quench the Thirst of the American West," *Milken Institute Review*, Fourth Quarter, 2010, p. 60.

[202] Libecap, 2010, p. 60.

long as they do not infringe upon the senior claimants' appropriation. Since senior rights always take priority over junior rights, the reduced risk makes them more valuable.[203]

Rights of appropriation can be leased and sold like property. In fact, during times of drought, senior rights holders can (and some often do) lease water to junior parties. With prior appropriation, claimants can lose their rights if they alter the place of use or point of diversion, if the usage falls outside the standards of beneficial use, or if they stop appropriating their allotment of water for a specified period. Forfeited rights either revert to the state or can fall to other claimants.[204]

Historically, an unfortunate consequence of the "use it or lose it" rule was that it discouraged conservation and promoted inefficient water utilization. One might think that the "beneficial use" rule would have dissuaded accessors from wasting water, but the scope of the rule is limited: "Beneficial use" applies only to the purpose of the usage, not the manner in which it is used. As a result, accessors often went to great lengths to exhaust their allotment. For example, to avoid forfeiture, farmers would commit portions of their property to water-intensive crops such as alfalfa that would soak up any excess allotment.[205]

As water resources were depleted, states sought to protect the public good by rectifying some of the perverse incentives introduced by the "use it or lose it" rule. Most states now designate restoring in-stream flows for ecological or conservation purposes as a "beneficial use."[206] They introduced water boards or other oversight agencies and established permit systems to attempt to direct allocation rights to uses that could best serve the public good. Some localities have encouraged more responsible water use by issuing water credits that can be banked or sold, which in turn facilitates the migration of water rights to those with the greatest need.[207]

For prior appropriation states, the development of water markets also may promote more efficient water use. As the supply dwindles and prices rise, it creates pressure for using water more efficiently so that the excess can be freed up for sale. This "creates more water" to be used for new uses, thereby also improving the ability to distribute water to where it is most "valued." But transforming water into a tradable commodity has its drawbacks. The most significant is that water will go to the highest bidder, not necessarily to where it will best serve the public interest. Regulation and oversight can alleviate some of the negative consequences of water markets and keep water from becoming a purely private good.[208]

---

[203] Libecap, 2010.

[204] Libecap, 2010, p. 60.

[205] Libecap, 2010, p. 62.

[206] Janet C. Neuman, "Have We Got a Deal for You: Can the East Borrow From the Western Water Marketing Experience?" *Georgia State University Law Review*, Winter 2004, p. 453.

[207] American Water Works Association, "White Paper: Water Rights and Allocations for Sound Resource Management," June 28, 1995.

[208] Neuman, 2004.

Hybrid Systems

Some states have hybrid surface water law systems that preserve riparian rights while recognizing prior appropriation rights. Each of the states with hybrid or "mixed" systems—California, Hawaii, Mississippi, and Florida—has adopted different aspects of riparian and prior appropriation systems. The combination of both systems adds greater complexity to determining which claimant has priority over the other. To reduce the number of legal challenges, states have adopted additional statutes to clarify the order of usage priority. For instance, in California, riparian owners have the greatest priority (with the greatest allotment of water going to the landowner with largest amount of waterfront property), followed by senior appropriators, and finally junior appropriators. To reduce water waste, California has extended the usage requirements associated with prior appropriation to riparian users as well, thus restricting both types of users to "reasonable and beneficial" use restrictions.[209]

Groundwater Rights

The laws governing groundwater rights use are very complex, in part because they developed when the relationship between groundwater and surface water was not well understood and because early water law systems assumed that groundwater was immeasurable and its movements unknowable.[210] We now know that groundwater interacts with surface water in a hydraulic system: Surface water (including precipitation) percolates below the water table where it flows through the groundwater system and discharges into streams or wetlands.[211] Hydrogeology is the field of study that measures groundwater supplies. Yet, proper groundwater measurements are still complicated enough to prohibit adequate monitoring and enforcement of usage regulations, and the burden typically falls on those who are harmed to litigate usage disputes. The result is that it usually is costly to define and enforce groundwater rights.

Adding to the complexity is the patchwork of legal approaches used to determine groundwater rights, making them even less well defined than surface rights.[212] The United States currently has 50 different types of groundwater doctrines, with aspects determined by local, state, and federal laws and authorities. Most, however, can be grouped in four categories:

1. **Absolute Ownership.** Owners of property sitting above a groundwater supply can pump as much as they want, for whatever purpose, regardless of the impact on adjacent property owners. Since pumping is unrestricted, it can result in a significant depletion of neighbors' groundwater supplies, resulting in disputes that are typically settled by

---

[209] California Environmental Protection Agency, State Water Resources Control Board, "The Water Rights Process," undated.

[210] Charles A. Job, *Groundwater Economics*, Boca Raton, Fla.: CRC Press, 2010, p. 142.

[211] Thomas Winter, Judson Harvey, O. Lehn Franke, and William Alley, "Ground Water and Surface Water: A Single Resource," U.S. Geological Survey Circular 1139, 1998, pp. 2–3.

[212] U.S. Army Environmental Policy Institute, "Army Water Security Strategy," December 2011, p. A-26; Job, 2010.

litigation. Due to these negative effects, most states have adopted more restrictive groundwater rights regimes, and today only Indiana and Texas continue to apply the absolute ownership doctrine.[213]

2. **Reasonable Use.** As with absolute ownership, under the reasonable use doctrine groundwater use is considered a property right and landowners can pump as much as they like. The only difference is that the pumped water must be used on the overlying land without waste. Pumped water can be transported away from the overlying land only if doing so does not adversely affect the groundwater use of adjacent overlying landowners. Because of these extra conditions, the reasonable use doctrine protects the rights of overlying landowners and promotes conservation. Arizona and eleven eastern states apply the reasonable use doctrine.[214]

3. **Correlative Rights.** This allows overlying landowners to access a "reasonable" amount of groundwater, which is typically based on the percentage of their property that lies over the water source.[215] Variations of correlative rights are applied in California, Michigan, Ohio, Wisconsin, Arkansas, Florida, Nebraska, New Jersey, and Missouri.[216]

4. **Appropriation,** Several western states apply modified versions of prior appropriation rules to groundwater rights. As with prior appropriation rights for surface water, groundwater appropriation rights are not tied to property and can be transferred or sold. The right is tied to a specific amount of water and subject to the beneficial use rule—which is defined just as it is for surface water.[217]

With multiple landowners sitting atop of the same pool, many groundwater basins are subject to competitive withdrawal. Such competition causes classic tragedy-of-the-commons conditions in which users bear only a fraction of the full cost of their pumping decisions. The consequences include land subsidence, saltwater intrusion, quality degradation and higher pumping costs (withdrawals by one user will lower the water levels. so all users must spend more to bring up deeper water). Aquifers also may deplete at faster rates. Many landowners have inadequate incentive to conserve water, and the groundwater basins can become overused.[218]

Primary Ways That Military Installations Can Obtain Water Rights

What is clear from the general discussion of surface and groundwater rights above is that determining water rights for a specific piece of property involves extensive analysis to determine the application of local, state, federal, and even tribal laws and regulations. The same applies for Army installations. Even though Army installations are federal property, there are points at which state, tribal, and local considerations affect the determination of installation water rights.

---

[213] Job, 2010.

[214] Job, 2010, p. 143.

[215] U.S. Army Environmental Policy Institute, 2011, p. A-26.

[216] Job, 2010, p. 144.

[217] Job, 2010, p. 144.

[218] For more information see Libecap, 2010, and Robert Glennon, *Unquenchable: America's Water Crisis and What To Do About It*, Washington, D.C.: Island Press, 2009.

The extent depends in part on how the rights were obtained, which in turn depends on how the land was acquired. For larger military installations, which have many parcels of land that were acquired in different ways, understanding an installation's total amount of water rights can get complicated.

Next, we discuss the different ways in which military installations obtain water rights.

Ways That Installations Obtain Water Rights

If a military installation's property was withdrawn land as a military reserve, then the installation water rights for that property are determined by the doctrine of implied "Federal Reserved Water Rights."[219] Typically the federal government has deferred to states on water allocation issues. However, under the Supremacy, Property and Commerce clauses of the Constitution, Congress has the authority to preempt state rights and withdraw land from the public domain for a federal use.[220] The reserved rights doctrine holds that when the government reserves public land, the implied intent is that it also reserves water to fulfill the purpose for which the land was reserved. Conversely, states are denied the right to allocate water in a manner that deprives the federal government of an adequate water supply for fulfilling the purpose of the reserved property.[221]

The reserved rights doctrine developed as the federal government began setting aside public property for the establishment of Native American reservations. Over time it was extended to other federal reserved lands, including national forests, national wildlife refuges, and U.S. military installations.[222]

In addition to federal reserved water rights, military installations can obtain water rights in other ways—namely through cession, acquisition, and legislation (which could grant an explicit federal water right, in contrast to the doctrine of implied rights described above). If water is ceded to the installation, then the cession authorization would determine whether the associated water rights were ceded as well. Like states, the federal government also can acquire property for installations either by purchase or establishing eminent domain. The federal government then would acquire a state law-determined water right (in the absence of any special legislation granting a superior water right), governed by the water rights regime in the state where the

---

[219] We distinguish between the doctrine of implied water rights, which is a judicial creation, and explicit reservation of water rights by Congress when it creates military installations. The implied rights doctrine was created in the courts to grant federal reserved lands a sufficient supply of water to meet the purposes of the reservation—even when there was no explicit grant of water rights at the time of the creation of the reservation. Alternatively, Congress has the ability to explicitly claim water rights when it created military reservations.

[220] California Department of Water Resources, "California Water Plan Update 2009, Chapter 15: Federal Interests," Vol. 4, 2009, p. 15.

[221] California Department of Water Resources, 2009, p. 8.

[222] U.S. Department of Justice, Environment and Natural Resource Division, "Federal Reserved Water Rights and State Law Claims," undated.

property was acquired. And finally, Congress can pass legislation to explicitly grant water rights to an installation.[223]

Controversies can arise over whether the federal government has water rights, and subsequently how much water it is entitled to use. Adjudicating these claims raised a host of federalism issues—namely, were cases concerning federal and state water rights more appropriately heard in federal or state court? The McCarran [224]Amendment (1952) was a legislative enactment that began to put limits on U.S. sovereign immunity in cases concerning the management of water rights. Prior to enactment of the legislation, states and other affected parties could not force the federal government to join court cases and other adjudications of water rights. The amendment "waived federal sovereign immunity for the joinder of the United States as a defendant in general stream adjudication," which provided states an avenue for demanding that the U.S. Government quantify, assert, and define its federal reserve water rights.

Hence, Federal Reserve Water Rights would appear to make it seem that obtaining water rights for some installations would be relatively straightforward. However, persistent legal challenges and changes in states' positions have added complications and uncertainty. For instance, military use may not explicitly fall under state designations for "beneficial use," it can be hard to quantify water entitlements for installations, and states differ considerably in terms of how they manage water allocation and the legal doctrines that states apply.[225] Since many installations contain land that was acquired in different ways, such as property purchases spanning over decades, it complicates understanding an installation's water rights.

Summary

This chapter reviewed historical and current water management trends, persisting challenges, and U.S. water rights. Water management initially focused on accessing water sources, and this history is replete with issues that continue to exist today. Massive federal investment in water storage and conveyance infrastructure supported population growth and development. Significant infrastructure was built to distribute the water supplies to homes, businesses, industries, and other users. This long-lived built infrastructure continues to carry water today, supporting established practices for sharing water in many regions of the country. Not until the late 1970s, as federal investment in infrastructure declined, did financial and environmental accountability become major factors in federal investment decisions. As we have seen throughout the

---

[223] U.S. Army Environmental Policy Institute, 2011, p. A-27.

[224] "The McCarran Amendment," 66 Stat. 560 (1952), codified at 43 U.S.C. § 666, 1952. For an overview, see U.S. Department of Justice, "The McCarran Amendment," Washington, D.C.: Environment and Natural Resource Division, undated.

[225] Michael J. Cianci, Jr., James F. Williams, and Eric S. Binkley, "The New National Defense Water Right - An Alternative to Federal Reserved Water Rights for Military Installations," *Air Force Law Review*, Air Force Judge Advocate General School, 48 A.F. L. Rev. 159, 2000.

development of water, objections over the environmental, social and economic changes that resulted from sharing water have not been without conflict.

Water development also required complex interstate compacts and international agreements requiring congressional approval to resolve disputes among states (congressional legislation and court decisions are two other methods for determining water allocation). More recent efforts have focused on the ability of these compact agreements to adapt water allocation and management to changing demographics, environmental conditions, and political interests, particularly in light of the effects that climate change is predicted to have on water availability.

Rapid growth, development of new water, new agreements, and uncertain water laws also have spawned a complex patchwork of water rights and significant water conflicts among states. Many claims for water rights and sources ended up facing legal challenges. States have seen similarly massive legal battles, called general stream adjudications, in an effort to better manage water resources and resolve disputes over water rights, especially where those rights are over-appropriated or when droughts occur.

Traditional water demands driven by population growth and economic development continue to accelerate in many regions of the country. Simultaneously, other demands for water are increasing stress on water resources, including increasing concern over ecosystem health, increasing awareness of the value of recreational services, and growth in water intensive industries.

However, nearly all of the traditional water sources are already allocated or over-allocated, and past approaches to water development are rarely options for addressing today's needs. Moreover, water supply variability has exposed weaknesses in past approaches as states, communities, and regional entities look for alternatives.

Today, all federal, state, and local governments, water utilities, and others are implementing an array of activities geared toward extending the supply of water; improving our understanding of water supply and use; and re-allocating water supply as new demands arise and water supply variability becomes more frequent. Today's more common water management practices seek to encourage water conservation, develop alternative water sources, and implement state and regional water plans (which incorporate conjunctive planning of groundwater and surface water). Activities to encourage water conservation are present in technology rebates to encourage the adoption of water efficient technology, consumer education, building codes, state and regional planning, and water pricing, to name a few.

Non-traditional sources of water are also being developed as options for new water sources become more limited and expensive. These non-traditional sources include recycled water (or reclaimed water), stormwater, and desalinated water. Many cities have plans to employ all of these sources, and others already have infrastructure making it suitable for them to use reclaimed water. Issues related to cost, water quality, environmental impacts, and water rights must be addressed in each of these approaches.

Many state, regional, and local water managers have been developing plans to help different jurisdictions better coordinate water management activities over the long term. While these plans can vary in focus and implementation, comprehensive and effective planning processes will better assess water supply and uses (both in terms of quantity and quality); treat surface water and groundwater conjunctively; incorporate stakeholder input; and lead to actions to implement the plans. Collaboration and partnership are a major part of developing and implementing state, regional, and local water management plans.

Sustainability and watershed management are two frameworks that are being used to a larger extent. Sustainability focuses on a more comprehensive set of values—economic, environmental, and human—to guide resource use and development, while watershed management focuses on the natural hydrologic systems for water management and planning. These frameworks also explicitly recognize that long-term water resource availability must be maintained. As these frameworks gain traction throughout the country, water managers are becoming responsive to the priorities that are given to ecosystem restoration and overall health in addition to more traditional concerns.

As water management evolves to consider more comprehensive and efficient uses of the resource, a number of key challenges that may affect water availability remain in both the short term and long term. Water managers must grapple with the pressures of meeting new demands from population growth and the consequences of climate change. Issues such as legacy infrastructure replacement, depleting aquifers, and water quality degradation (resulting from development, water use, and agricultural and industrial practices) are still unresolved. However, public demands for low-cost water, due in part to conventional pricing practices, make meeting these challenges all the more difficult.

Water rights, or the legal right to use water, have developed differently in individual states, creating additional complexity when managing water resources. All of the measures previously discussed (such as conservation, alternative sources, watershed planning, and state water resources planning), have legal ramifications that both restrict and facilitate action.

All of these water issues are important for military installation water managers to understand, because such broader trends and activities impact the water that installations have. They also affect how installation managers can make investments in water infrastructure and partnerships as well as managers' understanding of how water market opportunities are evolving.

3. Army Installation Water Goals, Project Funding Sources, and Partnership Opportunities

In this chapter, we provide some useful background information about water partnerships and Army installation water programs, investments and partnerships. First, information is presented about Army installations' water goals, which are driven by an array of federal, Department of Defense (DoD) and Army policies. Then we discuss the ways in which installations fund investments in water and wastewater systems. They include some traditional partnerships activities, such as Energy Savings Performance Contracts (ESPCs) and Utility Energy Service Contracts (UESCs). A partnership is when two or more organizations agree to work together for mutual benefits and invest in the partnership relationship by sharing responsibilities, resources, and risks. Since partnerships are an important method to help fund and implement installation water investments, this section actually begins with some basic background definitions on partnerships before discussing military funding sources. After discussing military funding sources, we provide an overview of the different types of water partnerships, including public-private, public-to-public, and regional partnerships, in which state and local governments are involved to help manage water resources for long-term water and drought planning, and to more cost-effectively invest in water infrastructure. In this section, we also briefly provide an overview of how these partnerships offer opportunities to help fund installation water projects and enhance installations' long-term water security.

Army Installation Water Goals

From a practical perspective, according to the "Army Water Security Strategy," Army installations' primary water goals are "to ensure the integrity of the water and wastewater systems to ensure an uninterrupted water supply and comply with applicable health standards and environmental regulations."[1] However, because of the pressure on water resources and the broader water-management trends discussed in the previous chapter, the Army also has goals related to water conservation and increasing the use of reclaimed water.

Federal, DoD, and Army Water Policy

Army installations are subject to several executive orders, DoD policies, and Army policies and regulations that have established benchmarks and guidelines for water efficiency in recent years. Federal-level regulations are largely grounded in Executive Order 13423, "Strengthening Federal Environmental, Energy, and Transportation Management"; the "Energy Independence and

---

[1] U.S. Army Environmental Policy Institute, 2011, p. 30.

Security Act of 2007" Section 432, Executive Order 13514, "Federal Leadership in Environmental, Energy, and Economic Performance"; and the Department of Defense Strategic Sustainability Performance Plan.

Signed in 2007, Executive Order 13423 directs each agency beginning in FY 2008 to reduce water consumption intensity, relative to the baseline of the agency's water consumption in FY 2007, through life-cycle cost effective measures by 2 percent annually through the end of the fiscal year 2015 or 16 percent by the end of FY 2015.[2] The "Energy Independence and Security Act of 2007" Section 432 requires federal facility energy managers to complete energy and water evaluations, to implement energy and water efficiency measures, and to measure and verify energy and water savings for the implemented measures.

Executive Order 13514 seeks to improve water use efficiency and management in the following ways:

- Reducing potable water consumption intensity by 2 percent annually through fiscal year 2020, or 26 percent by the end of fiscal year 2020, relative to a baseline of the agency's water consumption in fiscal year 2007, by implementing water management strategies including water-efficient and low-flow fixtures, and efficient cooling towers.
- Reducing agency industrial, landscaping, and agricultural water consumption by 2 percent annually or 20 percent by the end of fiscal year 2020 relative to a baseline of the agency's industrial, landscaping, and agricultural water consumption in fiscal year 2010.
- Identifying, promoting, and implementing water reuse strategies that reduce potable water consumption.
- Implementing and achieving the objectives identified in the stormwater management guidance referenced in section 14 of the order.[3]

In addition, the December 5, 2013 "Presidential Memorandum—Federal Leadership on Energy Management," requires federal agencies to install water meters on their buildings "where cost-effective and appropriate."[4]

The 2010 and 2011 *Department of Defense Strategic Sustainability Performance Plan* also outlines annual targets for meeting the following goals:[5]

- Reduce potable water consumption intensity by 26 percent of FY 2007 levels by FY 2020.

---

[2] See White House Office of the Press Secretary, "Executive Order 13423—Strengthening Federal Environmental, Energy and Transportation Management," January 24, 2007.

[3] See White House Office of the Press Secretary, "Executive Order 13514—Planning for Federal Sustainability in the Next Decade," March 19, 2015.

[4] White House Office of the Press Secretary, "Presidential Memorandum—Federal Leadership on Energy Management," December 5, 2013.

[5] U.S. Department of Defense, "Department of Defense Strategic Sustainability Performance Plan: FY 2010," August 2010, pp. II-14-15;U.S. Department of Defense, "Department of Defense Strategic Sustainability Performance Plan: FY 2011, October 2011," pp. II-32-33.

- Reduce industrial and irrigation water consumption by 20 percent of FY 2010 levels by FY 2020
- To the maximum extent technically feasible, maintain pre-development hydrology for all development and redevelopment projects of at least 5,000 square feet.

Section 23-20 of Army Regulation 420-1 requires that a Water Resource Management Plan (WRMP) be prepared for each installation. It says that the WRMP will include a water conservation program that includes best management practices such as water re-use, water metering, and landscape management.[6] It also directs that IMCOM will maintain the documentation necessary to protect garrison water rights.[7] Section 22-12 calls for the establishment of installation energy and water management plans and policies, and the following items:[8]

- Conducting surveys to check water safety and quality.[9]
- Conducting leak-detection tests to identify and repair leaks.[10]
- Increasing water efficiency for domestic water consumption by using water-saving fixtures and appliances.[11]
- Using reclaimed or recycled water for landscape irrigation.[12]
- Developing water management plans to implement best practices for water conservation.[13]

Army Water Priorities

Issues of water conservation and water efficiency increasingly have become priorities for the Army as well.[14] The 2006 *Army Energy and Water Campaign Plan for Installations: FY 2008–2013* made water conservation a priority on Army installations. It identified water scarcity as one of the most underestimated resource issues[15] and security of the water system supply "as important as energy security in order to maintain functioning installations."[16] The 2009 *Army*

---

[6] See Army Regulation 420-1, Section 23-20, par. a.

[7] See Army Regulation 420-1, Section 23-20, par. 5e.

[8] See Army Regulation 420-1, Section 22-12, par. a1.

[9] See Army Regulation 420-1, Section 22-12, par. e1.

[10] See Army Regulation 420-1, Section 22-12, par. e2.

[11] See Army Regulation 420-1, Section 22-12, par. e3.

[12] See Army Regulation 420-1, Section 22-12, par. e4.

[13] See Army Regulation 420-1, Section 22-12, par. e5.

[14] Katherine Hammack, "Energy and Sustainability Priorities and Opportunities," *U.S. Army Journal of Installation Management*, Spring 2011, p. 1.

[15] U.S. Department of the Army, *Army Energy and Water Campaign Plan for Installations: FY2008–2013*, 2006, p. ii.

[16] U.S. Department of the Army, 2006, p. iii.

Energy Security Implementation Strategy also specifically identified water conservation as a related priority.[17]

Army water security also is an important priority. The 2011 *Army Water Security Strategy* defines what water security means to the Army:

> Army water security is the assurance that water (potable and non-potable) of suitable quality will be provided at rates sufficient to fully support the Army wherever it has, or anticipates having, a mission in the future.[18]

This strategy document also lays out six overarching factors that require attention to achieve Army water security, see Table 3.1.

Table 3.1. Factors That Require Attention to Achieve Army Water Security

Sources	The quantity and quality of natural, raw water (surface and groundwater) available to the region.
Supply	The Army's entitlement and access to the raw water and means of distributing it to Army users.
Sustainable Practices	Net Zero water use efficiency concepts
Survivability	Treating raw water to federal drinking water standards and preventing and recovering from water supply disruption or contamination.
Sponsorship	Identification and alignment of Army water management responsibilities.
Stakeholders	Constructive engagement of other regional water users.

The Army *Installation Management Community Campaign Plan: 2010–2017* highlights energy and water efficiency and security as "key elements in supporting installation readiness"[19] and calls for installations to develop a Garrison Energy and Water Management Program (GEWMP) as well as other actions related to water conservation.[20] These have been reiterated in the latest version of the Army's *Installation Management Community Campaign Plan: 2012–2017*.[21]

The Army *Installation Management Water Portfolio: 2011–2017* provides an overview of the Army's water-management capabilities to eliminate unnecessary consumption, increase efficiency, and expand the use of recycled/reclaimed water. It also embraces the Army vision of a "Net Zero Water Installation." The Army defines a Net Zero Water Installation as "an

---

[17] U.S. Department of the Army, *Army Energy Security Implementation Strategy*, 2009, p. 12.

[18] U.S. Army Environmental Policy Institute, 2011, p. 2.

[19] U.S. Department of the Army, "Installation Management Community Campaign Plan: 2010–2017," March 2010, p. F-1.

[20] See Annex F of U.S. Department of the Army, March 2010.

[21] See Annex F of U.S. Department of the Army, "Installation Management Community Campaign Plan: 2012–2017," November 2011.

installation that reduces overall water use, regardless of the source, increases use of technology that uses water more efficiently; recycles and reuses water, shifting from the use of potable water to non-potable sources as much as possible; and minimizing inter-basin transfers of any type of water, potable or non-potable, so that a Net Zero water installation recharges as much water back into the aquifer as it withdraws."[22]

In April 2011, the Army identified the following six installations to participate in a pilot program to become net zero water installations by 2020:

- Aberdeen Proving Ground, Maryland
- Camp Rilea, Oregon
- Fort Buchanan, Puerto Rico
- Fort Riley, Kansas
- Joint Base Lewis-McChord, Washington
- Tobyhanna Army Depot, Pennsylvania.

The Army also has a broader goal of Net Zero Installations that includes net zero energy, water, and waste. These installations will be designed to conserve natural resources and to address the interrelationships between such resources. Fort Bliss, Texas, and Fort Carson, Colorado, were selected to participate in a pilot project aiming to be integrated net zero installations by 2020. Thus, eight installation pilots actually are focusing on net zero water.

As discussed later in this report, several Army installations have become leaders in water conservation efforts. For instance, over the past 15 years, Fort Huachuca has reduced groundwater pumping by 60 percent through projects such as installing artificial turf on physical training fields, waterless urinals, installing water-efficient irrigation, and exploiting rainwater for irrigation of grassy areas.[23]

Fort Bragg has reduced its water consumption by four to five million gallons between 2002 and 2010. This reduction was obtained through measures such as the installation of low-flow toilets and showerheads in buildings. In addition, an odd-even watering schedule was adopted for outside watering, since usage peaks during summer months.[24]

In 2009, the Tooele Army Depot was awarded the 31st Annual Secretary of the Army Energy and Water Management Award and the 2009 Federal Energy and Water Management Award because water conservation efforts (mainly fixing underground leaks in water pipes) saved the depot more than $60,000 and nearly 100 million gallons of water per year.[25]

---

[22] U.S. Department of the Army, "Army Directive 2014-02 (Net Zero Installations Policy)," Washington, D.C.: Secretary of the Army, January 28, 2014.

[23] Jennifer M. Caprioli, "Water Resources Management at Fort Huachuca Continues 15 Years Later," U.S. Army news article, July 26, 2010.

[24] Tina Ray, "Fort Bragg Focuses on Energy, Water Conservation," U.S. Department of the Army, September 16, 2010.

[25] Ely Trapp, "TEAD Team Wins Federal Energy Conservation Award," U.S. Army, news article, November 9, 2009.

Next, we discuss the different funding sources that installations use to fund investments in water conservation, water and sewer infrastructure, and other water projects.

Funding Sources for Installation Water Investments

Installations have different ways in which they can fund investments in water efficiency technologies as well as water and wastewater infrastructure. Such funding sources range from military construction and operations and maintenance funds to specialized one-time appropriations and ESPCs, UESCs, and other partnerships. The funding often is limited and competitive for the more traditional funding sources. Because of declining federal budgets, partnerships offer greater opportunities than some of the traditional sources. Since many different types of partnerships exist, we begin this section by defining partnerships. Then we briefly summarize some of the different funding sources and the challenges and opportunities in using them for installation water investments. For discussion purposes we have grouped them into the following categories:

- Traditional Military Funding Sources
- Army Water Utility Privatization
- Partnerships with Industry and Utilities
- Government Water Partnerships

Background on Partnerships

To understand the role that partnerships have in helping to fund installation water investments, we begin this discussion by defining partnership and explaining the different types of partnerships. The term *partnership* often is used many different ways, so here is our definition: A partnership is when two or more organizations agree to work together for mutual benefits and invest in the partnership relationship by sharing responsibilities, resources, and risks. Resources can include financial, technical skills, personnel, and equipment. Some organizational relationships that are called partnerships actually are closer to being collaborations because they do not involve the same level of commitment, organizational integration, interdependence, and sharing of risks, responsibilities, and resources. A true partnership involves a high level of commitment and a closer working relationship. Actually, partnerships can be viewed on a continuum from a loose partnership, which is more like a collaboration, to a more-interdependent relationship with more shared resources and risks, which is a more in-depth partnership. Table 3.2 shows eight criteria associated with successful partnerships, which helps define a more in-depth partnership.

Table 3.2. Criteria for Successful Partnerships

Criteria Name	Brief Description of the Criteria
Individual excellence	Both partners are strong and have something of value to contribute to the relationship. Their motives for entering into the relationship are positive (to pursue future opportunities), not negative (to mask weaknesses or escape a difficult situation).
Importance	The relationship fits the major strategic objectives of the partners, so they want to make it work. Partners have long-term goals in which the relationship plays a key role.
Interdependence	The partners need each other. They have complementary assets and skills. Neither can accomplish alone what both can together.
Investment	The partners invest in each other (for example, through equity swaps, cross-ownership, or mutual board service) to demonstrate their respective stakes in the relationship and each other. They show tangible signs of long-term commitment by devoting financial resources.
Information	Communication is reasonably open. Partners share information required to make the relationship work, including their objectives and goals, technical data, and knowledge of conflicts, trouble spots, or changing situations.
Integration	The partners develop linkages and shared ways of operating so they can work together smoothly. They build broad connections between many people at many organizational levels. Partners become both teachers and learners.
Institutionalization	The relationship is given formal status, with clear responsibilities and decision processes. It extends beyond the particular people who formed it, and it cannot be broken on a whim.
Integrity	The partners behave toward each other in honorable ways that justify and enhance mutual trust. They do not abuse the information they gain, nor do they undermine each other.

SOURCE: Rosabeth Moss Kanter, "Eight I's That Create Successful We's," *Harvard Business Review*, July/August 1994.

Water partnerships focus on water conservation, wastewater treatment, use of reclaimed water, water and wastewater infrastructure, water supply, or water efficiency technologies.

Installations can enter into many different types of partnerships. An installation partnership involves an installation and other organization (or multiple organizations) agreeing to work together for mutual benefits, and usually involves a long-term interdependent relationship. Installations can enter into public-private partnerships (PPPs), public-to-public partnerships (PuPs), and regional partnerships.

An installation public-private partnership (PPP) is a collaborative arrangement between an installation and a private entity, which could be a for-profit company or a non-governmental organization (NGO). "PPPs specify joint risks and responsibilities, which implies some sharing of risks, costs, and assets."[26] A typical installation public-private partnership involves the private company providing a service to the installation over the long term in exchange for ongoing services. For water and energy investments, an Energy Savings Performance Contract (ESPC), discussed further below, is an example of this type of partnership. Another example would be a Utility Energy Service Contract (UESC), also discussed further below, that involves a for-profit company.

A public-to-public partnership (PuP) is a partnership agreement between an installation and a local, state, or other federal agency, usually over the long term. PuPs allow public utilities to pool resources, therefore buying power and technical expertise. The benefits of a PuP are scale, public efficiencies, and lower costs. The two government agencies often bring different expertise to the table, such as financial knowledge, technical skills, or facilities to be shared. A UESC with a public utility is a traditional example of this type of partnership. We will discuss some other less traditional examples later.

Lastly, a regional partnership is when an installation has an agreement with multiple entities, which may be public and private, within a region for mutual benefit. Army installations have entered into regional partnerships to address encroachment concerns through the Army Compatible Use Buffer (ACUB) program[27] and environmental concerns through regional ecosystem partnerships. One of the best examples of a regional ecosystem management partnership that has involved an Army installation is the Central Shortgrass Prairie (CSP) Ecoregion Partnership, in which Fort Carson has played an active role.[28] The CSP partnership is a collaboration of federal, state, NGO, and private landowners to study, manage, and preserve the CSP ecoregion by protecting key ecological patches and conservation corridors so that managers can maintain a healthy, viable ecosystem. The CSP ecoregion encompasses approximately 56 million acres and includes parts of Colorado, Kansas, Nebraska, New Mexico, Oklahoma, Texas, and Wyoming. CSP partners include The Nature Conservancy, Colorado Association of Conservation Districts, Colorado state natural heritage programs, Fort Carson, Department of Defense, Colorado Division of Wildlife, U.S. Fish and Wildlife Service, U.S. Forest Service (FS), and other federal, state, and non-governmental agencies and organizations. Many of these regional partnerships tend to be more like collaborations than an in-depth partnership, because the partners invest and commit less to the partnership relationship. A

---

[26] Ike Chang, Steven Galing, Carolyn Wong, Howell Yee, Elliot Axelband, Mark Onesi, and Kenneth P. Horn, *Use of Public-Private Partnerships to Meet Future Army Needs*, Santa Monica, Calif.: RAND Corporation, 1999, p. 2.

[27] ACUB allows Army installations to use funds to enter into partnership agreements with county, state, or municipal governments, as well as with nonprofit organizations. The agreements enable the partner to purchase tracts of land or easements on lands from willing sellers to establish buffers around installations to maintain current land uses or to protect habitat. Buffer areas are established around Army installations to limit the effects of encroachment by preventing commercial and residential activities along installation boundaries and to maximize use of land inside the installation to support the training and testing mission of installations.

[28] See Betsy Neely et al., "Central Shortgrass Prairie Ecoregional Assessment and Partnership Initiative: Final Report," Nature Conservancy of Colorado and the Shortgrass Prairie Partnership, November 2006.

regional partnership also often takes more time to develop and implement, especially when many diverse partners are involved and the focus is on a large region, as installations have learned from their ACUB partnership experiences. Smaller regional partnerships involve fewer entities. Regional partnerships that focus on water are discussed below.

Traditional Military Funding Sources

Historically, Army installations have had two consistent Army sources of funding for some water investments: Army Military Construction (MILCON) funds which pay for construction of new buildings and other installation facilities and their associated infrastructure; and operations and maintenance (O&M) funds. A Department of Defense MILCON-funded program called the Energy Conservation Investment Program (ECIP) also can be used for some water conservation investments, but all military installations compete for these limited funds. These three funding sources are appropriated funds that are authorized and appropriated by Congress for specific purposes.

We describe each of these three funding sources below. However, we first briefly mention how we found that some installation personnel are creative and have been able to tap into other Army, Office of the Secretary of Defense (OSD), or congressional specialized funds for water project investments. How installations are able to acquire funding from these and other sources is discussed further in the Fort Carson and Fort Huachuca case studies in Chapters Five and Six. However, we briefly provide an example here for Fort Carson. It has acquired some water project funding through the Army's Utility Modernization Program and special congressionally appropriated funds for water utility upgrades.

Military Construction

Limited Military Construction, Army (MCA) funds are available for water efficiency projects, including planning, programming, designing, budgeting, and executing new construction. Congress authorizes and appropriates MILCON funds for the specific purpose of constructing new military buildings. These funds are regulated by AR 420-1, which outlines policies, procedures, and responsibilities for the Department of the Army MILCON budget.[29] Some MILCON funds are used for water efficiency investments, often because of the Army's Leadership in Energy and Environmental Design (LEED) Silver requirement.

LEED is a voluntary, consensus-based national rating system for developing high-performance, sustainable buildings and was developed by the U.S. Green Building Council (USGBC). In 2008, Army sustainability policies required that new buildings should achieve LEED for New Construction Silver standards at a minimum and that major renovation and repair projects should achieve LEED for Existing Buildings Certified standards when life-cycle cost-

---

[29] For more information, see U.S. Department of the Army Energy Program, "Appropriated Funds–MCA," updated January 4, 2010.

effective. In 2011, Army policy also required that new homes achieve LEED Silver standards. The LEED rating is based on a point system in which more points indicate a higher level of sustainability achievements. Buildings receive points for meeting requirements in a wide range of areas, including sustainable siting, water efficiency, energy and atmosphere, materials and resources, indoor environmental quality, innovation and design processes, awareness in education, and locations. LEED version 3 (v3), developed in 2009, uses a point-based accounting system, which awards up to 100 points for buildings that meet environmental sustainability requirements. The LEED process also awards additional points for meeting regional priorities, such as water efficiency in the West. Buildings that meet different point thresholds are awarded different levels of LEED certification—basic certification (40–49 points), silver (50–59), gold (60–79), or platinum (80+).[30]

Many installations have been constructing LEED buildings at the Silver or higher levels and have been including some water efficiency projects. For example, at Joint Base Lewis-McChord (JBLM), a LEED Gold FY05 barracks building included a range of different water projects. Water features included non-potable water toilet flushing systems (using water from roof drain systems), waterless urinals, and a stormwater management system. In the latter, runoff from the barracks parking lot is collected and separated into separate oil and water tanks; the water is routed to a nearby runoff pond rather than into the sewage system.[31]

One of the challenges with LEED, which has drawn criticism, is that the point system is so flexible that some people do not put in the most sustainable building features, especially because of costs.[32] For example, developers can get the same number of points for installing a bike rack as they can for an expensive water recycling system. In other cases, a developer may put in features that are inappropriate for the regional conditions. Given this issue and budget pressures, it can be difficult for Army installations to use the LEED standard to install more expensive water conservation measures, such as a water recycling system. However, the Army sustainability policy that requires LEED for Existing Buildings when it is cost-effective over their life cycle can be used to help focus on more sustainable water conservation measures.

In addition, the Army has faced other difficulties in achieving higher building performance and actually implementing more sustainable buildings:

---

[30] LEED 3.0 was introduced in spring 2009. For more details on LEED standard, see U.S. Green Building Council, LEED 2009: Technical Advancements to the LEED Rating System, 2011c; U.S. Green Building Council, LEED 2009 for New Construction and Major Renovations, 2008.

[31] Beth E. Lachman, Kimberly Curry Hall, Aimee E. Curtright, and Kimberly M. Colloton, *Making the Connection: Beneficial Collaboration between Army Installations and Energy Utility Companies*, Santa Monica, Calif.: RAND Corporation, MG-1126-A, 2011, p. 140.

[32] Some people have criticized LEED, saying it allows developers to reap the public relations benefits of building "green" without ensuring sustainability. For example, see Abby Leonard, "Architect Frank Gehry Talks LEED and the Future of Green Building," PBS, June 14, 2010.

- "Color of money" problems often arise when sustainable building investments, such as higher military construction costs to meet LEED Gold or Platinum standards, require additional up-front costs that are not funded because the savings from such investments occur in a different budget account. For example, MILCON dollars pay for the new LEED building and savings occur in reduced operations and maintenance costs for energy, water, or wastewater disposal. Fort Carson has found that the difficulty of being able to find and invest "capital to support higher first cost of sustainable construction in order to achieve life-cycle savings" is an impediment to erecting more sustainable buildings.[33]

- The National Defense Authorization Act (NDAA) of 2011 (Section 2830) prohibits the use of military funds for the LEED Gold and Platinum certification. Waivers are allowed if the Secretary of Defense submits notification that includes a cost-benefit analysis of the decision and demonstrates payback for the energy improvements or sustainable design features. This legislation has made it more difficult for installations to put up LEED Gold and Platinum buildings, which include more advanced water conservation features like those described in the JBLM example.

- Because of the additional cost of certification, the Army has not required the USGBC to officially certify any of its projects, requiring only that they be built to "certifiable" standards. LEED construction and certification usually adds less than 3 percent to the overall building construction costs.[34] When a building is certified, the USGBC requires extensive documentation to verify that the building meets the LEED requirements. Most of the certification costs come from preparing the certification documentation that must go to the USGBC, which includes extensive receipts to verify that green building features—such as energy-efficient light bulbs—were purchased and installed. For example, in 2004, Silver certification documentation costs for a 49,500-square-foot, $11.5-million Consolidated Support Facility (CSF) at Edwards Air Force Base were around 1 percent of the building's construction cost, or about $115,000.[35]

Since the buildings are not officially certified, even though this is an Army standard, energy- and water-efficient projects and other parts of the LEED standards are not always achieved. For example, it usually requires higher military construction costs to meet LEED standards, but since savings tend to occur in other military accounts, such as reduced O&M costs for energy, water, or wastewater disposal, LEED features, such as capital-intensive water-efficient equipment, are sometimes cut to save money as construction starts, particularly if costs are rising, so as not to overrun the MILCON budget.[36] To help deal with this problem, the Army issued a policy that

---

[33] Fort Carson, "Fort Carson 2006 Sustainability Report," Fort Carson, Colo., 2006, p. 8.

[34] Greg Kats, Leon Alevantis, Adam Berman, Evan Mills and Jeff Perlman, "The Costs and Financial Benefits of Green Buildings: A Report to California's Sustainable Building Task Force, U.S. Green Building Council, October 2003; Steven Winter Associates, "GSA LEED Cost Study," October 2004.

[35] For more details about the cost of LEED certification, see Beth E. Lachman, Ellen M. Pint, Gary Cecchine and Kimberly Colloton, *Developing Headquarters Guidance for Army Installations Sustainability Plans in 2007*, Santa Monica, Calif.: RAND Corporation, MG-837-A, 2009, p. 91.

[36] For more details about problems with buildings not achieving the LEED standard, see Lachman, et al., 2009, p. 69.

starting in FY13, 5 percent of new buildings will require third-party LEED certification. However, given current budget reductions, such reductions in LEED features likely are to continue to be a problem. Actually, given all the challenges just discussed, it is difficult and will likely become increasingly difficult for Army installations to use LEED and MILCON dollars to invest in water efficiency, especially for the more expensive projects.

Operations and Maintenance Funds

Installations also have Sustainment Restoration and Modernization (SRM) funds, more commonly known as operations and maintenance (O&M) funds, which are used to help operate, maintain, and upgrade buildings and infrastructure. Installation O&M funds can be used to pay water bills, fund repairs, and replace equipment. They also are used to comply with applicable health standards and environmental regulations. Insufficiencies in water systems' maintenance often waste significant water on installations. Therefore, some installations use these funds to invest in water efficiency for cost savings.

However, installations often find it difficult to use O&M funds for water projects. This is largely due to time constraints and the competition for scarce resources. O&M funds are intended to fund building or equipment maintenance and day-to-day operations rather than large-scale capital investments. O&M funds must be spent in one fiscal year, unlike MILCON funds, which can be used for up to seven years. By law, all construction must be 100 percent funded, so using O&M funds for water efficiency projects can limit the size of a project since the entire project has to be funded in one year. By contrast, MILCON dollars can be spent for larger water efficiency projects. In addition, O&M funds are limited, and water projects must compete with other installation projects such as fixing leaking roofs and other maintenance issues. Often, pressing short-term fixes receive priority over water efficiency projects. Again, given the current budget situation and the competition for O&M funds, it likely will be more difficult for installations to acquire O&M funding for water conservation and infrastructure projects.

Energy Conservation Investment Program

The Energy Conservation Investment Program (ECIP) is a Department of Defense MILCON-funded program. It is used to execute energy and water efficiency projects on DoD facilities that reduce associated utility energy, water, and other related costs. The ECIP program focuses on energy and water savings, implementation of renewable energy, and converting systems to cleaner energy sources. Army installations compete with other Services' installations to obtain these funds. In FY 2008, the Army was awarded 18 ECIP projects, which totaled approximately $24 million, and in FY 2009, it was awarded 15 projects, which totaled about $27 million.[37] Note that in FY 2009, the American Recovery and Reinvestment Act (ARRA) provided an additional

---

[37] U.S. Department of the Army Energy Program, "Appropriated Funds-ECIP," updated January 4, 2010.

17 ECIP projects to the Army, totaling more than $32 million.[38] In FY 2012, the Army was awarded 13 ECIP projects totaling more than $51 million.[39] In FY 2013, the Army was awarded 16 ECIP projects totaling almost $50 million. Army ECIP projects have mostly focused on energy efficiency and renewable energy projects. However, a few have focused on water conservation or on projects that involve both energy and water interests, such as solar hot water heaters. For instance, in FY08 Fort Rucker was awarded an ECIP project for water conservation activities costing $480,000, and having a savings-to- investment ratio (SIR)[40] of 1.91 and a payback of 6.7 years.[41] Since ECIP funds are appropriate for system expansion, Fort Carson used a FY13 ECIP project for $4 million to help expand its non-potable water system at a savings-to-investment ratio of 1.34. White Sands Missile Range was awarded an ECIP in FY13 for solar hot water and direct digital controls for $1.2 million, with an SIR of 2.14.[42] In desert areas, some installations even have been able to use ECIP dollars to install water-efficient landscaping, such as Nellis AFB in a FY 2007 ECIP, which cost $2.82 million with an SIR of 1.25 and a payback of 8.76 years. Because installation water costs tend to be so low, it is difficult to meet the required investment rate of return for water projects. However, as the above examples illustrate, opportunities can arise to use ECIP funding for some limited water conservation and infrastructure investments. Yet, given the payback requirement, it is difficult to use such funds for more expensive large-scale water infrastructure investments, such as replacing a wastewater treatment plant.

Army Water Utility Privatization

By the 1990s, the Army realized that these traditional sources for funding water investments, especially the needed large capital investments, were not enough. SRM and MILCON funding only were covering about 80 percent of the amount required to fully revitalize water and wastewater systems. It found "systematic underfunding of Army-owned water and wastewater systems."[43] Many installation water and wastewater systems were badly degraded because the Army was not investing enough in system maintenance and needed upgrades. Many in the Army, OSD and other Services also felt that the Army's and other Services' installations should focus

---

[38] This number can be compared with ECIP projects awarded to other services in ARRA: the Navy received 13 ECIP projects, the Marine Corps received 10 ECIP projects, the Air Force got four ECIPs, and four ECIPs were awarded to defense-wide projects.

[39] U.S. Department of the Defense, "FY2012 Energy Conservation Investment Program (ECIP)," Washington, D.C.: Office of the Under Secretary of Defense (Comptroller), undated.

[40] Savings-to-investment ratio (SIR) is $ saved / $ invested.

[41] U.S. Department of Defense, "FY 2008 Energy Conservation Investment Program (ECIP) Projects," Washington, D.C.: Office of the Under Secretary of Defense for Acquisition, Technology and Logistics, 2008.

[42] U.S. Department of Defense, "FY2013 Energy Conservation Investment Program (ECIP)," Washington, D.C.: Office of the Under Secretary of Defense (Comptroller), undated.

[43] U.S. Army Environmental Policy Institute, December 2011.

on core defense missions instead of the utility infrastructure management business and that installation energy, water, and wastewater systems would be better off being owned, managed, and maintained by private or public organizations with utility-management expertise.

As a result, DoD decided in 1997 that privatization was the preferred method for improving utility systems, and Congress approved legislative authority for privatizing DoD's utility systems with Public Law 105-85. The Department of Defense issued Defense Reform Initiative Directive No. 9 ("DRID No. 9"), "Privatizing Utility Systems," in December 1997, which started DoD's installation Utility Privatization initiative. It directed the military departments to develop plans for privatizing all of their utility systems (electric, water, wastewater, and natural gas) except those needed for security reasons, or when privatization was uneconomical. In 1998, DoD issued Defense Reform Initiative Directive No. 49, "Privatizing Utility Systems" ("DRID No. 49") to provide more specific guidance on utility privatization. Over the years, DoD has updated such guidance.

Utility privatization (UP) means that the Army actually sells the government-owned on-installation utility distribution systems to a private or public entity (such as a municipal utility) that will then operate the systems and provide utility services to the installation's buildings and activities.[44] Utilities privatization is a method by which military installations can obtain safe, technologically current, and environmentally sound utility systems, ideally at a relatively lower cost than they would under continued government ownership. By September 2012, 66 water or wastewater systems had been privatized at 35 Army installations.[45] For example, Forts Belvoir, Benning, Bragg, Hood, Irwin, and Wainwright have had two water privatizations, for both their water and wastewater systems, while Fort Pickett only has privatized its water system and Oahu/Schofield Barracks has privatized only its wastewater system.

Many of these more successful utility privatization deals also can be considered public-private partnerships or public-to-public partnerships because of the long-term relationships that develop. Some Army installations work with the privatized utility as a partner, such as at Fort Gordon. In 2008, Augusta and Fort Gordon signed a $290 million deal for the city to provide water to the post at wholesale cost, and the deal soon was expanded to include sewage treatment. Fort Gordon and City of Augusta view this relationship as a public-to-public partnership.

Many UP projects have "largely been successful in recapitalizing and upgrading Army water systems," according to the "Army Water Security Strategy." "UP provides a stabilized utility rate platform by amortization of project costs and the accumulation of Repair and Restoration (R&R) reinvestment funds."[46] Privatization of water utilities has ensured better compliance with

---

[44] For a good overview of what utility privatization means, see Jeffrey A. Renshaw, "Utility Privatization in the Military Services: Issues, Problems, and Potential Solutions," *Air Force Law Review*, January 1, 2002.

[45] Curt Wexel, "Army Utilities Privatization (UP) Program Primer," Washington, D.C.: U.S. Department of the Army Office of the Assistant Chief of Staff for Installation Management, Privatization and Partnerships Division, September 2012.

[46] U.S. Army Environmental Policy Institute, 2011, p. 30.

applicable health standards and environmental regulations, helped reduce safety and environmental risk and liability, improved system reliability, and provided cost stability while leveraging private capital and technical capability.[47]

It is important to note that because they are "addressing years of previously deferred infrastructure maintenance," according to the "Army Water Security Strategy," UP contracts "can result in substantial increases in the cost of water."[48] Installations need to plan their budgets accordingly. However, because utility bills are considered "must pay" items from a budget standpoint, the problems afflicting Army-owned water system improvements have not afflicted paying for UP projects.

Despite the many advantages, utility privatization also led to some problems with utility privatization. For instance, a 2005 GAO study found:

> Utility privatization can provide for quicker system improvements than otherwise might be available; however, there are questions about program savings. Although the services' economic analyses estimate that utility systems privatized to date will reduce the government's costs for utility services, GAO questions the estimates because they give an unrealistic sense of savings to a program that increases ongoing government utility costs in order to pay contractors for enhanced utility services and capital improvements. Other base support services could suffer unless budgets are adjusted to reflect these increased costs. Moreover, GAO found that long-term cost comparisons did not depict actual expected costs of continued government ownership in the event that systems were not privatized and DOD had not taken steps to ensure that the estimates were otherwise reliable. ... In some cases, contractors also include additional amounts in the contracts to cover costs associated with the fair market value payment. Thus, implementing the fair market value requirement in such cases results in higher contract costs because the government will pay back more than it will receive for conveying the systems. Two additional issues of concern identified by GAO related to limited oversight of privatization contracts and DOD's preferred practice of permanently conveying utility systems to contractors rather than using more limited arrangements which, according to DOD consultant reports, is a more prevalent private sector practice and one which may offer greater safeguards to the government.[49]

Each of the Services, including the Army, has been working to address such issues, such as improving the economic analyses, but other issues remain. For instance, the 2011 "Army Water Security Strategy" identified a range of security concerns with the privatization of water infrastructure, including:

- Potential risks of accident and malicious tampering with the water assets

---

[47] For more on the benefits of utilities privatization, see Wexel, 2012.

[48] U.S. Army Environmental Policy Institute, 2011, p. 31.

[49] U.S. Government Accountability Office, "Defense Infrastructure: Management Issue Requiring Attention in Utility Privatization," GAO-05-433, May 12, 2005.

- Relying on connections to external water utilities exposes the installation to the vulnerabilities associated with those utilities
- Whether the installation can be self-sufficient for an extended period of time in an emergency situation.

To help deal with these concerns, this same strategic document recommends an Army objective to "Provide Advance Planning, Contractual Flexibility, and Adequate Staff Support to Implement and Administer Army Water Privatization Contracts."[50]

In our interviews with installation and privatized utility personnel, we identified some other issues with utility privatization. First, at some installations, the privatized utility has little to no incentive to engage in water conservation activities, especially when the utility is selling the installation water and that commodity income is needed to maintain the water system and provide profit to the company.[51] Second, some installations may no longer appear to own their effluent, and it is difficult for them to use an asset that they used to own for installation irrigation and other non-potable water purposes. Current privatization contracts include a clause that protects installations' rights for using wastewater; however, some early privatization contracts may lack this. This issue points out a third, broader issue, which is that some installations are losing some control over and flexibility with water assets, which could make it more difficult to participate in water market mechanisms. Fourth, installations encounter unexpected consequences with UP projects. At one installation, management was concerned about possible community growth causing encroachment problems near the installation. The encroachment became possible because the UP water utility that had expanded sewer and water services for the installation now also was able to extend these services into new rural areas at a cheaper rate because it had acquired the installation's water infrastructure. The lack of nearby water and sewer infrastructure previously had limited such growth. At a different installation, which had expanded significantly because of gaining units from Base Realignment and Closure (BRAC) 2005, legal challenges arose over the new water system infrastructure, and negotiations as well as contract modifications were needed. The new infrastructure was not part of the privatized utility's original contract.

Lastly, we found that some installations may be at risk of losing water staff expertise, which is potentially needed to deal with long-term water scarcity, rights, security, and market opportunities. Many installations already face personnel shortages and do not have enough qualified personnel to deal with water management concerns, such as being able to fully document and maintain all water rights information. Installation water management personnel, facing declining personnel and budget pressures, were starting to increasingly rely on the privatized utility to deal with water issues, even for issues such as maintaining installation water

---

[50] U.S. Army Environmental Policy Institute, 2011, pp. 30–31.

[51] We should note that, at first, privatization deals conserve water because the systems are initially upgraded which helps fix leaks and improves system efficiency, however, the privatized contractor may not be interested in water conservation measures in the future because of the potentially negative impact to the company's profits.

rights and long-term water planning. Obviously, one of the benefits of water utility privatization is relying on those water experts for technical help; however, some issues, such as long-term water security and rights, are important installation functions.

Many of these issues could also be dealt within the privatization contracts, such as including monetary incentives for water conservation investments, and policies regarding installation staff, issues that are discussed further in the last chapter.

To summarize, UP has some advantages and disadvantages when it comes to funding installation water investments and meeting installations' long-term water management goals. All these different issues need to be assessed and addressed, according to the "Army Water Security Strategy," "to ensure that long-term privatization commitments deliver the greatest value to the Army and ensure the greatest level of security."[52]

Partnerships with Industry and Utilities

Army installations also can draw on some non-military sources for water conservation and infrastructure investments by partnering with industry and utilities. The two main types of partnership opportunities have been ESPCs and UESCs, two very similar programs. DoD has embraced using ESPCs and UESCs at military installations. As stated in a 2008 OSD memorandum on ESPCs and UESCs, the Army and each of the other Services "shall endeavor" to annually invest in ESPCs and UESCs at a value equivalent to at least 10 percent of the Service's annual energy consumption.[53] Other federal guidance, such as the December 2, 2011 "Presidential Memorandum—Implementation of Energy Savings Projects and Performance-Based Contracting for Energy Savings,"[54] also mentions the importance of using such tools.

Energy Savings Performance Contract

As mentioned earlier, the Energy Savings Performance Contract (ESPC) is a partnership between a military installation (or another federal facility) and a private company, which is called an Energy Service Company (ESCO). With this funding mechanism, installations can implement energy and water efficiency projects without upfront capital costs or special congressional appropriations. The private company pays for the energy and water efficiency investments, and the installation pays it back from its energy and water savings over time. The ESCO conducts a comprehensive energy or water audit to identify potential Energy Conservation Measures (ECMs)—activities that could be implemented that save energy or water and guarantees that the anticipated savings from the ECMs will cover the costs. After the ESPC project costs are repaid

---

[52] U.S. Army Environmental Policy Institute, 2011, p. 31.

[53] U.S. Department of Defense, "Energy Savings Performance Contracts and Utility Energy Service Contracts Memorandum," Washington, D.C.: Office of the Under Secretary of Defense, January 24, 2008.

[54] This memorandum discusses performance-based contracts, which includes ESPCs. For more information see The White House, "Presidential Memorandum—Implementation of Energy Savings Projects and Performance-Based Contracting for Energy Savings," December 2, 2011.

over a 10- to 25-year period, all additional savings accrue to the installation. From the late 1990s until 2009, approximately $2.3 billion was invested in U.S. federal facilities through more than 460 ESPCs.[55] Then, from 2009 to 2011, about $1.2 billion was invested in ESPC projects at federal facilities that saved more than $3.5 billion in energy and water costs.[56]

We illustrate ESPCs related to water for two different military installations: Fort Bliss and Dyess Air Force Base, both in West Texas, where drought and water scarcity have made water a key concern for local governments and military installations. Fort Bliss ESPC Project Number 5 along with energy ECMs, included a water ECM for retrofitting sinks, showers, urinals, and toilets in 400 buildings. The total implementation costs of this ECM was $2.09 million, with a simple payback of 8.64 years and an annual savings of $307,994 in natural gas, water, and wastewater costs. Having a simple payback of fewer than 25 years is a requirement for ESPC projects, but most installations want a even shorter payback, such as 10 to 15 years or fewer, which can be challenging for water projects given the low cost of water at most installations. In 2013, draft ESPC Project 7 at Fort Bliss focused only on water ECMs, including a reclaimed water piping solution for the parade ground, youth center, Finney Field, and Omar Bradley Complex; Xeriscape landscaping of the parade ground and two other locations; and a new irrigation systems for the parade ground and seven other locations. It also corrects inaccurate sewage billing and improves the future monitoring of the sewage utility. This ESPC project's total cost is $18.93 million, with an annual estimated cost savings of $1.29 million and the estimated annual reduction of 136,455 kgal potable water usage and use of 139,000 kgal of reclaimed water. The project has a simple payback of 14.7 years and would be financed over 23 years.[57]

Similarly, Dyess AFB, because of ongoing drought problem and concerns about the potable water supply in nearby Abilene, Texas, began using Abilene's effluent water for irrigation. An ESPC was used to add two 11-million-gallon holding reservoirs, two pump stations, and three miles of distribution piping to connect the effluent irrigation system. This ESPC reduces annual potable water consumption by 160 million gallons and saves $300,000 a year. It also saves 2 percent of Abilene's potable water supply.

The experiences at Fort Bliss and Dyess AFB, with ESPCs focused so extensively on water, especially recycled water, are not currently the norm. More ESPCs have focused more on energy than on water because of water's lower costs. Of the 278 ESPC projects awarded at federal facilities between FY 1998 and FY 2012, about 38 percent included some sort of water and

---

[55] U.S. Department of Energy, "Financing Mechanisms: Energy Savings Performance Contracts," updated September 3, 2009.

[56] These data include ESPC projects awarded under the DOE indefinite-delivery, indefinite-quantity (IDIQ) and the USACE Huntsville ESPC Contract. See U.S. Department of Energy, "Energy Savings Performance Contracts (ESPCs)," September 2012.

[57] Johnson Controls, "Energy Savings Performance Contract Feasibility Study Report: Fort Bliss, Texas Project 7," June 6, 2013.

sewer system tasks, but most of these ESPCs also included many more energy tasks.[58] In many of these ESPC projects, energy and water efficiency projects are combined to make the economics work. However, the Fort Bliss and Dyess AFB experiences show that opportunities exist to do more extensive water ESPCs as water costs rise. What helped in both of these installation circumstances was that local communities had the processes and infrastructure in place to use treated wastewater for irrigation, which helped lower the costs.

In addition to cost concerns, we found that some ESPCs do not focus on water as much as on energy because of the interests, skills, and backgrounds of ESCO and installation personnel. Most ESPCs will consider low-flow water fixture improvements, but some do not always consider more innovative water conservation projects, assuming they will not be as cost-effective.[59] For instance, in developing an ESPC proposal, an ESCO may focus more on an energy audit and not consider many ECMs for water projects because their technical personnel lack knowledge about more advanced water efficiency technologies or because they assume water ECMs will not be cost-effective. More emphasis needs to be placed on water and on developing water ECMs in ESPCs, such as by educating installations' personnel and by ensuring that ESCOs address water issues in their proposals.

Utility Energy Service Contract

The Utility Energy Service Contract (UESC) is a partnership between an installation (or other federal customer agency) and a utility company that enables the implementation of energy and water-efficiency projects. It allows the utility company to provide the installation comprehensive water and energy-efficiency improvements and demand-reduction services. As is the case with the ESPC, UESC projects are funded through their anticipated savings. Similar to ESCOs, the utility company will conduct an energy and water audit, identify cost-effective ECMs, arrange financing, and cover the upfront capital costs of the project. The utility company assesses, designs, and implements energy and water efficiency projects with optional O&M and optional measurement and verification (M&V). The ECMs are intended to reduce energy and water costs sufficiently to allow for repayment to the utility company over a period of ten years.[60] Namely, unlike an ESPC, a UESC has a 10-year simple payback requirement, which makes it harder to invest in some water ECMs.

Between 1994, when USECs began, through 2008, more than 45 gas and electric utilities had completed about $2 billion in energy projects at federal facilities through UESCs, with

---

[58] U.S. Department of Energy, "DOE IDIQ ESPC Awarded Projects," January 18, 2013.

[59] Note that, as discussed earlier, EISA 2007 Section 432 requirements are similar to ESPCs, which increases the motivation to include water measures in ESPC activities.

[60] For more information on UESCs, see Army Energy Program, "Alternative Financing-UESC," updated January 4, 2010; U.S. Department of Energy, "Financing Mechanisms: Utility Energy Services Contracts," updated October 28, 2009; and Lachman, Hall, Curtright, and Colloton, 2011.

approximately $135 million reported in FY 2007.[61] By 2011, about $2.3 billion had been invested in 1,680 UESC projects, which upgraded infrastructure and equipment at a range of different federal facilities to reduce water and energy consumption, saving more than 14 trillion BTU.[62]

Again, UESC energy and water efficiency ECMs are combined to create the needed rate of return for the overall UESC project and often have combined energy and water savings. For example, through a $17.3 million UESC project, Naval Air Station (NAS) Jacksonville, Florida, reduced its energy intensity by 4 percent (34 billion BTU of energy) and water consumption by 24 percent (79 million gallons of water) compared with the prior year during only four months of operation in FY 2011. The project examined more than 30 facilities and incorporated a range of ECMs, including air-handler unit ultraviolet lights, motor-variable frequency drives, direct digital controls, heating, ventilation, and air conditioning (HVAC) upgrades, fuel conversions, chiller retrofits and replacements, roof-mounted solar water heating systems, and boiler replacements. The UESC project's annual savings is expected to be more than 65 billion BTU and $3.3 million.[63] Some Army installations, such as Fort Bragg and Fort Huachuca, have developed and implemented UESCs that include both energy and water ECMs.

As with ESPCs, UESCs mostly have focused on energy-efficiency projects rather than water because of the lower costs of water. However, low-flow water fixtures are a standard ECM considered in most audits. But as with ESPCs, sometimes UESCs also do not consider the more innovative water ECMs because utility companies and installations' personnel have more interests, skills, and backgrounds related to energy. In addition, most UESCs have been with natural gas and electricity companies, rather than with a water utility.

As with ESPCs, more emphasis needs to be placed on water and developing water ECMs in installation UESCs. Despite the challenges, opportunities exist for more water projects in UESCs, especially as water costs rise. For example, the Army could reach out more to water utilities about the possibility of doing a UESC and try to educate both Army and utility company personnel about the opportunities.

Opportunities for Government Water Partnerships

Installations can learn a great deal from some of their counterparts in state and local governments on managing and planning for future water supply and shortages and finding funding for water and sewer system investments. As was discussed in the previous chapter, public water utilities are facing many of the same challenges as Army installations, including declining budgets and

---

[61] Richard Kidd, "Comments at the Federal Utility Partnership Working Group (FUPWG) meeting, November 19–20, 2008, Williamsburg, Va.," FUPWG Fall 2008 Report, Fall 2008.

[62] U.S. Department of Energy, "Utility Partnerships Overview," July 2011.

[63] U.S. Department of Energy, "2012 Federal Energy and Water Management Award Winners," February 7, 2013.

the need to find funding to upgrade aging and degrading water infrastructure. Some state and local governments also face potential uncertainties about long-term water supplies. Some of these public utilities are taking advantage of public-private partnerships, public-to-public partnerships, or regional partnerships to help fund water projects. In the latter two cases, water partnerships that involve different government organizations are being used to share water supply and infrastructure, including pipes, reservoirs, or treatment facilities, or to collaborate in regional water-supply planning. The key benefits to such government water partnerships are lower costs by achieving economies of scale, sharing infrastructure, and helping to ensure long-term water supplies. A main challenge is reaching agreement on sharing the costs, risks, and use of the facilities. We elaborate below on each of these three types of partnership opportunities.

Public-Private Partnerships

Public-private partnership agreements currently serve more than 2,000 North American water facilities.[64] However, owing to the age of the U.S.'s water infrastructure, the strained financial resources in the public sector, and cuts to government services, demand for public-private partnerships has been rising,[65] due to the potential for those partnerships to "reduce overall development risk and capital investment, improve efficiencies and cost effectiveness, and maximize the respective strengths of the public and private sectors."[66] Benefits often "include more efficient operations, guaranteed performance, strict environmental and safety compliance, increased training opportunities for employees, and shared risk liability."[67] In most of these water PPPs, the municipality still owns the potable water utility or wastewater treatment facilities, but the private company partner runs all or part of these facilities. Such partnerships are different from a typical industry fee-for-service contracting relationship because of the close mutually beneficial working relationship between the municipality and private company as well as a contract that includes community oversight and a termination clause if performance criteria for cost, quality, and customer service are not met.[68] For example, the City of Schenectady, New York, entered in a partnership with Veolia Water North America (Veolia Water) in 1991 for Veolia Water to run the city's biosolids program and wastewater treatment plant. Through this partnership, Schenectady has saved money and experienced significant operational improvements at its wastewater treatment plant, including reductions in odor complaints. In fact,

---

[64] National Association of Water Companies, "Quick Facts of U.S. Water Public-Private Partnerships," undated.

[65] Richard Norment, The National Council for Public-Private Partnerships, "The Framework for Public-Private Partnerships," undated.

[66] Western States Water Council, "Western Water Resources Infrastructure Strategies: Identifying, Prioritizing and Financing Needs," June 2011.

[67] National Association of Water Companies, "Quick Facts of U.S. Water Public-Private Partnerships," undated.

[68] National Association of Water Companies, undated.

in 2011, this partnership won a public-private partnership award from the U.S. Conference of Mayors. In addition,

> Outside the scope of the wastewater contract, the partnership has helped extend community involvement beyond the plant. Working together, Veolia Water and Schenectady have helped to beautify the city by cleaning up parks, landscaping various neighborhoods and preserving a nationally recognized historic pump station.[69]

Such public-private partnerships are another opportunity for installations to leverage industry technical expertise and save money. Such partnerships are an alternative to installation UP projects. In fact, they are very similar to UP, but without selling off the assets.

Public-to-Public Partnerships

Some public water utilities also are taking advantage of public-to-public partnerships to help with water investments. In fact, in 2007 the International City/County Management Association (ICMA) surveyed municipalities about what types of organization provides services for them, whether they were in-house, or another government agency, private for profit, or private non-profit organization. ICMA received 164 responses.[70] The results showed that for water, the services mostly were provided in-house; however, for sewage collection and treatment, a PuP with another government agency accounted for 27 percent. Similarly, water treatment was 24 percent PuPs, and water distribution was 16 percent PuPs.[71] Many of these involve having one municipality partner provide the services at the second partner's facility and taking over operating and maintaining any existing second-partner facilities. One interesting example involves a Native American tribe, the Chickasaw Nation, in Duncan, Oklahoma. The Chickasaw Tribal Utilities Authority has partnered with the City of Duncan for water services at its Chisholm Trail Casino. Under this agreement, in 2013 it was planned that the city would take over existing water and sewage services at the casino, as well as providing any additional electricity needed, which the tribe will pay for their utility bills. In addition, the Chickasaw Tribal Utilities Authority would build a wastewater lift station to hook up with the city's wastewater system.[72]

Other partnership arrangements actually involve a close working relationship for sharing water infrastructure to achieve economies-of-scale savings. For example, in August 2008 the

---

[69] U.S. Conference of Mayors, "Executive Summary: Filling the Void," 2011. For more information, also see U.S. Conference of Mayors, "Four Cities Honored for Excellence & Innovation in Public-Private Partnerships," January 21, 2011.

[70] ICMA actually sent the survey to 2,207 municipalities.

[71] Mildred E. Warner and Amir Hefetz, "Service Characteristics and Contracting: The Importance of Citizen Interest and Competition," *Municipal Year Book 2010*, International City/County Management Association, Washington, D.C., 2010.

[72] Mike Smith, "City, Casino Partner on Water Services," *The Duncan Banner*, August 14, 2013.

cities of Lake Oswego and Tigard, Oregon, signed a partnership agreement for sharing drinking water resources, infrastructure, and costs. Lake Oswego's water supply system was near capacity and needed major improvements. The City of Tigard wanted its own water supply so that its residents could have a secure, dependable water source. The partners agreed to create a joint water supply system to serve the two cities. A joint Oversight Committee with representatives from both city councils provides leadership and guidance. The City of Lake Oswego manages and builds the water system improvements. Project improvements over the years have focused on upgrading and expanding a range of existing facilities, including the Clackamas River intake; the water treatment facility in West Linn; the pipes that convey finished water to both communities; Bonita Road pumping station; and the Waluga Reservoir.[73] Costs are allocated 47 percent to Lake Oswego and 53 percent to Tigard because Lake Oswego owns the original facilities.[74] The main funding sources have been both cities' issuance of revenue bonds, which are repaid by water customers' monthly water bills. The partnership has saved significant costs for both cities and their water customers.[75]

Public-to-public partnerships also can focus on water supply and even water-rights issues. For instance, in Texas the Lower Colorado River Authority (LCRA) and the City of Austin "use a new partnering relationship to plan for future water needs and jointly manage their water rights."[76] LCRA is a Texas conservation and reclamation district charged with managing the water supply and environment of the lower Colorado River basin, overseeing control, storing, preservation, and distribution of the river and its tributaries.[77] Austin holds "significant run-of-river water rights to divert and use water from the Colorado River for municipal and steam electric purposes."[78] In 2007, Austin estimated that by about 2050, the city will "need more water than it will have available from the Existing Water Sale Agreements and Austin's Existing Water Rights.[79] In 2007, LCRA and the City of Austin signed a collaborative agreement, called the "Supplemental Water Supply Agreement" and also called the Water Partnership, "to collaboratively manage water supplies and evaluate and implement strategies designed to optimize water supplies to meet the needs of" Austin and LCRA, as well as their customers, and

---

[73] For more information about these improvements, see City of Lake Oswego and City of Tigard, "Intergovernmental Agreement Regarding Water Supply Facilities, Design, Construction, and Operation," August 2008; Lake Oswego Tigard Water Partnership, "About the Partnership," undated.

[74] Lake Oswego Tigard Water Partnership, "Lake Oswego Tigard Water Partnership," factsheet, undated.

[75] City of Lake Oswego and City of Tigard, 2008; Lake Oswego Tigard Water Partnership, factsheet, undated; Lake Oswego Tigard Water Partnership, undated.

[76] Lower Colorado River Authority, "Water Partnership with City of Austin," 2013.

[77] LCRA also provides public recreation areas, and supports community and economic development.

[78] City of Austin and Lower Colorado River Authority, "Supplemental Water Supply Agreement," November 14, 2007, p. 1.

[79] City of Austin and Lower Colorado River Authority, 2007, p. 1.

the environment.[80] The Water Partnership evaluates various water supply alternatives to meet estimates of Austin's future water demand and makes recommendations to the Austin City Council and the LCRA Board.

PuP partnerships for sharing water infrastructure and supply also can occur among more than two government entities. For example, Aurora Water, Denver Water, and South Metro Water Supply Authority have worked for several years to create the Colorado Water Infrastructure and Supply Efficiency Partnership (WISE) to share excess water infrastructure capacities and unused reusable water. WISE could be considered a regional water-sharing partnership for the Denver region.

The WISE agreement means 72,250 acre-feet of treated water from Denver and Aurora would be permanently delivered to members of the South Metro Water Supply Authority (SMWSA), which consists of 13 smaller water utilities in south Denver. Some of the water that flows down the South Platte River and out of Colorado would be recaptured by Aurora's 34-mile Prairie Waters Pipeline, which has excess capacity, and pumped back to a water-purification facility near the Aurora Reservoir to be treated and piped to Denver southern suburbs. The water delivery will begin in 2016. Members of the SMWSA are required to have the distribution infrastructure in place to move the water from the purification plant. The cost of the water and this infrastructure is estimated at $250 million over the next 10 years. Individual members have to finance their own shares of the project.[81]

Ten members of SMWSA have signed the intergovernmental agreement (IGA) for WISE: the town of Castle Rock; Stonegate Village Metropolitan District; Dominion Water and Sanitation District; Cottonwood Water and Sanitation District; Pinery Water and Wastewater District; Centennial Water and Sanitation District; Rangeview Metropolitan District; Parker Water and Sanitation District; Meridian Metropolitan District; and Inverness Water and Sanitation District.

The WISE partnership has benefits for all of the partners. Denver Water gains access to unused water supplies and access to water from the South Platte River downstream of Denver. WISE water also has the potential to replace Denver Water's "Strategic Water Reserve" in mountain reservoirs. For SMWSA members, WISE reduces their reliance on groundwater, helps minimize their need to purchase new water rights and efficiently uses one of their system pipelines (which has excess capacity), and other regional infrastructure. For Aurora Water, WISE means more efficient use of their Prairie Waters Project system (which has excess capacity), and the ability to offset the cost of Prairie Waters and future water rights purchases.[82]

[80] City of Austin and Lower Colorado River Authority, 2007, p. 2.

[81] "Reuse: The WISE Partnership Gets Approval from the Denver Water Board," *Denver Business Journal*, August 20, 2013.

[82] Tracy Kosloff, "A WISE Project for the Denver Metro Area," American Water Resource Association Colorado, March 30, 2010.

Although WISE is beneficial for the different partners, it has had challenges. One was obtaining agreement and commitment from so many different partners. According to one analysis, the WISE partners had to "agree on each of their water demands and structure an equitable payment system for the joint project." The partners also needed "a firm grasp of how much water each entity will use and under what conditions they will want to use it." They also needed "to analyze the pros and cons of any available supply options."[83] Even though the partners have worked through these issues and expect cost savings over the long term, some concerns remain about the costs of the agreement and impact on consumer water utility rates.[84]

Army installations already are engaged in some diverse water public-to-public partnerships. Most of them focus on water infrastructure issues, but some also involve small water-supply issues. We present four different examples for three installations: Forts Carson, Huachuca, and Bliss. Fort Carson and the City of Fountain have a partnership to share water resources from Keeton Reservoir, a small on-post reservoir (explained more in Chapter Six). Fort Carson also partnered with the Colorado Department of Corrections to share in the replacement cost of a 30-mile water line near the Piñon Canyon Maneuver Site (PCMS). The City of Trinidad supplies the water, but Fort Carson owns the water line, and a state prison taps into this line. When the line was leaking and needed replacement, the prison and Fort Carson shared the cost.

Fort Huachuca is partnering with Huachuca City to make use of nearby Huachuca City's water effluent at the post's wastewater treatment facility to provide more reclaimed water for Fort Huachuca's purposes. The lagoons of the Huachuca City's wastewater treatment facility were located in the flood plains of the Babocomari River, and a consent decree was issued for these to be decommissioned.[85] Fort Huachuca negotiated a deal to bring in the effluent from the city. As of summer 2012, Fort Huachuca was working to have the final piping installed for transferring wastewater from the City to the post's facility (as is explained more in Chapter Six). This multipurpose public-to-public partnership has multiple benefits for both partners: (1) Fort Huachuca will improve the operations of its own wastewater treatment facility (by making use of excess capacity), (2) Fort Huachuca will increase reuse or recharge with extra reclaimed water, and (3) Huachuca City obviates its need for a new wastewater treatment facility saving significant costs. This example, and the Fort Carson water line example, show how creative partnerships with local governments for sharing water infrastructure are doable and can have benefits for both the installations and local governments.

Fort Bliss and El Paso Water Utilities (EPWU) partnered in the development and construction of a desalination plant located on Fort Bliss but that EPWU owns and operates. The

---

[83] Kosloff, 2010.

[84] Rhonda Moore, "Castle Rock Still Wants WISE Partnership Water but There are Worries about Rates," *Castle Rock News-Press*, February 28, 2013.

[85] The state of Arizona has issued a "Notice of Opportunity to Correct" to the city's wastewater facility, but this case was resolved and closed due to the plan to reroute the effluent to Fort Huachuca.

$91 million Kay Bailey Hutchison Desalination Plant,[86] completed in 2007, was funded by different federal sources, including a $3.3 million engineering and environmental analysis by Fort Bliss, and EPWU. The facility treats brackish groundwater and helps protect water quality in the aquifer, which helps prevents brackish water from contaminating Fort Bliss' potable well sources. The total plant capacity is 27.5 MGD of potable water; however, it averages about 3 MGD of RO water. It can make up to a total capacity of 15 MGD RO water, which is then blended with Fort Bliss well water to produce the potable water. EPWU gives Fort Bliss multiple credits on its monthly water bill for the easement for this plant, the use of Fort Bliss well water, and for the original engineering and environmental analysis.[87]

Army installations now have more opportunities to develop and implement more public-to-public partnerships because of the passage of NDAA 2013 Section 331 in January 2013. It authorized the military Service Secretaries to enter into Intergovernmental Support Agreements (IGSAs) with state or local governments to provide, receive, or share installation support services if the agreement will enhance mission effectiveness or create efficiencies (such as through economies of scale savings), including reducing installation costs. Such agreements can include water services and infrastructure partnerships with nearby cities and towns. In summer of 2013, OACSIM created the Intergovernmental Support Agreement Program to help installations implement more public-to-public partnerships with communities.

Regional Partnerships

Regional collaborations and partnerships among state, regional, and local governments as well as other relevant entities are becoming more important for water planning and management. This is because of the joint use of water supplies, the fact that water is a public good, multiple stakeholder interests, increasing water demands, and increasingly scarce supplies of water. Regional water partnerships often have focused on collaboration in regional water supply and long-term planning, drought management and planning, and/or watershed management and planning. Some regional partnerships are closer to collaborations than true partnerships. For example, the Upper San Pedro Partnership, which was discussed earlier, is more of a collaboration than an in-depth partnership. In fact, this is often true partly because when more entities are involved it complicates the process. Entities are less likely to invest as much in the relationships, it is more challenging, and the collaboration is less likely to develop into an in-depth partnership. But it does not matter where a water partnership falls on the partnership continuum, because all can be useful in reaching the partners' goals.

We next provide some examples of regional partnerships. Since we already discussed some regional watershed partnerships in Chapter Two, such as the Chesapeake Bay Program and the

---

[86] Kate Galbraith, "Texas' Water Woes Spark Interest in Desalination," *Texas Tribune*, June 10, 2012.

[87] For more information, see El Paso Water Utilities, "Desalination: Setting the Stage for the Future," 2007.

Puget Sound Partnership, we will not discuss them here. We present three diverse examples in depth to illustrate the difference and complexities of some of these partnerships:

- The Integrated Regional Water Management (IRWM) Planning in California for planning and managing regional water resources. This example is interesting because it involves state and regional water planning, partnerships within partnerships and implementing diverse projects including ones that focus on water supply, watershed management, water quality, natural resources and flood management.
- The Great Lakes Partnership involving governors and its efforts to manage water uses, address water quality concerns, improve maritime transportation, promote beach safety and implement water conservation projects.
- The Western Recycled Water Coalition (WRWC), an innovative regional partnership focused on implementing more recycled water infrastructure. The WRWC is also an interesting collaboration example because it has grown over time both in membership (from seven to 22) and in geographic area.

All of these examples illustrate the types of regional partnerships in which Army installations could participate. Such regional partnerships to help plan, manage, and/or share water resources are likely to grow and are an important opportunity for installations to increase their long-term water security.

Integrated Regional Water Management Planning in California and the Greater Monterey County

As was discussed on Chapter Two, California's Integrated Regional Water Management (IRWM) Planning is a collaborative effort to manage all of a region's water-resource aspects. IRWM crosses jurisdictional, watershed, and political boundaries; involves multiple agencies, stakeholders, individuals, and groups; and attempts to address the issues and differing perspectives of all the entities involved through mutually beneficial solutions.

IRWM regions in California are defined as a contiguous geographic area encompassing the service areas of multiple local agencies. They are tasked with maximizing the opportunities to integrate water management activities and integrating water-management programs and projects within a hydrologic region as defined in the California Water Plan, a Regional Water Quality Control Board (RWQCB) region, or subdivision or another area that has been specifically identified by the California Department of Water Resources. These regions cut across political, watershed, and jurisdictional boundaries to ensure that all aspects of an area's water management are addressed comprehensively. The IRWM management body within the region must file an application with the state and meet certain criteria to be eligible for grant funding. The state Department of Water Resources offers grants, funded through several California state propositions, for activities and projects that help develop and implement local regional plans. Projects that improve water quality, flood control, watershed management, or wastewater infrastructure are eligible. By summer 2013, $1.8 million has been authorized by the state alone.

As of summer 2013, around 50 IRWM regions encompass approximately 87 percent of California's geographic area and 99 percent of its population.[88]

Integrated water management through regional collaboration and coordination is viewed as a key component of future water management in California. With various officials recognizing the diversity that exists among the regions in their water issues, they are making efforts to incorporate the various regional IRWM plans into the statewide water plan.[89]

One of these regions, the Greater Monterey County IRWM in the Central Coast of California, began developing a plan around 2002 to 2004 (its predecessor was the Salinas Valley Region). The location of the Greater Monterey County IRWM is shown in Figure 3.1.

Figure 3.1. Greater Monterey County IRWM

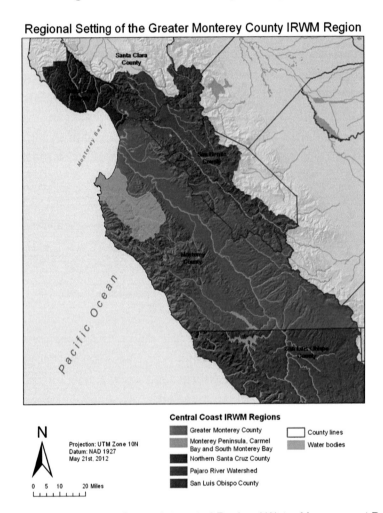

SOURCE: "Greater Monterey County Integrated Regional Water Management Program."

---

[88] Greater Monterey County Integrated Regional Water Management Program, website, undated; California Department of Water Resources, "Strategic Plan for the Future of Integrated Regional Water Management in California," undated.

[89] Paul Massera, "Integrated Water Management in the California Water Plan," IRWM Conference, May 24, 2011.

Around the time planning efforts for the region began, state funding for regional planning was unavailable because of severe budget issues. The Greater Monterey County Region's multi-year planning process was jointly funded with grants obtained by the Big Sur Land Trust and in-kind support from the Monterey Bay National Marine Sanctuary personnel as well as the Monterey County Water Resources Agency.[90]

At present, the partnership is operating under a memorandum of understanding (MOU) with associated bylaws, and includes representatives from 19 different entities that include government agencies, nonprofit organizations, educational organizations, water service districts, and private water companies, along with representatives from agricultural, environmental, and community interests. As stated in the MOU, its purpose is to:

> recognize a mutual understanding among entities in the greater Monterey County area regarding their joint efforts toward Integrated Regional Water Management (IRWM) planning. . . . Partners commit to participate in the ongoing process established pursuant to the Safe Drinking Water, Water Quality and Supply, Flood Control, River and Coastal Protection Act (also known as Proposition 84) and develop a comprehensive Integrated Regional Water Management Plan (IRWMP) for the Greater Monterey County IRWM Region.

The MOU establishes the mutual responsibilities of partners and the goals and objectives for the partnership. It also contains goals for the region's water planning; identifies strategies that must be considered within the integrated water management plan;[91] addresses the comprehensive scope of the integrated water management plan; and expresses the desire to work cooperatively and collaboratively among the public, private and nonprofit sectors. The MOU also requires the inclusion of all stakeholders in an open, collaborative process along with consideration of projects that may be too large for one entity (and therefore would require collaboration/partnership), and utilizing the most cost-effective means for advancing the plan's objectives.[92]

The implementing organization, the Regional Water Management Group, is comprised of members representing the above interests within the region. These organizations include the Big Sur Land Trust, California Water Service Company, Castroville Community Services District, City of Salinas, City of Soledad, Coastlands Mutual Water Company, Elkhorn Slough National Estuarine Research Reserve, Environmental Justice Coalition for Water, Garrapata Creek

---

[90] Monterey County Water Resources Agency Board Of Directors, "Greater Monterey County Regional Full Planning Packet," undated.

[91] These strategies are "ecosystem restoration, environmental and habitat protection and improvement, water supply reliability, flood management, groundwater management, recreation and public access, storm water capture and management, water conservation, water quality protection and improvement, water recycling, and wetlands enhancement and creation. Optional additional strategies that may be considered include: conjunctive use, desalination, imported water, land use planning, nonpoint source pollution control, promotion of the steelhead run, surface storage, watershed planning, water and wastewater treatment, and water transfers." See Monterey County Water Resources Agency Board Of Directors, undated.

[92] Monterey County Water Resources Agency Board Of Directors, undated.

Watershed Council, Marina Coast Water District, Monterey Bay National Marine Sanctuary, Monterey County Agricultural Commissioner's Office, Monterey County Water Resources Agency, Monterey Regional Water Pollution Control Agency, Moss Landing Marine Laboratories, Resource Conservation District of Monterey County, Rural Community Assistance Corporation, San Jerardo Co-operative, and Watershed Institute at California State University Monterey Bay.

The Greater Monterey County Regional Water Management Group is charged with developing and implementing the IRWM for the Greater Monterey County Region to include reviewing and selecting projects for funding. The regional plan was adopted in April 2013 after a public comment period, and seven projects were funded in 2011 through a competitive grant program. The projects being implemented range from stream restoration to water reclamation. Table 3.3 presents a list of the seven projects funded in 2011, totaling more than $4.1 million. Some actually are being performed through partnerships among several entities within the region. For example, the Integrated Ecosystem Restoration in Elkhorn Slough is a collaborative effort between the Elkhorn Slough Foundation in partnership with the Elkhorn Slough National Estuarine Research Reserve, the Moss Landing Harbor District, the Monterey County Water Resources Agency, and the County of Santa Cruz. Meanwhile, the Water Quality Enhancement of the Tembladero Slough is a collaborative effort among county planners, farmers, scientific researchers, and the community.

Table 3.3. Greater Monterey Region IRWM Projects Funded in 2011 ($ millions)

Project Title	Objective	Description	Funding
City of Soledad Water Recycling and Reclamation Project	Water Supply	Research the use of recycled water for agricultural and landscape purposes and build additional infrastructure for the use of recycled water.	$904.5
Castroville Well Treatment	Water Supply and Quality	Develop a deep well and treatment capability (for arsenic) to replace shallower wells experiencing seawater intrusion.	$581.0
San Jerardo Wastewater Project	Water Quality	Upgrade wastewater treatment to prevent nitrate contamination of groundwater used for drinking purposes.	$924.5
Integrated Ecosystem Restoration in Elkhorn Slough	Natural Resource, Flood Management, and Water Quality	Restore tidal salt marsh and grassland buffer to provide habitat and reduce non-point source pollution.	$822.2
Water Quality Enhancement of the Tembladero Slough and Coastal Access for the Community of Castroville	Natural Resource, Flood /Watershed Management, and Water Quality	Enhance the degraded water body, which has 14 listed pollutants that flow untreated into the Monterey Bay National Marine Sanctuary, through a variety of methods.	$341.7
Santa Rita Creek Watershed Approach to Water Quality Solutions	Flood/Watershed Management and Water Quality	Restoration of the creek degraded from agricultural and urban runoff through irrigation and nutrient management practices and revegetation.	$372.4
City of Salinas Stormwater Toxicity Reduction through Low Impact Development	Water Quality	Evaluation of the efficacy of bioswales in reducing contaminant concentrations in stormwater from the city.	$192.7

SOURCE: Greater Monterey County IRWMP, "Current Projects."

As these projects are completed, they will be evaluated on an individual basis as well as on their contribution toward the plan's goals.

The Presidio of Monterey's personnel should try to participate in the Greater Monterey Peninsula Integrated Regional Water Management Planning process even though the installation relies on water from California American Water Utility, because it would help the installation strategically understand where its water comes from and what is being done to help ensure this supply.

Great Lakes Partnership

The Great Lakes contain 20 percent of the world's fresh water, while the region generates nearly 30 percent of the U.S. gross domestic product and hosts about 60 percent of all U.S. manufacturing. The Council of Great Lakes Governors (CGLG) is an international partnership that seeks to protect the water in the Great Lakes Basin while enhancing economic growth. Governors of Illinois, Indiana, Michigan, Minnesota, New York, Ohio, Pennsylvania and Wisconsin have joined the premiers of Ontario and Quebec to formally cooperate on issues of mutual benefit since the early- to mid-1980s (Canadian provinces became associate members later).

Early on, the governors recognized that many water policy, environmental, and economic issues could be more effectively addressed on a regional basis and formed the council as a result. The first agreement entered into was the Great Lakes Charter, which created a mechanism for regional collaboration along with a guide for protective measures to be taken by each jurisdiction, and information-sharing practices to improve decision-making. In 1989, the governors formed the Great Lakes Protection Fund, which the states fund to support collaborative efforts that maintain and improve the Great Lakes ecosystem.

In the early 1990s, the CGLG also coordinated a response to the U.S. EPA water-quality regulations to ensure they provided the desired levels of flexibility and effectiveness. And more recently, the governors and premiers signed the Great Lakes-St. Lawrence River Basin Sustainable Water Resources Agreement (a good faith agreement between the states and provinces) and endorsed the Great Lakes-St. Lawrence River Basin Water Resources Compact (which became law on December 8, 2008, and legally binds the states). An MOU establishes the Council of Great Lakes Governors as the secretariat to the compact's council. Through these agreements, the states and provinces will develop and implement water conservation programs and use a consistent standard to review proposed uses of Great Lakes Basin waters (new diversions are banned for the most part).

The Council of Great Lakes Governors is engaged in several areas related to the management of the Great Lakes water resources and the promotion of economic development. Activities and projects have supported water management (implementing the compact, collecting and sharing water-use information); water quality protection and restoration; addressing aquatic invasive species; and the promotion of international trade as well as tourism. The most recent summit identified the following areas for the Council's focus:

- Seek improvements to maritime transportation through the Great Lakes-St. Lawrence River Maritime Initiative
- Expand international trade and export opportunities through passage of a resolution
- Promote the region's water-oriented companies worldwide with the Great Lakes-St. Lawrence Water Partnership
- Coordinate regulatory efforts for unwanted aquatic invasive species using a priority list
- Continue to enhance economic cooperation between Canada and the United States
- Promote beach safety through awareness and a beach safety application
- Prioritize water quality issues related to nutrient enrichment and harmful algal blooms
- More effectively coordinate regional efforts to monitor water quality and quantity.

Past projects have addressed pollution prevention, brownfields, water quality, spill protection, biomass use, recycled materials, and workforce issues; oftentimes in partnership with other organizations such as the Environmental Defense Fund, U.S. automobile companies, and the Printing Industries of America.[93]

---

[93] Council of Great Lakes Governors, "Overview," undated; Council of Great Lakes Governors, "Council of Great Lakes Governors Projects," undated.

The Western Recycled Water Coalition

The Western Recycled Water Coalition (WRWC) is a collaboration among cities, water and wastewater districts, and water utilities to develop and implement locally managed projects to help ensure water supplies in the western United States. Seven Northern California government agencies originally started it in 2009 as the San Francisco Bay Area Recycled Water Coalition. This collaboration focuses on acquiring funds from different sources for implementing recycled water infrastructure. Implementation projects are part of integrated regional water-management solutions. Members work together under a Memorandum of Agreement. Collaboration helps gives the members more visibility, unity, and "clout" when lobbying for state and federal funds. WRWC helps its members make a better case for why the funding is important. It also helps with educational and media outreach and to gain more public acceptance for the use of recycled water.[94] Other membership benefits include organized lobby trips to Washington D.C., the opportunity to increase the visibility of recycled water projects with the USBR, and monthly meetings to share information and collaborate on federal funding strategies.[95]

Since its start, the coalition said it has made significant progress in meeting members' goals:

> From 2009–2012, the Bay Area Recycled Water Coalition secured $38.1 million in federal Title XVI funds,[96] which were combined with $114 million in state and local resources to build eight recycled water projects and prepare feasibility studies for 14 more. The constructed projects now produce over 35,000 acre-feet of recycled water annually for irrigation and industry.[97]

As of March 2013, the WRWC had 22 members (shown in Table 3.4), having expanded beyond the San Francisco Bay Area to include the California Delta region and Central Valley.[98] At that time, the members were planning 20 new recycled-water projects. When funded and built, these projects were expected to annually produce 82,000 acre-feet of water that does not rely on limited surface and groundwater sources.[99] WRWC projects[100] have or will supply water

---

[94] Paul Burgarino, "Use of Recycled Water Trickles into Delta Region," *Contra Costa Times*, March 8, 2013.

[95] Jane Strommer, "Successful Bay Area Recycled Water Coalition is Expanding Across Mid-Pacific Region and Open to New Member Agencies," Central Valley/Sierra Foothills Water Reuse California, Central Valley/Sierra Foothills Chapter Newsletter, Vol. 1 No. 2, December 2012.

[96] Title XVI of P.L. 102-575, as amended (Title XVI), provides authority for USBR's water recycling and reuse program. Through the Title XVI program, USBR helps fund efforts to reclaim and reuse wastewaters in the 17 Western States and Hawaii. The Title XVI Program includes local project funding for planning studies and the construction of water recycling projects in partnership with local governmental entities. For more information, see U.S. Bureau of Reclamation, "Title XVI—Water Reclamation & Reuse Program," June 17, 2013.

[97] Dan Oney, "Recycled Water Coalition Expands Membership to Delta, Central Valley; Plans 20 New Projects," PUBLICCEO.com, February 22, 2013.

[98] Western Recycled Water Coalition, "Western Recycled Water Coalition," factsheet, March 14, 1013.

[99] Oney, 2013.

[100] For more information about the projects, see the project factsheets at WRWC, "Projects," 2013.

for irrigation of parks, school grounds, agriculture, and government facilities; for cooling processes for power plants; for manufacturing facilities; and for environmental restoration.

We illustrate three sample projects:

- The Redwood City Recycled Water Project consists of constructing tertiary treatment facilities, two 2.2 million gallon storage tanks, and a distribution pump station and pipelines to convey recycled water to the Redwood Shores, Greater Bayfront and Seaport areas. This project replaces the use of potable water with recycled water in a range of applications, including landscape irrigation, dust control, cooling, commercial vehicle washing, commercial window washing, and commercial laundry. The estimated cost is $72 million with part of it already built and paid for; Redwood City is seeking $3.5 million from local and state sources and $1.1 million from federal sources. Project benefits include reducing the city's use of potable water from the Hetch Hetchy water system, expanding the region's use of recycled water, and improving the reliability and conservation of the local water supply.[101]

- Palo Alto has been implementing the three-phase Palo Alto Recycled Water Pipeline Project for years and is seeking help through the WRWC with the third phase, which will expand the regional recycled water system to 50 additional large water-using customers in Palo Alto. The project includes about 5 miles of transmission pipeline and distribution pipes, a pump station, and a booster station. This project is expected to cost $33 million; Palo Alto is pursuing about $4 million in state grants, $8 million in federal funds and $21 million in local and state loans. This project's benefits include reducing dependency on state potable water supplies, reducing wastewater discharges to the South San Francisco Bay, improving water supply reliability for irrigation customers, and increasing water use efficiency.[102]

- The Ironhouse Sanitary District Recycled Water Project involves constructing a recycled-water distribution system for current and future landscape (such as parks and schools), agricultural, commercial, and industrial users in and near the City of Oakley, California. Proposed facilities include a distribution pump station, storage, and about 15 miles of transmission and distribution pipelines. Total estimated cost is $29 million, with plans to acquire $7 million in federal funds and $21 million through local or state grants and/or loans. Project benefits include reducing the dependence on Delta water supplies, reducing the stress on sensitive Delta fish species, improving water supply reliability, and increasing water use efficiency.[103]

---

[101] City of Redwood City, "Redwood City Recycled Water Project: Seaport," factsheet, undated.

[102] City of Palo Alto, "Palo Alto Recycled Water Pipeline Project," factsheet, undated.

[103] Ironhouse Sanitary District, "Ironhouse Sanitary District Recycled Water Project," factsheet, undated.

Table 3.4. Western Recycled Water Coalition Members

Central Contra Costa Sanitary District	City of Brentwood	City of Ceres
City of Fresno	City of Hayward	City of Modesto
City of Mountain View	City of Palo Alto	City of Pleasanton
City of Redwood City	City of San Jose, South Bay Water Recycling	City of Sunnyvale
City of Turlock	Del Puerto Water District	Delta Diablo Sanitation District
Dublin San Ramon Services District	Ironhouse Sanitary District	Monterey Regional Water Pollution Control Agency
San Jose Water Company	Santa Clara Valley Water District	Town of Yountville
Zone 7 Water Agency		

As the sample projects illustrate, the main benefits include reducing diversions from natural watercourses and aquifers; reducing the pumping of fresh water from the California Delta; improving surface water quality; attracting green infrastructure; increasing economic development; and providing sustainable, drought-resistant, and affordable water supplies for industrial, agricultural, government, and municipal uses.[104]

The WRWC may be expanding, with possible new members from as far away as Oregon and Nevada. We should note that this partnership would be considered a collaboration because members do not invest as many resources or time in the relationship. It could also be used as a model for Army installations to develop regional collaborations for recycled-water collaboration with state and local governments in other parts of the United States.

As mentioned above, most of the regional partnerships and collaborations involving installations have focused on environmental and/or encroachment issues rather than water. These regional partnerships and collaborations provide useful lessons for installations pursuing regional water partnerships. For instance, lessons learned include the fact that such partnerships take more time to develop, and it is important to involve all of the relevant stakeholders early in the process. In addition, opportunities are available for more regional water partnerships with communities because of Section 331 in the 2013 NDAA and because so many municipal water utilities are looking for cost savings with economy of scale opportunities. Participating in regional partnerships that help manage water supplies and plan for the future, including drought and other water uncertainties, can help installations ensure long-term access to water and improve installation water security. The state and local government regional partnership examples presented here help to illustrate some of the possibilities. Lastly, regional partnerships to help plan, manage, or share water resources are likely to grow and are an important opportunity for installations now and in the future.

---

[104] Western Recycled Water Coalition, "Meeting the Water Challenge in the West," undated.

123

Summary About Partnerships

Because traditional funding sources for installation water investments, such as MILCON, O&M, and ECIP, have become more challenging to obtain, especially with declining military budgets, Army installations have been using other alternatives to fund installation water investments. Such alternatives include privatizing installation potable water and wastewater systems and partnering with industry and utilities in ESPCs and UESCs. ESPC and UESC partnerships have become important methods to help fund and implement installation water investments, especially for providing capital, installing, and operating water efficiency technologies. However, since installation water costs tend to be so low, it is difficult to meet the required investment rate of return for such water projects. In addition, other types of installation partnership opportunities exist for Army installations to help fund water investments. An installation partnership is when an installation and other organization (or multiple organizations) agree to work together for mutual benefits, and it usually involves a long-term interdependent relationship.

Installations have started implementing other types of water partnerships, such as Fort Bliss partnering with EPWU in the development and operation of a desalination plant. However, installations can learn a lot from what some of their counterparts in state and local governments are doing to manage and plan for future water supply and shortages, and to find funding for investing in water and sewer systems. Public water utilities are facing many of the same challenges as Army installations, including declining budgets and the need to find funding to upgrade aging and degrading water infrastructure and ensuring access to long-term water supplies. Some of these public utilities are taking advantage of different types of partnerships to help fund water projects. Water partnerships have focused on sharing water supply and infrastructure, including pipes, reservoirs, or treatment facilities; collaborating in regional water supply planning; and water conservation. The key benefits to such water partnerships are lower costs by achieving economies of scale and sharing infrastructure and/or helping to ensure long-term water supplies. A main challenge is reaching agreement on sharing the costs, risks, and use of the facilities.

These partnerships can be grouped into three main types: public-private partnerships (PPPs), public-to-public partnerships (PuPs), and regional partnerships. An installation public-private partnership (PPP) is a collaborative arrangement between an installation and a private entity, which could be a for-profit company or a non-governmental organization (NGO). Local governments are partnering with private companies to run their water and wastewater infrastructure, such as Schenectady, N.Y., which has partnered with a private company to run its wastewater treatment plant. Such public-private partnerships are another opportunity for installations to leverage industry technical expertise and save money. Such partnerships also can be an alternative to installation UP projects. In fact, they often are similar to UP, but without selling off the assets.

An installation PuP is a partnership agreement between an installation and a local, state, or other federal agency, usually over the long term. PuPs allow public utilities to pool resources, and therefore buying power and technical expertise. The benefits of a PuP are scale, public efficiencies, and lower costs. The two government agencies often bring different expertise to the table, such as financial, technical, or shared facilities. For example, in August 2008 the cities of Lake Oswego and Tigard, Oregon, signed a partnership agreement for sharing drinking water resources, infrastructure, and costs, which has saved both cities significant costs in operating their water supply systems. Similarly, installations such as Fort Carson and Fort Huachuca have taken advantage of PuPs with local governments to save costs on water infrastructure. Greater opportunities now are available for such partnerships between installations and communities because of Section 331 in the 2013 NDAA, which provided additional authority for such installation PuPs and because so many municipal water utilities are looking for cost savings with economy-of-scale opportunities.

An installation regional partnership is when a military installation has an agreement with multiple entities, which may be public and private, within a region for mutual benefit. Many Army installations have experience with regional partnerships to address encroachment concerns through the Army Compatible Use Buffer (ACUB) program, although fewer have been involved in regional water partnerships. Because of pressures from over-allocated water resources, droughts, and competing demands for finite water sources, state and local governments are developing and implementing more regional water planning, supply, management and infrastructure collaborations and partnerships.

These water partnerships, which are more complex to develop and implement (as installations have learned from their regional ACUB partnerships), also are opportunities for Army installations. For example, consider the Integrated Regional Water Management (IRWM) Planning in California, which involves state and regional water planning, partnerships within partnerships and implementing diverse projects including ones that focus on water supply, watershed management, water quality, natural resources and flood management. This partnership process provides an opportunity for Army installations within different regions of California to learn more about these issues and to help ensure installations' water supplies. Second, consider the Western Recycled Water Coalition (WRWC), an innovative regional partnership focused on investing in implementing more recycled water infrastructure, which could be used as a model for Army installations to develop regional collaborations for recycled water collaboration with state and local governments elsewhere.

All of these PPPs, PuPs, and regional partnerships to help plan, manage, develop, and/or share water infrastructure and resources are likely to grow and offer an important opportunity for installations to increase water investments and long-term water security.

4. Water Market Mechanisms

Water market mechanisms are approaches that treat water as a commodity and that can be used to transfer water among users, reallocating water-using price.[1] Because they are voluntary and have the potential to move water efficiently, water market mechanisms are viewed in policy discussions as one possible approach for more effectively managing water resources. As described in Chapter Two, the evolution of water management policies and approaches (from an emphasis on developing and conveying water supplies to encouraging more efficient water use and reallocating water among users) are complemented by market approaches. As new demands for water use arise, and water resources approach their limits, market approaches may offer an efficient and effective means for sharing this finite resource. When the price of water reflects its true economic value, markets can be used to encourage conservation and direct water to high-value uses.

However, water markets are controversial, in large part because water is essential for life and as a result does not fit neatly into the category of a private good to be traded. In fact, various states' water rights laws treat water as a private, common, or public good, determining the viability of using water market mechanisms. In addition, the transfer of water from one user to another may have third-party effects (in part because water rights are usufructuary).[2]

This chapter provides an overview of several of the more common market mechanisms used to allocate water within the United States. The approaches included in this chapter are quite varied in respect to the specific actors and purposes, but central to each is the use of price to manage water. (One approach, water quality trading, facilitates the exchange of discharge credits among polluters in order to reduce overall pollution.) Many facilitate trade because they have procedures for establishing a price and making the transaction occur more smoothly and quickly. Some actually move water (transfer wet water[3]), while others trade paper rights. Water market mechanisms included in this discussion are basic leasing and selling, water auctions, water banks, block pricing, and water quality trading. Table 4.1 provides an overview of the market approaches used for various aspects of water management described in this chapter.

---

[1] In contrast, some argue that access to water is a human right that should be provided regardless of ability to pay, and resist treating it in the same way as other commodities. See, for example, Holly Young, "Live Q&A: water, public good or private commodity?" *The Guardian Global Development Professionals Network*, September 1, 2014.

[2] Meaning the right is to use the water, not for the water itself.

[3] Wet water refers to water rights for actual water, in other words, water that a rights holder would have used, or stored, as opposed to simply the paper right to water, since this water may or may not be available. When purchasing a water right, it is critical to ensure that the water right is more than a paper right; in other words, that the water will actually be available to the purchaser and is not simply a certificate. Wet water has four potential sources: stored water, excess water that cannot be stored, water available because of conservation measures, and groundwater.

Table 4.1. Market Approaches to Water Management

Market Approach	Actors	General Purpose	Price Determination
Water market transfers	Water rights sellers and buyers; may include government or water authority	Transfer water rights either temporarily or permanently	Negotiated among buyers and sellers
Water banks	Water rights sellers and buyers, Bank as administering authority	Transfer water rights to an intermediary (the bank)	Can be established by the bank or negotiated
Water auctions	One buyer (or seller) and multiple bidders	Minimize the cost of procuring water (or maximize revenue from sale of water)	Established through competition among bidders
Block pricing	Utility	Encourage conservation while covering costs	Determined by utility rate structure
Water quality trading	Point and nonpoint dischargers	Reduce the cost of complying with water quality regulations	Negotiated by parties

Some water market mechanisms are well established and routine in select regions of the country (such as the water markets in California, Colorado, and Texas). However, in other areas, more limited pilots have been initiated that seek to either expand the use of these approaches or apply novel market approaches. Overall, the use of market-based approaches is not widespread, and therefore they are likely a longer-term issue for installation personnel. As these approaches grow and develop, they may affect the price of water (and therefore the value of existing water rights), create new opportunities for installation water resources, and alter the financials of water-related investments. For example, market approaches could potentially be used to expand water supply should growth on the installation occur, sell water to finance new water infrastructure or to increase water supply reliability during dry years. Developing an understanding of market approaches, and their applicability to Army installations, will improve long-term planning and investments. This section discusses the potential benefits and challenges of using water market mechanisms in general, followed by a discussion of several specific water market mechanisms.

Water Market Mechanisms Can Be an Effective Way to Allocate Water

Water market mechanisms are attractive as a means of allocating water because they are economically efficient, responsive to changing conditions, and are voluntary. And since decisions are decentralized (the transfer participants determine the criteria for water transfers to occur), localized conditions and considerations are more readily incorporated into the process. The need to transfer water may arise because of over-appropriation of water rights, historic aquifer overdraft, drought, habitat and environmental concerns, population growth, or industrial development of water-intensive industries. All of these situations are likely in the future to affect a wide range of regions in both the west and in the east. Because of the flexible, responsive, and voluntary nature of water market mechanisms, these approaches can be preferable to more traditional ways of getting water to new users. For example, building new supply infrastructure (such as dams and reservoirs or desalination plants) is costly, can take years, and has significant environmental effects. Lawsuits, another conventional option for allocating existing supplies when competition occurs for these resources, also are costly and can take years to resolve.

Water market mechanisms use price to distribute water among users. In an efficient market, the last unit of water sold will be sold at the price that equals the long-run cost of providing the last unit of water (marginal revenue equals marginal cost). It is at this point in which the amount supplied and demanded for a given price yields the greatest value (for example, when social welfare has been maximized). When water prices reflect its true value (and if water can legally and physically be transferred), the opportunity cost of using water creates an incentive for users to ensure that water is being put to its best use. Water users will be more likely to monitor water use, use water more efficiently, and apply it to the most productive uses. Basically, the price of water contains a lot of information about water availability, quality, reliability, infrastructure, individual actors, and the resource's uses in a market.

In contrast, it is universally noted that in the United States, water prices do not reflect the full cost of providing water. For example, most utility prices, which are regulated, do not include long-term infrastructure costs required for future supply, nor do they fully reflect water scarcity.[4] Nor does irrigation water supplied to farmers typically include the costs associated with federally subsidized infrastructure. Artificially low prices make it economical for a greater range of water uses, and those uses may not include the most valuable ones. Nor do artificially low prices provide the proper incentive to conserve water or invest in water projects. As a result, there is some loss in overall welfare. Buyers and sellers do not establish these prices, so they do not contain all the information on water supplies, costs, potential uses, and their values that market-driven prices contain. Price, in these contexts, is not a reliable indicator of economic value.[5]

---

[4] Utilities are regulated to ensure that reasonable cost water is available to consumers and that utilities do not misuse their monopoly powers.

[5] For a more detailed discussion of the economics of water markets, including overall welfare, see Sheila M. Olmstead, "The Economics of Managing Scarce Water Resources," *Review of Environmental Economics and Policy*, Vol. 4, No. 2, 2010, pp. 179–198; Terry Anderson, Brandon Scarborough, and Lawrence Watson, "Tapping Water Markets," Washington. D.C.: Resources for the Future Press, 2012; W.M. Hanemann, "The economic conception of water," in *Water Crisis: Myth or Reality?* (Edited by P.P. Rogers, M.R. Llamas, L. Martinez-Cortina), London: Taylor & Francis, 2005; and U.S. Environmental Protection Agency, "The Importance of Water to the U.S. Economy Part 1: Background Report," September 2012; and Glennon, 2009.

Another benefit to water market mechanisms is that they are voluntary and flexible in response to changing conditions such as the needs of new uses, temporary fluctuations in supply, or longer-term water scarcity. These changes ultimately will be reflected in the price of water, and buyers and sellers will respond according to their individual preferences. This is in contrast to more prescriptive approaches to allocating water that rely on judgments or assessments of value for classes of users (such as rationing during drought, prioritization by first-use or type of use) and those which may not be as readily altered (utility rates that are established annually and therefore are not responsive to short-term supply fluctuations). Furthermore, these agreements between the seller and the buyer can be tailored to each one's specific needs, resolving potential conflicts a priori through the negotiation and review processes. This is in contrast to many riparian states where the courts are the only option for clarifying water allocation and disputes are settled through litigation, which may take years to resolve and which frequently leads to unsatisfactory or inconclusive results. For example, Georgia, Alabama, and Florida have been taking court actions over the Appalachicola-Chattahoochee-Flint and Alabama-Coosa-Tallapoosa River Systems on and off since 1990. Scarce water supplies and competing demands among municipal users, agriculture, hydropower, species protection, and navigation, fishing and recreation have led to productivity losses in these states.[6] Therefore, because water market mechanisms are *voluntary* transfers among parties, they may avoid some of the political animosity over more arbitrary approaches (such as legal or regulatory) that either can slow or block the reallocation of water.

However, allowing price alone to determine water allocation within the United States has not been practiced. This is true primarily because water itself is considered a common or an essential commodity. It is necessary for growing crops, developing cities, sustaining ecosystems, providing energy and many other purposes that may not be readily characterized in economic terms. As water flows, multiple users can use it. As a result, water rights and water transfers are regulated. However, states vary in their approach to allocating and regulating water *use*, which in turn will affect the likelihood that a specific market mechanism can be used. Before we describe specific water market mechanisms, we must briefly discuss how water rights vary across the United States and how that relates to water market mechanisms (for more details on water rights, see Chapter Two). Water rights, which are determined at the state level, are the critical determinant as to whether or not water market mechanisms even are feasible, since they define the terms under which water can be used.

---

[6] Water Policy Institute, "Water Wars: Conflicts over Shared Waters, A Case Study of the Apalachicola-Chattahoochee-Flint and Alabama-Coosa-Tallapoosa River Systems in the Southeastern United States and the Broader Implications of the Conflicts," Hunton & Williams LLP, March 2009; and Dave Williams, "Georgia Wins Tri-state Water Ruling," *Atlanta Business Chronicle*, June 28, 2011.

Water Rights Regimes Determine Market Feasibility

In eastern states, water is allocated according to the riparian doctrine. It links water use to land ownership; in other words, landowners have rights to the water that runs through or adjacent to their property. Allowable water use is not quantified, but subject to reasonable use without interfering with other riparian users. Non-appurtenant[7] water use is rare, and in times of low supply there is no definitive means to reallocate the reduced supply of water. In the strictest sense, water use is treated as a common good and any disputes regarding its use are settled using tort law. Since water rights are not quantified and are limited to appurtenant users, no real opportunity for market mechanisms is available in these states.

A subset of eastern states follow regulated riparianism[8] in which the state has some ability to manage water supplies through the permitting process and state water planning. One can argue that in these states, water is treated more as a public good. Disputes regarding water use are managed through the state permitting agency, typically following a priority-based system of reasonable water use. The result is greater certainty regarding the likely supply of water to a particular permit holder, at least for the duration of the permit. Additionally, in theory—should states allow it—water permits, which quantify allowable water use, can be transferred. In practice, this has been more difficult to do for a variety of reasons discussed in later sections.

Western states allocate water using the prior-appropriation doctrine, allocating water rights permits to first-in-time users for beneficial uses. Since water permits quantify the allowable diversion, water trading is feasible.[9] Indeed, water permits are tradable assuming certain conditions are met such as beneficial use (which can include maintaining in-stream flows), maintaining the point of diversion, and the quantity diverted. By maintaining these conditions, regulators seek to reduce third-party effects. Should water availability decline, however, senior rights holders have first priority and can use the total amount of water identified in the permit, regardless of effects on downstream users. Thus senior rights are more valuable than junior rights. Therefore, water permits within prior appropriation are the most akin to private goods (with some limitations) that are more amenable to trading.

Therefore, western states using the prior appropriation doctrine are much more likely to have experience with water market mechanisms. Under regulated riparianism, water trading has the

---

[7] *Non-appurtenant* refers to a water right that is not attached to land ownership; it is separable and usually refers to a right to water that is used some distance from the source.

[8] See Joseph W. Dellapenna, "Special Challenges to Water Markets in Riparian States," *Georgia State University Law Review*, Vol. 21, No. 2, Article 9, Winter 2004. Dellapenna coined this term.

[9] The quantification of water rights in the West is performed through an adjudication process. While these processes vary from state-to-state, some can take decades. Often historic uses are not well quantified or legally defined, as is the case with Native American water rights as well as with military installations' water rights. Trading rights then becomes more difficult in these situations, since without legally defined water rights trading cannot occur. See David Brookshire, Philip Ganderton, Mary Ewers, Bonnie Colby, and Steve Stewart, "Water Markets in the Southwest: Why and Where?" *Southwest Hydrology*, March/April 2004, pp. 14–15.

potential to occur, while under traditional riparianism water trading is much more difficult (if not impossible). As a result, most of the subsequent discussion of water market mechanisms is based on the experiences in the western states. The application of these approaches in the East is rare.

Challenges in Implementing Water Market Mechanisms

Market mechanisms can be an effective and efficient way to allocate water. However, the existence of externalities to water use, water rights heterogeneities, and transaction costs create challenges for more widespread use of these approaches.[10] State laws determine how water rights are determined, quantified, and transferred. First and foremost, every state has a requirement for beneficial or reasonable use of water rights. In western states, if the standards for beneficial use are not met, water rights can be forfeited, usually for non-use. In eastern states water withdrawal permits may be terminated. Water banks and leasing are examples of water market mechanisms that only are viable if forfeiture does not occur with rights transfers (Utah, for example, was one of the first states to make changes to allow a water bank, the origins of which date to 1932).[11] State water law also will determine the reliability of the water right—in other words, how likely the water right actually will access wet water. As explained in the previous section, in pure riparian states water rights are not well quantified, making widespread transfers infeasible (and illegal).

If states allow transfers through markets, there may be limitations on the purposes for which the water may be used and when and where they can occur because of the externalities of water use. Referred to as third-party effects, transfers that change the point of diversion, or the way in which water is used, are assessed for effects on others. For example, water taken out of agricultural use for another purpose will negatively affect the agricultural suppliers and employees, as well as the tax base in the region, but may positively affect water quality when fewer pesticides enter waterways. Because of concerns over third-party effects in particular, inter-basin and interstate transfers are especially problematic. Oregon and many other states do not allow inter-basin transfers.[12]

Additionally, state agencies must approve all transfers before they can occur. These procedures vary from state to state. Typically the burden of proof is on the seller. In this regard, California has a particularly cumbersome and complex process for approving water rights transfers involving multiple agencies at all levels of government (such as the California Water

---

[10] Economic efficiency is when the price of water is at the point at which the marginal costs of providing the water are equal to the marginal benefits gained from its use. It is at this point at which social welfare is maximized.

[11] Sandra Zellmer, "The Anti-Speculation Doctrine in Water Law: Ghost-busting, Trust-busting, or Ensuring Reasonable, Beneficial Use?" American Bar Association Section of Environment, Energy, and Resources, 26th Annual Water Law Conference, San Diego, Calif., February 21–22, 2008, p. 13.

[12] Ray Hartwell, and Bruce Aylward, "Auctions and the Reallocation of Water Rights in Central Oregon," Deschutes River Conservancy, River Paper Series No. 1, April 2007, p. 2.

Resources Control Board, Department of Water Resources, and the USBR) to determine if the transfer would significantly affect the overall economy.[13] While these laws and procedures in theory ensure that water allocation is equitable and serves the public values, there is a tradeoff with creating uncertainty and adding time and costs to using water market mechanisms.[14] These approval processes and procedures create inefficiencies, or "transaction costs" in the economics nomenclature. Thus, state water rights laws and administrative procedures are important factors for determining the viability and scope of water market mechanisms. For example, an analysis of transfers in California found that for those counties with a review process, fewer transfers occur overall. A greater percentage of the remaining transfers also are in-county (and yield lower prices in general), likely due to the up-front costs and potential for negative public opinion.[15]

Other transaction costs that influence the attractiveness of water markets are those associated with such issues as:

- Identifying and attracting buyers and sellers
- Water rights verification and substantiation
- Administrative procedures and negotiations
- Water conveyance infrastructure.

The relative importance of these costs will affect the numbers of buyers and sellers and ultimately the observed market price for water. For example, where markets are thin (fewer buyers or sellers), an individual may skew price away from the most efficient in an economic sense. The success of the water market mechanisms will depend on these factors, and on having clearly defined, verifiable water rights that can be transferred.

The applicability of water market mechanisms also will depend on the specific characteristics of the location, such as the reason for needing to transfer water, available infrastructure, hydrological conditions, soils, microclimate, competing demands for water, availability of water, and political climate. For example, if no physical way is available to transfer water, or if transporting water is expensive (due to the cost of capital, energy, and conveyance losses) relative to the value that can be derived from it, then water market mechanisms will not be cost-effective. And in the Columbia River Basin in Washington state, attaining economies of scale in market transactions for improving in-stream flows has been hampered by the unique nature of each transaction, as noted below:

> Many Qualified Local Entities (QLEs) are of the opinion that this [economies of scale] is unlikely because the context of each transaction is unique, and so it is difficult or impossible to turn water for in-stream flow into a commodity. The particular hydrology, groundwater surface water interactions, and landowner

---

[13] Libecap, 2010, p. 63.

[14] Some argue that the state laws and procedures primarily preserve the status quo. See Libecap, 2010.

[15] Ellen Hanek, "Who Should be Allowed To Sell Water in California? Third Party Issues and the Water Market," Public Policy Institute of California, July 2003, p. viii.

context varies from reach to reach, and as a result, requires a significant investment to understand and deal with. Information and experience gained in one reach may not be directly applicable to other reaches.[16]

Indeed, many water transfers have occurred in areas where water rights are relatively homogenous and infrastructure exists, such as the Colorado–Big Thompson Project (C-BT), California's Central Valley Project and Imperial Irrigation District, or Arizona's Central Arizona Project (CAP).

Even though water market mechanisms are regulated, concerns over equity, maintaining social values, and access to water can contribute to a political environment that limits participation. For example, some areas fear that those with the monetary resources can trump water uses that may have poorly defined economic values (such as environmental or other social values, which are difficult to put in economic terms) or lead to an inequitable distribution of resources. For example, the perceptions are stark surrounding a potentially high-bidder, such as an industrial farm or an oil and gas company, that can afford to pay higher prices for water than a socially valued family farm. However, the flip side also is true: Farmers are able to gain income by selling water rights that are temporarily in excess or may not generate as much income if used for low-value crops. Political conflicts can arise when traditional water users have to compete in markets with newer demands for water. An example would be when municipalities or environmental trusts look to use water that agriculture communities typically use. Either way, expanding the use of water market mechanisms necessarily will involve engaging individuals who may be unaware of how they work, or are skeptical of the process. This will be costly and time-consuming. Concerns over water speculation also surround expanding the use of water market mechanisms. There have been occasional examples of speculation as a possibility, such as the selling of Great Lakes water or Ogallala Aquifer water. However, state water laws requiring beneficial use and forfeiture limit speculation, and the laws of all western states specifically ban speculation.[17]

Finally, all of the challenges described above—the heterogeneity of water rights (in terms of priority, reliability of supply, quantification, location, etc.), information asymmetries, regulatory policies and procedures, and the highly localized (and in some cases politicized) nature of water markets—can yield water prices that are not completely efficient in an economic sense. Any further development and expansion of markets must continue to address these challenges in order to gain the full benefits of markets. Therefore, while transfers have been occurring for decades in some states, many of these mechanisms are emergent and have not been used on large scales.

---

[16] Jared Hardner, and R. E. Gullison, "Independent External Evaluation of The Columbia Basin Water Transactions Program (2003–2006)," Hardner and Gullison Associates LLC, October 7, 2007, p. 35.

[17] Three important exceptions to water anti-speculation law are municipal water supplies, foreign water (water not part of the natural stream either from other basins, demining water or treated wastewater), and in-stream flows. For more on the issue of water speculation, see Zellmer, 2008.

The success of these approaches continues to be on a trial-and-error basis. So while in theory such transfers may be promising, in practice details matter.

The following sections describe some of the more common water market mechanisms that have been discussed in the literature. These include water transfers through sales and leases, water auctions, water banks, block pricing, and water quality trading. What these market mechanisms have in common is that they use price to affect behavior. However, these mechanisms serve very different purposes. Some are used to transfer water between buyers and sellers, while utilities use another to encourage conservation. The last is used to offer alternative approaches to complying with water quality regulations. A brief overview of each approach is provided along with illustrative examples. Some of these approaches have been used successfully, while others have not advanced beyond the proposal phase. Overall, the use of water market mechanisms is still an evolving and nascent concept for water allocation.

Water Transfers: Leasing and Selling Water[18]

Water rights (or entitlements) are a legal determination giving the right to use water. Water rights are established and managed by state laws and regulations that give the owner the right to use the water in perpetuity, as long as it is put to beneficial use on a continuous basis. In those states where the water right is not tied to land ownership (the prior appropriation states in the west), the legal documents establishing water rights typically specify the amount of water, diversion point location, and date of priority. These are the critical characteristics of a water right. In these states, water rights can be leased, sold and exchanged (provided that certain conditions on location of diversion, type of use and third party affects conform with the particular state's laws). The Western Governors' Association (WGA) and WSWC define a water transfer as follows:

> A water transfer is a voluntary agreement that results in a temporary or permanent change in the type, time or place of use of water and/or a water right. Water transfers can be local or distant; they can be a sale, lease or donation; and they can move water among agricultural, municipal, industrial and environmental uses.[19]

Most water transfers involve leasing or selling water or a water right. This can be done in an informal market or a formal one that a government entity or water authority administers.

---

[18] Material from this section is drawn from Elizabeth Basta and Bonnie Colby, "Water Market Trends: Transactions, Quantities, and Prices," *Appraisal Journal*, January 1, 2010; Gregory W. Characklis, Brian R. Kirsch, Jocelyn Ramsey, Karen E. M. Dillard, and C. T. Kelley, "Developing Portfolios of Water Supply Transfers," *Water Resources Research*, July 11, 2005; Zack Donohew, "Water Transfer Level Dataset," University of California-Santa Barbara document, undated; Terence R. Lee and Andrei S. Jouravlev, "Prices, Property and Markets in Water Allocation," Santiago, Chile: United Nations Economic Commission for Latin America and The Caribbean, 1998; and Michael O'Donnell, and Bonnie Colby, "Dry-Year Supply Reliability Contracts: A Tool for Water Managers," Department of Agricultural and Resource Economics, Tucson, Ariz: University of Arizona, October 2009.

[19] Western Governors' Association and Western States Water Council, *Water Transfers in the West: Projects, Trends and Leading Practices in Voluntary Water Trading*, Denver, Colo., and Murray, Utah, December 2012, p. 8.

Several types of water transactions may occur between a demander and a provider when water is traded separately from the land. Within this list, leases and permanent sales are the most common type of transfer used.

- **Permanent sales**: Such sales are the "permanent transfers of water rights from one party to another. Water rights are typically defined on an annual basis. Therefore, an entity holds the right to use a specified amount of water each year for perpetuity."[20] Sales are useful when a permanent shift in water use has occurred.
- **Leases for a specified term**: Leases "involve the transfer of a specified amount of water annually for a certain time period. Thus, water rights are not exchanged, only the water associated with the right."[21] The lessor can use leases to earn revenue while not giving up water rights. Used for spot-markets (short-term immediate transfer of water), leases of one year or less are common, but leases with terms of five or 10 years occur as well. Leases frequently are used among farmers within an irrigation district and are useful when a temporary change in water supply or demand occurs. They also are used to reallocate water during peaks and valleys when supply in the long term is variable or if a permanent transfer is not desirable. Leases avoid some of the political concerns related to a permanent transfer of water rights that occur with sales. Under a typical lease contract, the lessee pays the water right owner (the lessor) periodic payments, but there can also be an up-front payment to initiate the lease. "In the short-run, leases are usually a cheaper source of water than permanent transfers, but water can and will fluctuate in price, water supplies are only secure until the contract expires, there is also an expense in the constant renewal costs for those who depend on short-term leases, and there is a risk of default, for legally the lessee has only a contract as protection, not a property right."[22]
- **Contingency or option contracts**: A special type of lease is a contingency or option contract. These contracts allow for the temporary right to use water if specified conditions occur in the future, but the permanent right remains. There are many types of contingency or option contracts, but in a typical option contract, the purchaser provides the supplier an upfront payment to reserve the right to lease water at a later date at an agreed-upon "exercise" price. Among other benefits, these types of water transfers are useful for hedging against short-term price volatility as well as for improving the reliability of water availability during dry years.[23]
- **Exchanges:** Exchanges cannot be classified as sales or leases. These are a small percentage of transfers, the most common type is when "a developer who gives a water right to a city in exchange for the city allowing the developer to connect a development to city utilities."[24]

---

[20] Donohew, undated.

[21] Donohew, undated.

[22] Lee and Jouravlev, 1998.

[23] Dry-year reliability contracts are contracts entered into to establish firm water supplies during low-supply periods. Option-contracts are a primary type of contract used for dry-year reliability, but other contract types can be used for this purpose as well. For more on this topic, see Michael O'Donnell and Bonnie Colby, "Dry-Year Supply Reliability Contracts: A Tool for Water Managers," Tucson, Ariz.: University of Arizona Department of Agricultural and Resource Economics, October 2009.

[24] Donohew, undated.

- **Wheeling transfers:** These involve the exchange of water with different levels of water quality so that water supply is better aligned with its use. For example, farmers can transfer high-quality water to municipalities for use as potable water while municipalities can send farmers lower-quality water for irrigation use.

These types of transactions may occur between a single provider and demander. Alternatively, a centralized authority may use them to transfer water among a group of providers and demanders. Each transaction will include price and volume information.

> The annual flow amount is the quantity of water that is transferred in the first year of a contract and only in the first year. This is the typical way to quantify a water transaction, by the annual acre-feet assigned to the water right. However, not all transactions are for set amounts of water. For instance, a transaction might be the purchase of a share in a ditch company, which provides an expected or average flow of water but whose actual yearly flow depends on the ditch company's allocation of water. It is often possible to determine from the details of a transaction what type of flow is expected, and what are the maximum and minimum quantities associated with the right.[25]

In addition to price and volume information, such agreements contain details on the location of water; the timing of the transfer; the likely water quality; demander and provider responsibilities; and liabilities for failure to compete the transfer. A demander also may have to enter into additional agreements to physically transfer the water using canals, aqueducts, or piping.

> Researchers note the following potential complications to water transfers: a) the financial and environmental costs of moving the water from one place to another may exceed the benefits gained from trading water;
>
> b) water rights can be difficult to measure or are vague;
>
> c) geographical boundaries and legal restrictions may limit water trading—for instance, state law may not permit inter-basin transfers or interstate transfers;
>
> d) statutory protection and political considerations may require consideration of environmental or third party impacts resulting from the transfer;
>
> e) there may be loss due to evaporation or seepage during conveyance.[26]

States can play a role in facilitating transfers with assistance, guidelines, and policies that reduce the time and risk associated with these complications. For example, one of the biggest difficulties for a state when reviewing water transfers is that water rights are, in effect, paper rights. The actual amount of water that can be transferred is determined through an analysis of the amount of water that actually had been diverted and used historically. To facilitate transfers, states could develop and provide consumptive use and return flow factors that may be readily modified for site-specific conditions. Additionally, state guidelines or assistance could be used to

---

[25] Donohew, undated.

[26] O'Donnell and Colby, 2009, p. 5; Western Governors' Association and Western States Water Council, 2012, p. 35.

help suppliers and demanders navigate the multiple federal, state, and local agencies that are frequently involved in the transfer process. And a variety of approaches may help address third-party effects, which can be the most complex and prolonged part of the process. Streamlining these complications will help reduce cost and uncertainty and extend the use of water transfers.[27]

The other primary challenge for water transfers is negotiating a fair price when information on similar transfers is not readily available *a priori*. Perhaps the simplest approach to support active and efficient market transfers is to have many buyers and sellers and publish past transactions so that price information is easily obtained. As mentioned earlier, water rights are not homogeneous (in terms of priority, reliability of supply, quantification, location, etc.) and depend on conveyance infrastructure, water quality, and other factors, making it difficult for buyers and sellers to compare prices. In many cases water markets are thin—there are not many buyers and sellers, so individual participants may have market power to influence prices. Therefore, ensuring that good price information is readily available may help facilitate more economically efficient market transfers. To the extent this information is not available to buyers and sellers, or costly to attain, the market may fail to reallocate water to higher-value uses.

Overview of Water Leasing and Selling in the West

Water has been leased and sold in the western United States for decades. Historically these transfers were common within the agricultural sector, but in the last couple of decades or so agriculture-to-municipal transfers have expanded as population growth and development occurred (power companies also have been buyers of water). More recently, environmental concerns have created a market for acquiring and retiring water rights as well as biofuel and bio-ethanol production. The main sellers of water rights include farmers; municipalities with surplus entitlements; Native American tribes; bankrupt businesses; and decommissioned industrial users, such as power plants and mines.

The active states for water transfers include Colorado, Arizona, California, Idaho and Texas. While Colorado dominates in terms of *numbers of transactions,* Arizona, California, and Texas are the top three in terms of the *amount of water committed,* while Colorado and Idaho are ranked fourth depending on the specific measure used.[28] An analysis of the water transfers from 1987 to 2005 found that agriculture is the primary source of water transactions and that the number of market transactions increased over that period, primarily due to agriculture-to-urban

---

[27] Western Governors' Association and Western States Water Council, 2012, pp. 56–59.

[28] This analysis looked at annual leases, multi-year leases and sales. Two different measures for the amount of water transferred are used because different mechanisms, annual leases, multi-year leases and sales are employed to transfer water. While annual leases commit water volumes for one year, sales and multi-year leases commit water volumes several years. So in this analysis, the volume of water transferred is measured in two ways: the annual volume transferred and the total committed volume to illuminate underlying trends in water transfers. See Jedidiah Brewer, Robert Glennon, Alan Ker, and Gary Libecap, "2006 Presidential Address Water Markets in the West: Prices, Trading and Contractual Forms," *Economic Inquiry*, Vol. 46, No. 2, April 2008, pp. 91–112.

trades.[29] While the annual flow of water transferred is relatively constant over this period, the amount of water committed is increasing as long-term leases and permanent sales are becoming more common. Figures 4.1 and 4.2 show the total quantity transferred by state (in terms of annual flow, not the total committed for a period of time) and the number of transactions within each state by transaction type (lease or sale) from 1988 to 2009.

Figure 4.1. Total Volume of Water Transferred by Transaction Type and State (1988–2009)

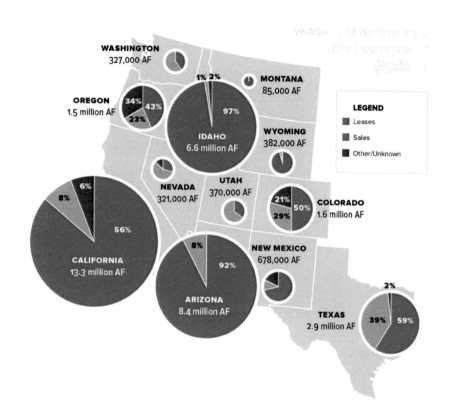

SOURCE: Tom Iseman, "Innovative Water Transfers: West-Wide Practice and Potential," American Water Resources Association, April 27, 2011.

---

[29] This was identified in the trade publication *Water Strategist,* whose data on water transfers was the only publically available information on water transfers and was widely used. However, the dataset had gaps; only 12 of the 17 states in the WGA are included, some water transfers are missing and the final year of publication, 2009, is incomplete. *Water Strategist* data no is longer openly available (as of 2010), but had been used in many studies of water market activity. See also Western Governors' Association and Western States Water Council, 2012, p. 14.

Figure 4.2. Number of Water Transfers by Transaction Type and State (1988–2009)

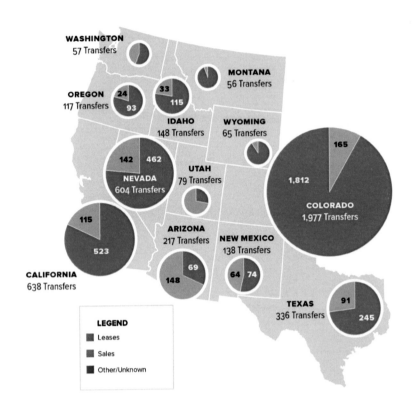

SOURCE: Tom Iseman, "Innovative Water Transfers: West-Wide Practice and Potential," American Water Resources Association, April 27, 2011.

The differences in the types of transactions that dominate within each state are due in large part to each state's unique market, regulatory, and infrastructure characteristics. For example, the Colorado–Big Thompson project on the eastern slope of the Rockies has a significant established infrastructure to transport water. Not only that, but water rights are well-defined and homogeneous (each right holder owns units of water within the system that can fluctuate proportionately with the available water supply). A streamlined process exists to transfer water; since the water is within the same management area and no requirements are in place for downstream return flows, third-party effects are minimal. As a result, transfers are straightforward and transaction costs are minimized, so the large numbers of transactions shown in Figure 4.2 are not surprising. In other states, such as California, the infrastructure to move water exists (for example, the Central Valley Project), but the water rights are not homogeneous. Nor is the process as streamlined, and this affects the numbers of leases and sales. As the figure indicates, other active market states in terms of the numbers of transactions are Texas, Arizona, Nevada, and New Mexico.

In terms of trading between sectors, by the measure of committed water volume, agriculture-to-urban trades account for the majority of water transferred, as shown in Table 4.2. Patterns of committed flow indicate that municipalities have a need for longer-term sources of water,

140

reflected in the greater quantities of water acquired through sales and multi-year leases and also reflected in higher prices (agriculture-to-urban trades demand higher prices than for those solely within agriculture). If annual volume (which does not provide the long-term volume of water acquired) is used, then agriculture-to-agriculture trades transfer the largest amount of water.[30] Table 4.3 presents the distribution of transfers by sector as a percentage of each state's total transferred quantity.

Table 4.2. Share of the Transfers Classified by Sector for All States (1987–2007)

	nnual Flow acૄ aaૄ			oc c aૄad Flow acૄ aaૄ		
	nૄculૄૄ ૄan	nૄculૄૄ nૄculૄૄ	ૄan ૄan	nૄculૄૄ ૄan	nૄculૄૄ nૄculૄૄ	ૄan ૄan
alaa	,29 ,066	2 , 3	602, 3	22,32 ,323	3, 22,632	2, 09,620
na aaul aaaa	3,22 , 23	6,322,333	,2 7,633	3,22 , 23	6,322,333	,2 7,633
ulૄ aaul aaaa	729,732	90,629	679,996	3,79 , 23	990,393	9,96 ,0 9
oૄal	2,27 ,92 302	6,367, 3 392	2,233, 62 3 2	37,366,279 72	2,763, 6 202	26,3 3,327 332

NOTE: The annual flow is the annual amount of water committed applied to the year the lease or sale was initiated. Committed flow is a measure of the total amount of water that is transferred over the term of the lease or sale (discounted using a 5% rate).
SOURCE: Jedidiah Brewer, Robert Glennon, Alan Ker, and Gary Libecap, "2006 Presidential Address Water Markets in the West: Prices, Trading and Contractual Forms," in *Economic Inquiry*, Vol. 46, No. 2, April 2008, pp. 91–112.

---

[30] Brewer, Glennon, Ker, and Libecap, 2008, pp. 91–112.

Table 4.3. Share of the Transfers Classified by Sector to Each State's Total Quantity Transferred (1987–2007)

	Annual Flow				Committed			
	Agriculture-to-Urban (%)	Agriculture-to-Agriculture (%)	Urban-to-Urban (%)	Total (million af)	Agriculture-to-Urban (%)	Agriculture-to-Agriculture (%)	Urban-to-Urban (%)	Total (million af)
Arizona	15	46	39	8.34	31	37	32	21.72
California	41	32	27	5.04	37	32	31	12.6
Colorado	51	29	20	0.59	75	8	17	5.88
Idaho	39	55	6	1.59	29	67	5	2.36
Montana	55	45	0	0.02	95	5	0	0.22
New Mexico	15	78	7	0.10	36	55	10	0.91
Nevada	84	0	16	0.22	72	0	28	2.39
Oregon	0	100	0	0.10	0	100	0	0.29
Texas	48	15	37	1.75	50	3	47	25.3
Utah	38	32	29	0.31	53	3	44	4.05
Washington	49	36	15	0.16	79	3	18	1.93
Wyoming	37	63	0	0.10	38	62	0	0.41

NOTE: The annual flow is the annual amount of water committed applied to the year the lease or sale was initiated. Committed flow is a measure of the total amount of water that is transferred over the term of the lease or sale (discounted using a 5% rate).
SOURCE: Jedidiah Brewer, Robert Glennon, Alan Ker, and Gary Libecap, "2006 Presidential Address Water Markets in the West: Prices, Trading and Contractual Forms," *Economic Inquiry*, Vol. 46, No. 2, April 2008, pp. 91–112.

Although millions of acre-feet of water have been transferred within select states, the Western Governors' Association (WGA) assessment shown in Figure 4.3 shows that only 2 percent of total surface water withdrawals has been transferred annually.[31] Over time, the total annual water transfers have fluctuated from a low of 600,000 acre-feet per year (which is about what it takes to meet the needs of Los Angeles) to 2.6 million acre-feet within a year (which is about what the state of Nevada annually withdraws).[32] And while they are a small percentage of

---

[31] Note that the amount transferred is reported as annual flows, not committed amounts, so it is consistent with the other study. If committed flows were reported, we should expect the amount of water traded to be increasing slightly over time. The WGA analysis employed the University of California Santa Barbara Bren School database of *Water Strategist* transfers data and USGS data of water withdrawals.

[32] Western States Water Council and Western Governors' Association, "Innovative Water Transfers Overview," 2008.

overall water use, others have noted that they are an important source of water for new uses, particularly as monies for new infrastructure projects are more difficult to obtain.[33]

**Figure 4.3. Annual Water Transfers in the Western United States
Total Transactions and Annual Volume Traded (1988–2009)**

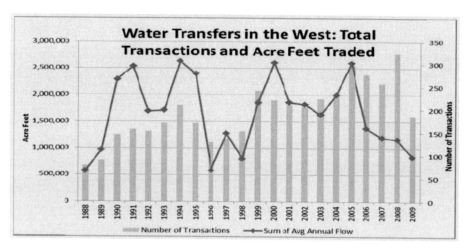

SOURCE: Western States Water Council, Western Governors' Association, "Innovative Water Transfers Overview," 2008.

The Future of Water Leases and Sales

The United States has the largest water market in the world. This market, especially water leases and sales, is likely to grow. As evidence, several investment firms have formed in response to the new demands for water and the development of water rights transfer markets. As these markets for water and water-rights transfers develop, despite their complexity and fragmentation,[34] opportunities open up for a broader range of firms interested in water rights. These firms believe money can be made in investing in water rights as new demands for water such as development, environmental concerns and regulations, as well as increasing biofuel and ethanol production, in areas where water supplies are limited or restricted and tradable. As a result, they invest in water companies, lands with groundwater resources, working farms, and other water rights investments. These firms may specialize in assisting others in the legal and logistical aspects of water-rights trades. Still others actually develop and invest in water-related businesses such as pipelines and other infrastructure; water treatment, desalination, or conservation technologies.

---

[33] Clay Landry, "The Wet Water Market—The Case for Water Rights," *Global Water Intelligence*, Vol. 1, No. 1, October 2010.

[34] The water rights market and the value of water rights are complex in large part because water as a commodity is not easily transported; therefore, in investment terms the market is not liquid. Not only can the costs to convey water be high, but politics, government policies, agricultural commodity prices, and demographics also play an important role in shaping the market. Fragmentation occurs because of the hydrological conditions and because state and local level regulatory requirements can limit water-rights transfers.

Because it is illuminating to see how valuable water rights are, here are some examples of the firms active in this area and their activities.

Companies with significant investments in water rights include J.G. Boswell Company (grower of cotton and other crops), Limoneira Company (a lemon grower), and PICO Holdings Inc. (a holding company in the insurance industry that is the largest water-rights holder in Nevada and once was the largest in Arizona). A subsidiary of PICO Holdings is Vidler Water Company. Considered the most established in the market of water rights, Vidler Water Company completed a 35-mile pipeline capable of delivering 8,000 acre-feet of groundwater from the Fish Springs Ranch to the northern edge of Reno, Nevada at a time when the city was growing rapidly. In the Reno area, water rights prices reached some of the highest levels in the United States, peaking at $40,000 per acre-foot in the 2005–2006 timeframe. This project has been marginalized after the real estate market faltered and prices for water rights fell.

T. Boone Pickens, a well-known businessman, also owns the largest amount of water rights of any individual in the United States. Pickens, who once had planned to build pipelines to supply municipalities with his water, owns rights in the largest underground aquifer in the world, the Ogallala Aquifer, which supplies 27 percent of all U.S. water used for irrigation and which is the source for 70 percent to 90 percent of the irrigated water in Kansas, Texas, and Nebraska.

Cities also are realizing the value of their water rights. In 2012, the city of Aurora, Colorado, entered into a five-year, $9.5 million lease to deliver 2.4 billion gallons of treated wastewater to the Anadarko Petroleum Company for use in oil and gas drilling.[35] The city of Greeley, Colorado, sold 1,575 acre-feet of water in 2012 for about $4.1 million to companies that supply oil drillers. In comparison, the city also sold nearly 157,500 acre-feet of water to farmers, but made much less, around $396,000.[36] In Arizona, the township of Prescott Valley successfully auctioned off water rights to treated wastewater to the investment firm Water Asset Management to raise funds for infrastructure development (which is discussed later in greater detail).

And Pure Cycle Corporation, which purchased water and land interests in the Arkansas River Valley from another water rights investment firm,[37] is planning a long-term conveyance project that could potentially bring Arkansas River water to Denver-area buyers. Because many interests compete for water in the Arkansas River Basin—agricultural, urban, recreational, and environmental—Pure Cycle Corporation is working with natural resource, governmental, and

---

[35] Justin Dove, "Three Easy Ways to Invest in Water," *Investment U*, No. 1742, April 2, 2012; "Across the USA: Colorado," *USA Today*, July 11, 2012, p. 9A.

[36] Garance Burke, "Fracking Fuels Water Fights in Nation's Dry Spots," *Associated Press*, June 17, 2013.

[37] Southwestern Investments Group purchased very senior water rights dating back to 1883 in the Arkansas River basin as part of its High Plains LLC project. This purchase is notable because farmers and others became concerned about the permanent loss of water rights to nonagricultural interests and changes to the overall community. One outgrowth of that concern was the creation of the Arkansas River Water Bank, which is discussed in a later section.

agricultural stakeholders to seek ways of maintaining the agricultural community[38] in the basin while allowing some of the water to be sent to urban centers.[39]

A few firms specialize exclusively in water investments (as distinguished from the examples above, which generally include other businesses)—water rights, water-related technologies and infrastructure described earlier. Summit Global Management, considered the largest in this area, has about $500 million in assets and $250 million in water rights within the United States and Australia; Water Asset Management, around since 2005, has $300 million in water-related assets and was the successful bidder at a water rights auction held by Prescott Valley, Arizona; and Aqua Capital Management LP been operating since 2006.[40]

In regions where water is over- or fully-appropriated, and as the pressure on water supplies continues to grow (because of drought, population growth, and environmental priorities), it is expected that the use of market mechanisms to exchange water will expand in response. Three-quarters of the 17 states in the Western States Water Council (within the Western Governors' Association) responded that water transfers likely will be used to meet demand.[41] Another market-based tool used for transferring water is the auction. The structure of auctions and some examples are discussed in the next section.

Water Auctions

Water auctions can be used to buy or sell water, using price to allocate or attract water supplies. In a traditional auction, multiple buyers bid on an item (water) that is being sold by a single seller who is trying to obtain the highest possible price. In a procurement auction, a single buyer seeks bids from multiple sellers and is trying to buy the item (water) at the lowest possible price. Water auctions can be of either type. Auctions are a useful mechanism for water transfers because typically ex ante price information on water is limited and difficult to ascertain. Through the bidding process in an auction, better information on price becomes available, resulting in a more beneficial allocation of water resources. Auctions also are a way to facilitate water transfers by bringing together buyers and sellers.

Ensuring that participants bid according to their true value is the key to structuring an auction successfully. If a large number of bidders take part in a competitive auction, success—as defined in a procurement auction by bidders using their true prices and the auctioneer buying at the lowest cost—is relatively easy to achieve. Structuring an auction appropriately is not clear-cut when the information is poor on the characteristics of water rights, when the values of alternative

---

[38] These ideas include those that improve farm operation efficiencies along with annual crop rotations.

[39] Landry, 2010; Southwestern Investor Group, "High Plains LLC Water Rights Project," undated.

[40] Manuela Badawy, "Looking for Gold in Water Investments," Reuters, December 8, 2011; "The West's New Gold Rush for Water Rights," *Global Water Intelligence*, Vol. 8, No. 12, December 2007.

[41] Western Governors' Association and Western States Water Council, 2012.

users are unknown, or when the auction draws a limited number of bidders. These are conditions are likely in water auctions. For example, farmers may not have a sense of the true value that municipal or industrial users place on water, nor may they even have a sense of each other's uses because of soil conditions, microclimates, crops, or other issues. As a result, they may intentionally or unintentionally game the system. Water auctions also are complicated by the fact that water rights may not always be homogeneous (in terms of conveyance costs or reliability), so that when holding an auction all units purchased (or sold) may not represent the same value. In this case, determining the winning bids, and the ultimate success of the auction, is much more challenging.

Key Features of Procurement Auctions

A common auction type used for buying water is the procurement auction. Procurement auctions can be run in three primary ways—as an ascending, descending, or sealed-bid auction. An ascending auction in one in which the starting price is relatively low and the winning bidder ultimately offers the highest price. Descending auctions work the other way, starting with a relatively high price and with bids falling. The third type of auction, sealed-bid auctions, bidders submit confidential bids and winners are selected using certain criteria that include such characteristics as price, quantity, or location. Most water procurement auctions are sealed-bid auctions. If the procurement auction targets farmers as bidders, then the timing of the auction becomes important to account for seasonal plans for planting crops.

The key features of procurement auctions are as follows:[42]

Who is eligible to bid: To limit third-party effects and uncertainty regarding conveyance costs, it may be best to limit the auction to a specific watershed or geographic area. It is possible it may not be legal to include water from out of state. In the case of the Flint River auction, the Georgia Environmental Protection Division prequalified bidders.

What kind of water rights are eligible, and what units of volume to use: Water is not homogeneous and can vary in terms of quality or availability (seniority of rights) or location and conveyance costs. The more that an auction includes homogeneous and well-defined water rights, the easier it is to structure and administer. However, there is a tradeoff between limiting the auction for ease and clarity and the level of competition that is affected by reducing the potential number of participants. To reduce administrative costs, it may be best to include a minimum quantity for eligible water rights. Or if maintaining in-stream flows to downstream users is of concern, then the auction may be limited to active permit users and not take the risk

---

[42] This discussion is taken largely from Michael O'Donnell and Bonnie Colby, "Water Auction Design for Reliability: Design, Implementation and Evaluation," Tucson, Ariz: University of Arizona Department of Agricultural and Resource Economics, May 27, 2009; and Ray Hartwell and Bruce Aylward, "Auctions and the Reallocation of Water Rights in Central Oregon," Deschutes River Conservancy, River Paper Series No. 1, April 2007.

that flows would be reduced or otherwise disrupted. And if the purpose of the auction is to ensure that reliable water is available, it may limit participation to those with more senior rights.

How much is included: Another consideration is whether the entire quantify specified in the water right is included—just the consumptive amount or some other portion of the right. Requiring that the entire amount be auctioned simplifies verification afterward, but allowing bidders to select the amount they wish to put up may attract more bidders. And in some states, the beneficial-use requirement may apply to consumptive use, so this may be the only portion that can be legally transferred. Therefore, federal, state and local laws must be considered to determine how much of the water right can be legally *transferred* to another party. In terms of units of volume, acre-feet of water can be used, as can land acreage—if water rights are for land that will lie fallow or if water rights per acre are relatively uniform.

How bidding occurs: An auction can have either a single round of bidding or several rounds in which bids are provisionally accepted or rejected. Predictably, single-round auctions are easier to administer and are easier for participants. However, multiple-round auctions give flexibility to achieve the desired results and in theory will minimize the cost paid and maximize participation. By holding multiple rounds, the auctioneer can mitigate the risks associated with poor information availability. The 2001 Flint River Auction held eight rounds of bidding (the acceptable price was not revealed) before the bidding was suspended. However, with each round, bidders can gain more information on acceptable bids through the process and through communicating with one another, so collusion becomes a risk. Much of the success of these auction formats depends on the number of bidders and the accessibility of information.

Price structure: The price paid either can be uniform or discriminatory; that is, the price paid to the winner(s) of the auction can be at the same price (uniform) or the actual price contained in the bid (discriminatory). Experiments have found that uniform-price auctions tend to bring in water at a lower price, because bidders do not have a strong incentive to overstate the value of their water as much as they do in a discriminatory auction. However, some water agencies—for example, in the case of Oregon's Deschutes River Conservancy and the Flint River Irrigation Auctions—felt that bidders would be more likely to participate and so they used discriminatory auctions (and paid the bid-price). If the procurement auction has a fixed budget, the uniform-price approach would make it easier to make awards matching budget expectations. Discriminatory auctions can be especially problematic because of the incentives to overstate value when the number of bidders is small.

What information to provide to bidders: A trade-off exists between providing enough information about the auction to encourage a large number of bidders and putting so much information out that bidders will be able to determine acceptable offers or to collude, in which case their bids do not reflect their true price. For instance, in the case of the Flint River Auction, the auction's budget was public knowledge. Had the Georgia EPD revealed the goal for total acreage to lie fallow, bidders easily could have inferred a likely winning bid and adjusted their bids accordingly.

How winning bids are selected: Auctions often use what is called a reserve price to triage bids. Those bids below the reserve price are accepted, while those above are not. A second approach is possible when there is a fixed budget or a target amount of water desired. The auctioneer would rank bids based on price and accept bids starting from lowest to highest until the budget limit is reached or the amount of water sought is reached. Hartwell and Aylward stated: "Especially in thin markets for goods without well-established values, appropriate reserve prices are essential in avoiding embarrassing failure of an auction ... Of course, determining an appropriate reserve price can be difficult, especially for goods whose market values are unknown."[43] Academic researchers noted, "Using both a reserve price and a budget cap is generally preferable to using either method alone because it simultaneously minimizes the likelihood of overpayment per unit and ensures that the overall budget is not exceeded."[44] Should the bids be for heterogeneous items, or if the auction has more than one objective, indexing according to predetermined criteria is another option for ranking bids, but this procedure adds complications to the process. A somewhat informal indexing process was used in the Edwards Aquifer auction for irrigation rights in which bids were evaluated on crop type, irrigation system and commitment to dry-land farming in addition to price.[45]

Finally, since the goal of auctions is to allocate water according to a price that as closely represents the true value bidders and auctioneers place on the resource, Hartwell and Aylward noted that the perception of competition is critical.[46] If bidders and auctioneers believe the auction is operating in a competitive situation, they will be more likely to bid according to the true value of the resource. If they believe otherwise, they may bid in such a way as to increase their individual profits. Therefore, the amount of information available regarding price, bidders, budgets, and other factors is a critical determinant in the success of the bidding process and ultimately the auction itself.

For public entities involved in auctions, there is a policy trade-off between making the auction accessible and less complicated to administer and producing the maximum level of competition. Public entities may want to sacrifice a level of efficiency for a greater likelihood that the auction will be easy to administer, to ensure bids are not exploitive and that bidders trust the process.

Auctions have been proposed and implemented in several states for a variety of purposes ranging from selling off excess water rights to buying rights for fallowing irrigated croplands. The success of auctions has seen mixed results largely as a result of the localized nature of water markets. Some of the key features of auctions include who is eligible to bid (typically determined by geographic boundaries), the type of water right sought, and how bidding occurs (transparency

---

[43] Hartwell and Aylward, 2007, p. 8.

[44] O'Donnell and Colby, 2009, p. 8.

[45] O'Donnell and Colby, 2009, p. 11. See this source for more details on indexing.

[46] Hartwell and Aylward, 2007, p. 8.

and number of rounds). These features will have significant effects on the success of the auction. In the following subsections, we describe two auctions that were structured differently and served different purposes. These two auction examples are Prescott Valley, Arizona (an effluent rights auction) and Flint River, Georgia (an irrigation rights auction). In later subsections, we will summarize the advantages and disadvantages of auctions and provide a quick assessment of their future implementation.

Prescott Valley, Arizona

The Prescott Valley auction, held in 2007, is considered one of the most successful water-rights auctions. In the decade or so leading up to the auction (and since), Prescott Valley Town was in a high-growth area of Arizona, constrained by limited water-rights availability. To raise money for water infrastructure, the town held a traditional auction to sell groundwater credits that resulted from discharging newly available treated wastewater effluent into the aquifer.[47]

Prescott Valley Town is in the Prescott Active Management Area (AMA) of Arizona, an area in which groundwater resources have special regulatory and management controls (discussed more in Chapter Six). As of 1999, Arizona water law required new developments to secure access to reliable water sources for the next 100 years, without which developers could not build. The town of Prescott Valley was especially constrained because of the area's high growth and limited options for new supplies of water; the town does not have access to water from the Central Arizona Project (CAP)—while its larger high-growth neighbor Phoenix does.[48]

The town discharges its treated effluent into the dry bed of the Agua Fria River, where it helps to recharge the underground aquifer. A hydrological study has estimated it will take 20 years for the recharged effluent to travel underground to the nearest groundwater well.[49] Nevertheless, the town receives groundwater credits for effluent discharged into the recharging riverbed. Since the new water rights made available by the town are in effect paper rights in the form of groundwater credits, they are more flexible than wet water (that is, they are more easily transported). The amount of water offered was 2,724 acre-feet, enough to support an additional 12,000 homes or other industrial, commercial, and recreational uses. So the offering of these water rights was attractive to developers and other potential water users in the water-poor area. However, establishing a price for these water rights was not easy, since the value is in part a function of anticipated growth and development in the area. By employing an auction to sell

---

[47] The treated wastewater resulted from an upgrade to the town's wastewater treatment facility. An irony is that the needed infrastructure was a pipeline to carry groundwater from another aquifer (within the same management area) to support already-permitted development in Prescott Valley.

[48] The Central Arizona Project is a 336-mile long system of aqueducts, tunnels, pumping plants and pipelines that transports Colorado River water (as well as water from other sources) through several regions of the state and provides the single major source of water.

[49] Peter Friederici, "Making an Effluent Market," *High Country News*, September 17, 2007.

off the newly acquired water rights at "market" prices, the town hoped to generate enough income to fund its infrastructure requirements.

Initially few bids were made, because the terms of the sale required payment of the entire purchase price up-front. Developers were not prepared to finance the payment in this way. Eventually, the auction was restructured to allow payments to be timed with completion of certain permitting requirements. The price floor also was lowered from $30,000 per acre-foot to $19,500 per acre-foot. By establishing a price floor, which was guaranteed by a financial investor (Aqua Capital Management LP, based in Nebraska, who agreed to pay $53 million for the rights if there was not a higher bidder), the city ensured that the auction would yield an acceptable result. The restructured auction was held about a year later, in late 2007, and more bidders both locally and from out of state became interested. A New York-based financial firm (Water Asset Management's subsidiary Water Property Investors, LLC) ultimately offered the highest bid, successfully bidding for the water at the cost of $24,500 per acre-foot. According to one market analyst:

> The fact that the bidder providing the floor and the eventual winner of the auction were both financial investors reflects the development of the water rights market from an illiquid market with geographically limited buyers and sellers to a more liquid one, with its own specialist market-makers.[50]

The price paid by the investment firm reflects the higher-scarcity value of water in the Prescott Valley Town and region; at these prices, water conservation becomes more economically viable. For comparison, the cities in the Phoenix area negotiated deals with nearby Native American communities to buy the use of tribal water for the next 100 years for $1,500 to $1,800 per acre-foot (recall that they have access to the CAP infrastructure).[51]

The winner was required to put the water to beneficial use in the town. The investment firm subsequently has sold about one-third of the water rights to developers for undisclosed prices, despite the slowing in the residential market.[52]

At the time, this was the largest sale for the highest price paid for water rights, yielding more than $67 million for the town. The auction allowed the town to raise money from its water resource, a resource whose true market price was not well known before the event. An open process was used to derive the market price, which reflected the scarcity of water in the area and which also will motivate conservation. To hold this auction, the town had to attain regulatory

---

[50] "Arizona Water Rights Auction Tops $20m," *Global Water Intelligence*, Vol. 8, No. 11, November 2007.

[51] Friederici, 2007.

[52] In 2009, a developer from Denver purchased 200 acre-feet for an undisclosed price and a real estate investment firm from Scottsdale, Ariz., purchased 700 acre-feet for an undisclosed price. See Christopher Scott, "Case Study: Effluent Auction in Prescott Valley, Arizona," Tucson, Ariz.: University of Arizona, 2012.

approvals, including the clarification of water rights, and it benefitted from partnering with the private sector to sell these water rights.[53]

Flint River, Georgia, Irrigation Auction

In 2000, after Georgia had experienced several droughts, the legislature passed the Drought Protection Act. After it was determined that the Flint River could not sustain the totality of permits supported by it, provisions were made in the law for irrigation auctions should a severe drought occur. After a severe drought was declared in 2000, Georgia's Environmental Protection Division held two procurement auctions for irrigation rights in 2001 and 2002. The auctions were held in an effort to reduce water withdrawals from the Flint River after it was determined that these quantities could not be sustained during the drought if the river was to flow. While permits were issued for both surface and groundwater withdrawals, only surface water withdrawals were included in the auctions, because it was unknown exactly how groundwater withdrawals would affect river flow. Eligible permit holders were those in the Flint River basin and its tributaries and perennial streams, along with groundwater users within the area of the Floridian aquifer that affects the Flint River flow.

The bids were to be the price-per-acre of land upon which irrigation would be suspended for the remainder of the year. Both auctions required permit-holders to suspend irrigation on all acres covered by the particular permit to make enforcement easier. Because permit-holders may have had more than one permit for their lands, it was up to them to decide which permit to offer and at what price. In other words, not all of the permits held by an irrigator had to be bundled, and each permit held by an irrigator could be offered at a different price or not offered at all. So, were their bids accepted, permit holders would have received a payment calculated using the acres removed from irrigation multiplied by the price per acre in the bid.

Recall that one of the benefits of using auctions is that they reveal a more accurate price for water, since in general true price information is difficult to determine and obtain. Additionally, procurement auctions are designed to yield the lowest price for the water purchased. To ensure the best price, the 2001 and 2002 auctions were structured differently, and the information provided during each was designed to lower the likelihood that the bidders (irrigators) would be able to game the system so that their true value to the water would be revealed. Consequently, since the available budget for the auction was public knowledge, the Georgia Environmental Protection Division director's target for irrigated acreage reduction was *not* announced prior to either auction, as bidders easily could have calculated the average price from available funds and bid accordingly, defeating the purpose of the auction.

---

[53] See "Arizona Water Rights Auction Tops $20m," November 2007; "Arizona Town Holds Successful Effluent Water Auction," *Underground Infrastructure Management*, January/February 2008, p. 44; Friederici, 2007; Christia Gibbons, "Prescott Valley Water Auction Could Impact Phoenix," *Phoenix Business Journal*, October 29, 2006; and Scott, 2012.

The 2001 auction was an iterative one in which eight rounds of bidding were held in an attempt to drive down bid prices. As it turned out, in the last round the highest accepted offer price increased by more than 50 percent and more than twice as many offers were accepted than in any previous round. "While the participants did not know the exact cutoff used in any previous round, they spoke with one another during the auction and knew the approximate level of the cutoff prices," one study noted.[54]

The second auction, held in 2002, was structured differently. A cap was placed on the maximum bid[55] that would be accepted, and only one round of sealed bidding occurred. Conditions also differed in that EPD could use its authority under the Drought Protection Act to remove lands involuntarily from irrigation if the auction failed to attract enough bidders to reach their targets for removal of land from irrigation.[56] The auction achieved its goal for lands removed from irrigation at an average price that was lower than the 2001 auction. Because only one round was held, it was easier to administer and permit-holders preferred it. An analysis of the auction suggests that the bids reflected the crop value to permit holders, since those with higher value crops, such as peanuts, did not participate proportionately as much as those with lower-value crops such as corn or cotton. Those who could generate greater wealth valued water more highly. Winning bids also were proportionately distributed geographically among permit holders.[57]

In terms of procuring irrigation water rights at a fair price, both auctions were considered successful (although one analysis suggests that the same volume of water could have been purchased more cost-effectively).[58] The primary criticisms of the auctions were two-fold. First, funding of the process is considered unsustainable, since the legislature set aside money from a national tobacco settlement[59] for the irrigation auction and it was unclear where money for any future auctions would come from. Second, it is debated whether the auctions were successful in the critical outcome desired—maintaining stream flows—since some suspect that irrigation did not really subside because farmers substituted groundwater for the surface-water withdrawals they sold.[60] Because historical data is unavailable on how water was used in irrigation in

---

[54] Ragan Petrie, Susan K. Laury, and Stephanie Hill, "Crops, Water Usage, and Auction Experience in the 2002 Irrigation Reduction Auction," Atlanta, Ga.: Georgia State University Andrew Young School of Policy Studies, Water Policy Working Paper #2004-014, December 2004.

[55] The cap was set at $150 per acre, $50 below the highest price paid in the first auction.

[56] Involuntary takings would be compensated at the average price paid to bidders in the voluntary auction.

[57] Petrie, Laury, and Hill, 2004.

[58] Petrie, Laury, and Hill, 2004; Ronald G., Charles A. Holt, and Susan K. Laury, "Using Laboratory Experiments for Policy Making: An Example from the Georgia Irrigation Reduction Auction," Atlanta, Ga.: Georgia State University Andrew Young School of Policy Studies Research Paper Series, Working Paper 06-14, 2003.

[59] The Master Tobacco Agreement of 1998 required the tobacco industry to make payments to states to cover, in part, the Medicaid costs of those persons with smoking-related illnesses.

[60] Joseph W. Dellapenna, "Special Challenges to Water Markets in Riparian States," *Georgia State University Law Review*, 21 Ga. St. U.L. Rev. 305, Winter 2004.

previous years, there is no real way to know the ultimate effect of the auction. Finally, one farmer sued over the EPD's analysis of perennial streams, claiming that faulty analysis denied him the opportunity to bid.[61] And some newspapers reported glitches in payments, both to farmers who did not participate in the auction and lack of payments to those who did.[62] But overall, the auctions were considered a success in soliciting bids to allow land to lie fallow during the drought.

Other Auctions

Other auctions have been held in various states for a range of purposes. In 2009, the Chino Basin Watermaster intended to hold a traditional auction of excess water rights formerly used by a Sunkist plant in the area and improve the infrastructure with the proceeds. Bidders were concerned about the uncertainly regarding the local water agencies' fees for transporting the water. In response to these concerns, the Watermaster was seeking fixed-price agreements with the agencies for transporting the water before holding the auction. It is unclear whether the auction ultimately occurred.[63]

The Northern Water Conservancy District in Colorado routinely auctions off surplus Colorado River water. In one auction, it sold water to an oil and gas company for fracking, setting off concerns about maintaining the hydrological cycle in the region. Reverse auctions have been used to procure leases and sales of water for water banks (in Washington state, for example).

A proposal for auctioning additional water supplies provided through the CAP in Arizona met with resistance in 2011 from stakeholder groups who were uncomfortable with the new approach to distributing water. The proposed auction involved an iterative bidding process in which the CAP authorities would offer the new sources of water for a proposed price (starting with the costs to acquire, develop and deliver it) and potential buyers would bid for the desired quantity. Assuming the demand for water exceeds the supply at the proposed price, another round of bidding would occur for a higher proposed price and so forth until the amount demanded within the bids equaled supplies for a given price. A majority of stakeholders felt this process was unfair and that prices could exclude buyers. Their preference was to distribute water according to the quantity demanded by each buyer at a set price; should the total quantity be insufficient, by say 15 percent, all buyers' quantity received would be reduced equally—in this case by 15 percent—of what they desired. Needless to say, this approach does not take into

---

[61] Paul L. Hollis, "Farmer's Irrigation Lawsuit Dismissed," *Southeast Farm Press*, January 16, 2002.

[62] Dellapenna, 2004.

[63] Chino Basin Watermaster, "Working Collaboratively to Maximize the Benefits from the Chino Groundwater Basin," undated; Chino Basin Watermaster, "Water Auction Legal Summary," undated; and "Chino Basin Auction Postponed as Bidders Take Fright," *Global Water Market Intelligence*, November 5, 2009.

account the buyers' uses or the true value of the uses. It would not maximize overall welfare gains nor reveal accurate prices.[64]

Concerns regarding equity and access distinguish water from other types of commodities and must be addressed in any approach to allocating water. Local responses to these issues in regard to the use of market mechanisms for water allocation will vary; however, trust and transparency in the process will help to overcome any fears and will improve participation.

Advantages and Disadvantages of Auctions

Auctions are viewed as an effective way compared to other mechanisms to establish a price for transferring water rights that more accurately reflects the true market value. They also can be a means to facilitate the purchase or sale of water by attracting and connecting sellers and buyers more easily. And in periods of drought, auctions can provide a voluntary and less arbitrary way of purchasing irrigation water rights over mandatory fallowing. Overall they can be a more efficient (and depending on one's point of view, a fairer) way to allocate water than administrative processes, which must rely on subjective judgment when price information is not well known.

However, auctions are tricky to administer and structure if the desired optimal outcomes are to be attained. The auctioneer cannot take a cookie-cutter approach to structuring and administering an auction when it comes to water. Water, unlike other commodities, is subject to complex legal and geographic constraints, potentially high costs for transport or conveyance, environmental considerations, water rights verification and quantification, enforcement issues, and third party effects. All of these factors must be considered when structuring and administering an auction if it is to be successful.

General public acceptance of auctions is mixed, because water is typically viewed as a public good. As a result, selling it as a commodity to the highest bidder may appear to restrict access to water. And when prospective bidders do not trust the particular auctioneer or the specific auction process, they will not participate, as in the case of a Yakima River water auction in which the auction failed because it drew only one bidder.[65]

Moreover, the public is used to receiving water at artificially low prices, and so it is unaware of the true cost to provide it. Auctions, as other any market mechanisms, likely will lead to water reallocation that alters the status quo, prompting changes in investments and the current way of doing business. These kinds of changes are likely to create political forces that seek to influence the process, or at worst thwart it.[66]

---

[64] Taylor Shipman, "Why Not Auction Water? Perspectives on Auctioning the Next Bucket in Arizona," American Water Resources Association annual meeting presentation, Albuquerque, N.M., November 9, 2011.

[65] O'Donnell and Colby, 2009, p. 9.

[66] Shipman, 2011.

The Future of Water Auctions

Auctions are common in other commodity markets. Since auctions are a way to gather information on market prices, their use is likely to expand with the growth of non-regulatory approaches to transfer water.[67] Concerns regarding equity and access distinguish water from other types of commodities and must be addressed in any approach to allocating water, including auctions. Local responses to these issues in regard to the use of market mechanisms for water allocation will vary; however, trust and transparency in the process will help to overcome fears and will improve participation.

Water Banks

Water banks are another market mechanism used to transfer water or water rights. Operating within a designated geographic or hydrological region, the bank in simple terms is an administrative body that facilitates water transfers among multiple buyers and multiple sellers for current or future use. Water suppliers (or depositors) can donate, lease, or sell their water rights. Depending on how the bank operates, demanders (lessees or buyers) may lease or purchase these rights based on market prices, a fixed price, or an auction, including an administrative fee to the bank. The bank itself facilitates and oversees these transactions, through an established process for transferring water rights that reduces uncertainty, confusion, and transaction costs. Figure 4.4 presents a schematic of a water bank to include water or water rights providers, water or water rights demanders, and an administrative body. All operate within the regulatory system of a specific state and locality.

---

[67] Recall that the market price of water is difficult to discern, because local factors have such a major influence on prices and the true value of water is not always appreciated.

Figure 4.4. Water Bank Schematic

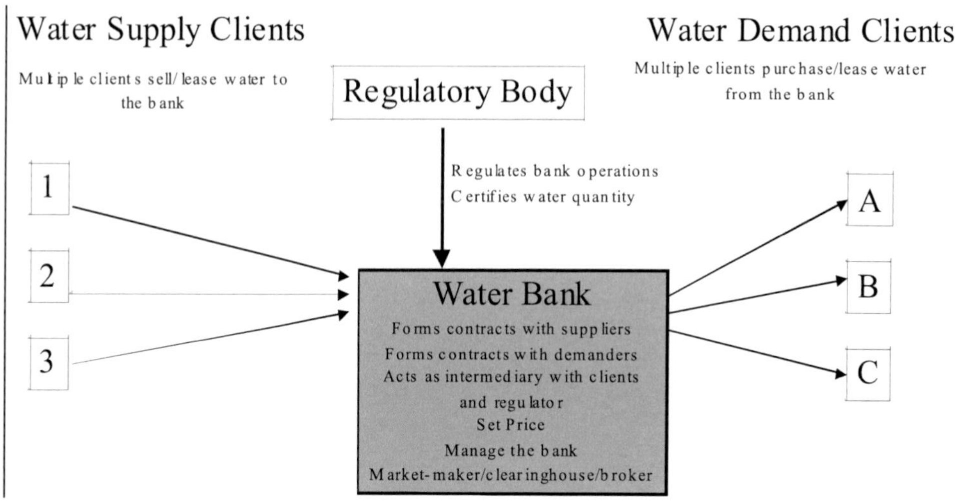

SOURCE: WestWater Research, 2004.

Water Bank Structures and Functions

Water banks are organized within a specific geographic region that generally is determined by storage and conveyance opportunities, numbers of buyers and sellers, and water uses.[68] Federal agencies, state agencies, water districts, non-profit organizations, and private for-profit firms can administer them. They employ a range of structures that include those listed below:

- **Surface storage banks physically store water in a reservoir for future use**. Water can be stored either by transferring the entitlement to the storage reservoir or by leaving a portion of the entitlement in the reservoir. As a result, infrastructure expenses, conveyance costs and transmission loss will occur. However, since the water is physically stored, supply is quite secure. Examples include the California Drought Water Bank and Dry Year Purchasing Programs.

- **Groundwater storage banks can work in a couple of different ways**. Water deposited into the bank can be used to recharge an aquifer, either through percolation in a recharge basin or by piping it directly. This is referred to as direct recharge. Another approach that groundwater banks use is to substitute surface water for groundwater use, thereby maintaining the water in the aquifer. This is referred to as indirect recharge. Groundwater banks have the advantage of having potentially low infrastructure costs and can reduce land subsidence. However, in areas where groundwater rights are poorly defined, it may be difficult to protect stored water from over-pumping, so water availability is less

---

[68] There is a trade-off in the size of the bank's geographic coverage. The geographic coverage should be large enough so that sufficient numbers of potential buyers and sellers are available for transactions. However, if the bank covers too large of an area, then conveyance costs will rise, buyers and sellers may not feel as confident in the bank. Or they may not be comfortable with potentially transferring water out-of-basin or district and participation can be reduced.

156

secure.[69] And when surface water is substituted for groundwater use, some question the effect these banks have on in-stream flows. Examples include Arizona Water Bank Authority, Edwards Aquifer Authority Groundwater Trust, Kansas Water Bank Association, and the Kern County (California), bank.

- **Institutional banks do not store water but rather facilitate the exchange of water rights by reducing the costs and uncertainties associated with water transfers.** A special kind of institutional water bank, a water trust, holds water entitlements for a predetermined amount of time (either limited or indefinitely) and is generally used to augment in-stream flows. Institutional banks avoid large infrastructure costs, but the security of water availability is diminished. Examples include the Arkansas River Water Bank, Idaho Water Supply Bank, and the Texas Water Bank.

The key market features of a water bank are what water rights are eligible to participate, what kinds of transactions take place, and how transfer price is determined. State water policies and laws, as well as the specific purpose for the bank, will drive the determination of these features (eligible water rights and the types of market transfers). At times, state legislation may be required to establish a bank. Banks can transfer either wet water or water rights using term leases or sales, which can be arranged bilaterally between the depositor and demander or with the bank when resources are pooled.

Just as with water auctions, determining the appropriate price for water is challenging, since water rights are not all the same. Water rights (or entitlements) are not uniform (junior rights are not as reliable as senior rights) and water volume, location, and quality can vary. As with other aspects of water banking, several approaches are taken to determine price. Institutional banks that facilitate transfers through bilateral trades generally will allow the seller and demander to determine price. Bilateral trades can be tricky, because the price and the quantity supplied must match the price and quantity demanded. On occasion, a major trader has skewed market prices as well. Other banks transfer water using a predetermined fixed price to facilitate trading. From a market-oriented point of view, this is not preferred because the bank is attempting to predict the true market value of water and may create inefficiencies if the predicted price is not close. Moreover, a larger bank may influence the market price in this way and cause additional distortions. Auctions can be used to buy water for the bank as well as to sell it. And finally, some banks that seek to provide water in drought conditions may enter into contingent contracts in which suppliers would provide water at a given price only in periods of drought (usually irrigators fallowing land to provide water to municipalities). Some banks, such as the Arizona Water Bank Authority, are not market-based in that a public entity annually determines the purchase price, and the groundwater credits are transferred to another public agency that either can sell them to their customers or retire the credits if the water needs to be recovered.

As the shaded box in the schematic of a water bank shown in Figure 4.4 indicates, water banks must have processes and procedures for verifying water rights; forming contracts with

---

[69] See O'Donnell and Colby, 2010 or Clifford, Laundry, and Larsen-Hayden, 2004 for solutions to protect over-pumping of groundwater.

suppliers; forming contracts with demanders; ensuring participants and transactions abide by legal and regulatory requirements; establishing transfer price; managing bank operations and obtaining funds; marketing its services; and resolving disputes. Verifying the water right is a critical function of the bank. Managed properly, this process can facilitate transactions, which is a primary benefit of water banks. If the bank is not structured properly, trades will not be made. For example, the Arkansas River Pilot failed after nine years in the pilot phase. The bank made no trades for a variety of reasons already enumerated, including the estimate that administrative paperwork and procedures were to take three months—proportionately a long time for a one-year lease. Since information on water providers and demanders were listed on a bulletin board, those parties easily could make the trades outside of the bank.[70]

Because water banks are quite diverse in terms of purpose and structure, they are challenging to characterize as a whole. Water banks, centered on a specific geographic area, will vary in terms of eligible suppliers and demanders,[71] their purpose, as well as the details of how the bank is administered (all the elements within the shaded box in Figure 4.4). And as with water auctions, the key to a water bank's success is meeting its intended purpose while ensuring the elements of the bank are structured properly to facilitate trade in a timely manner at a low cost within the overarching regulatory system.

Although farmers historically have used water banks to make temporary water trades within an irrigation district, the oldest of which dates back to 1932, the banks, like auctions, continue to be an emerging approach toward transferring water for a variety of purposes. The banks predominately occur in the West, which had an estimated 30 water banks in 10 western states as of 2004.[72] Since then, water banks have developed in Kansas and Nebraska.[73] Arizona, California, and Idaho are the states considered to have high levels of water bank activity, but overall results are mixed. For example, some banks have had few or no transactions.[74]

Water banks have been developed to serve the following objectives:

- Balancing available water supply during seasonal fluctuations or periods of drought. This also may include increasing water supply reliability to junior rights holders.
- Improving the conjunctive use of surface water and groundwater.

---

[70] O'Donnell and Colby, 2010.

[71] Water demanded can be leased or purchased depending on how the bank functions.

[72] This estimate includes active banks as well as pilots. Washington State Department of Ecology and WestWater Research, Inc., "Analysis of Water Banks in the Western States," Publication No. 04-11-011, July 2004, has a brief description of each bank.

[73] Kansas Department of Agriculture, Voluntary Incentive Program, undated; Central Platte Natural Resources Division, "Water Bank," Nebraska, undated.

[74] For example, the Arkansas River Bank in Colorado had no trades made after nine years. The Kansas Water Bank Association had three in its first five years. See WestWater Research and Washington Department of Ecology, "Water Banks in the United States," briefing, undated (data indicate likely 2003–2004).

- Facilitating the reallocation of over-appropriated water, or limited water supplies, to new uses (such as development or environmental purposes).
- Maintaining rights within a specific district or watershed as a hedge against permanent loss to other uses through sales or interstate agreements.
- Enabling water conservation.
- Ensuring in-stream flows are sufficient for environmental objectives, regulatory requirements, interstate agreements, or treaties.

While each water bank will have a primary purpose, any single water bank may address multiple objectives. For example, the California Drought Banks reallocate water during drought periods, but also ensure a minimum flow requirement is met in the delta area of Northern California. The Arizona Water Bank improves the conjunctive use of surface and groundwater, provides a source of water when supplies are low, and also provides another mechanism to manage Colorado River water among Arizona, California, and Nevada according to the interstate compact and other agreements. To illustrate this range of purposes and structures, several different water banks examples are described in the next subsection (more detailed descriptions are included in Appendix A).

Water Bank Examples

California Drought Water Banks: Balancing Water Supply

The California Drought Water Banks of 1991, 1992, and 1994 provide good examples of water banks that are used to balance water supply.[75] After years of repeated drought, a water bank was established in California in 1991. This was significant in that it was the first time voluntary, market-based water transfers occurred on a large scale in California.[76] The primary purpose of the bank was to enable voluntary transfers of water from suppliers (agriculture) in northern California to demanders (urban, municipal, and agriculture) in the central and southern parts of the state (within the State Water Project (SWP) customers and the Central Valley Project (CVP) service areas) at a time when traditional water supply sources were stressed. The bank activities served to protect existing water rights holders while providing water to critical needs without damaging fish and wildlife.

The California Drought Banks were activated in the 1991, 1992 and 1994 drought years. The California Department of Water Resources administered the banks and purchased water from:

---

[75] Peggy Clifford, Clay Laundry, and Andrea Larsen-Hayden "Analysis of Water Banks in the Western States," Water Resources Program, Washington Department of Ecology, Report 04-11-011, July 2004; Lloyd Dixon, Nancy Moore, and Susan Schechter, *California's 1991 Drought Water Bank,* Santa Monica, Calif.: RAND Corporation, MR-301-CDWR, 1993; "Innovations in American Government Awards: California Drought Water Bank," Cambridge, Mass.: Kennedy School of Public Policy, Harvard University, 1995; Michael O'Donnell, and Bonnie Colby, "Water Banks: A Tool for Enhancing Water Supply Reliability," Tucson, Arizona: University of Arizona, Department of Agricultural and Resource Economics, January 2010.

[76] Dixon, Moore, and Schechter, 1993, p. 1.

- Surface water made available by groundwater substitution
- Surface water stored in local reservoirs that was surplus to local needs
- Water made available by farmers fallowing or forgoing irrigation of designated farmland.

Demanders were required to submit bids quantifying critical needs for water. Priority was given to water needed for drinking, health, sanitation, fire protection, and critical agricultural needs. The California DWR negotiated contracts with suppliers at a range of prices that a committee of potential demanders had established. It then sold the water to demanders for a fixed price according to critical needs allocation rules that took into account economic, social or environmental losses.[77]

Activity levels of the 1991, 1992, and 1994 banks varied but were high at times. Fifty-one percent of the water came from no-irrigation contracts, 32 percent came from sources using groundwater substitution and 17 percent from stored water sources.[78] To supply this water, 166,000 acres of agricultural land were fallowed.[79] Analyses determined that statewide, the banks contributed jobs and economic opportunity, although some economic activity shifted away from the agricultural areas that supplied water to the southern part of the state. In terms of third-party effects from the water transfers, these were determined to be minimal.[80]

Arizona Water Bank: Balances Supply and Improves Conjunctive Use

The Arizona Water Bank provides an example of a water bank that balances water supplies improving the conjunctive use of surface and groundwater.[81] The Arizona Water Bank became active in 1997 to enable the storage of excess Colorado River Water within the Central Arizona Project system for future use during periods of low supply.[82] Until the Arizona Water Bank was created, Arizona did not use its full 2.8 million acre-feet allotment of Colorado River water

---

[77] For the 1991 bank, DWR negotiated a range of prices with sellers. Subsequent banks developed water pools that were price uniformly.

[78] Dixon, Moore, and Schechter, 1993, p. 5.

[79] David Zilberman, Ariel Dinar, Neal MacDougall, Madhu Khanna, Cheril Brown and Frederico Castillo, "Individual and Institutional Responses to the Drought: The Case of California Agriculture," *Journal of Contemporary Water Research and Education*, Vol. 121, No. 1, 2012.

[80] Dixon, Moore, and Schechter, 1993, p. 54.

[81] The Arizona Water Bank Authority was created and modified by two Arizona bills—House Bill 2492 in 1996 created the bank, while in 1999 the enactment of House Bill 2463 amended the earlier legislation. Creation of the bank was not controversial since it was created within the limits of existing laws and revenue sources (which were redirected to the bank). See Clifford, Laundry, and Larsen-Hayden, 2004, pp. 32–36; and O'Donnell and Colby, 2010.

[82] The AWBA procures excess CAP water supplies from the CAWCD. Water is allocated according to a priority system established by the U.S. Secretary of Interior as follows: municipal and industrial, Indian, non-Indian agriculture and miscellaneous. Therefore, excess CAP water is the most junior priority. The three Arizona counties within this system are Maricopa, Pinal, and Pima. See Arizona Senate Research Staff, "Arizona State Senate Background Brief: Arizona Water Banking Authority," November 15, 2010.

(excess Central Arizona Project (CAP) water was available in part because farmers were able to pump groundwater more cheaply and did not choose to purchase CAP supplied water).[83]

The Arizona Water Bank's administering authority is the Arizona Water Banking Authority (AWBA), which purchases excess water from the Central Arizona Project (CAP). The Central Arizona Water Conservation District (CAWCD) sets the price each year.[84] The purchase is considered consumptive use and therefore is not subject to non-use forfeiture. The AWBA is permitted to store this water within state-owned underground facilities and aquifers (direct recharge), or irrigation districts are allowed to use the water in lieu of pumping groundwater (indirect recharge). In exchange for storing the water, the AWBA receives credits that it can then transfer at some future time to either the Central Arizona Groundwater Replenishment District (CAGWRD) or the Arizona Department of Water Resources (ADWR) for use for their municipal and industrial customers along the Central Arizona Project canal should they require it during dry periods. The amount transferred is reduced by a 5 percent conservation requirement.[85] Therefore, the Arizona Water Bank enables the use of Arizona's allotment of Colorado River water to replenish groundwater aquifers, to implement Arizona's groundwater management plan, and to provide water supplies for municipal and industrial users during periods of low water supply. Figure 4.5 shows the rapid increase in water stored in Arizona's groundwater storage facilities due to the Arizona Water Bank Authority.

---

[83] Arizona Water Banking Authority, "Executive Summary," undated.

[84] Subsequent legislation in 1999 allows the AWBA to store effluent in addition to Colorado River water if there is excess capacity after the surplus Colorado River water is stored; to firm water supplies for non-CAP municipal and industrial users within the service area along the canal such as the Salt River Project, Maricopa Water District, and Roosevelt Water Conservation District; and to establish a lending process for water storage with other entities.

[85] The ADWR also could decide not to sell but to retire these credits.

Figure 4.5. Cumulative Storage in Arizona's Groundwater Storage Facilities, by Storer

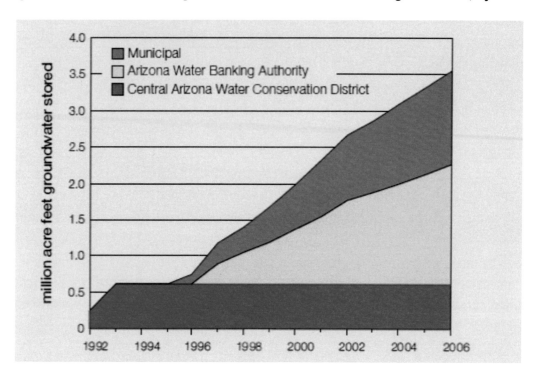

SOURCE: Taylor Shipman and Sharon B. Megdal, "Arizona's Groundwater Savings Program," *Southwest Hydrology*, May/June 2008, p. 11.

In 2005, the AWBA also became responsible for working with Native American tribes to fulfill the state's obligation in firming[86] their water supplies, which also provides drought protection. In addition, the AWBA can contract with authorized entities in California and Nevada[87] to store Colorado River water for a fee, thereby replenishing Arizona's groundwater aquifers. Under these agreements, when either California or Nevada would need their stored water, they would draw the stored amount from Lake Mead and the Colorado River. In 2005, the AWBA began storing water for the Southern Nevada Water Authority and suspended these operations in 2013.[88]

---

[86] *Firming* a water supply refers to making the supply more secure, or reliable. This is often done in periods when supplies are limited by supplementing water supplies with water stored in a reservoir or a water bank. For example, the Arizona Water Bank's unused Colorado River water is used to ensure that water is available for Arizona municipal uses, industrial uses and for Indian water rights settlements when other supplies are limited. See Arizona State Senate, "Fact Sheet For H.B. 2835: Arizona Water Settlements Act Implementation," undated. The Windy Gap Firming Project in Colorado is increasing reservoir capacity in order to make water supplies to municipalities on the Front Range more secure.

[87] Up to 1.25 million acre-feet for Nevada.

[88] Arizona Water Banking Authority, homepage, undated.

The Central Kansas Water Bank Association is an example of a water bank that enables conservation and redistributes water use away from sensitive, water-poor areas. It was the first water bank in Kansas and was established in 2005. Located in central Kansas within the Big Bend Groundwater Management District, the Central Kansas Water Bank Association is a groundwater bank created to alleviate stress on the Rattlesnake Creek Basin using voluntary means. The water bank makes it possible to reduce groundwater usage near the creek by "depositing" the water rights in the bank, and making it possible to lease water rights from other areas. It promotes water conservation and provides an alternative management approach for reallocating limited water resources away from supply-poor areas to improve streamflow (groundwater sources near the creek are hydrologically linked to the surface water) and can provide water resources to areas with growing demand.

The Central Kansas Water Bank Association is a private, not-for-profit groundwater bank that provides an electronic bulletin board to bring together depositors and lessees. The bank manages water deposits, leases, and safe deposits but does not own, buy, or sell water rights. Deposits can be made annually for up to five years, while leases can be for a maximum of ten years. Water rights deposited in the bank constitute "due and sufficient cause for non-use" and so therefore are not forfeited. A fee schedule is used to generate funds to operate the bank.

Water deposited in the bank can be used anywhere within the same hydrological unit covered by the geographic area of the bank for beneficial use, expanding both the location and uses of a water right. The actual amount of water leased is restricted to the quantity that had been used historically (rather than the authorized quantity) for the same level of consumptive use (calculated using standard factors). A conservation requirement that varies by area is then applied. The water can be leased for up to ten years at a price agreed upon by the depositor and the lessee. Because it is a term-permitted water right, it would be junior to an established water right in the area.[89]

After the initial pilot period, a review committee determined that low participation was due to several factors: cumbersome procedures; expensive fees; complex formulas for determining water quantity available for lease; large reductions taken on deposits and leases for conservation and consumptive use; and the uncertainty due to the temporary nature of the bank. Despite this, evaluators felt the bank could be a useful mechanism (with modifications) for conserving and transferring water and recommended making water banks a permanent water management option in Kansas.[90]

---

[89] Kansas Department of Agriculture, "Voluntary Incentive Programs," undated ; Tracy Streeter, "Testimony before the House Agriculture and Natural Resources Committee on HB 2516," Kansas Water Authority, January 31, 2012; Central Kansas Water Banking Association, "Kansas Water Banking Act 2001," undated.

[90] For more details on the recommendations, please see Stover, et al., 2011.

The first pilot of water banking in Washington state was the Kittitas Water Exchange in the Yakima Valley. It was developed to ensure stream flows as well as to provide a source of water for future development (note that 15 of 23 water banks surveyed in 2004 had some environmental objective). Lying within the Yakima River Basin, the area covered by the Kittitas Exchange has extensive irrigation using USBR reservoir water from the Yakima River. It also has problems stemming from threatened and endangered species issues (trout and salmon), congressionally required minimum streamflow requirements to sustain Yakima Nation fishing rights, and pressures from population growth. Compounding these demands, the river basin has experienced low total water supply in 2001, 2004, and 2005. Water rights in Kittitas County were fully appropriated by 1945. In dry years, even senior water rights holders (those who purchased their rights prior to 1905) did not get their entire share of water, and junior rights holders have not received their full allotment since 1990.[91] Because of these pressures and the associated uncertainties with the water supply, in 2009 Upper Kittitas County and the state issued a moratorium on new groundwater wells in the area. Later, the moratorium was lifted and new wells were allowed, but only if they could be water-neutral. In other words, any newly developed groundwater well in Kittitas County must purchase mitigation water to offset effects on streamflows and senior water-rights holders to receive a permit. About the same time a United States Geological Survey (USGS) study determined that wells affected surface water in the Yakima Basin.[92]

As envisioned, the water banks in Kittitas County are a way of providing water sources for new needs in addition to being a management tool to help the state meet minimum streamflow requirements resulting from the Endangered Species Act and fulfill Yakima Nation immemorial water rights. Senior rights holders who face rationing, and junior rights holders who face curtailment benefit as well. Currently there are seven water banks in the Kittitas Water Exchange: the Lamb and Anderson bank (within the Upper Kittitas Basin), the Roan bank

---

[91] All rights to Yakima Basin surface water have been adjudicated under the priority system: Those predating 1905 are considered senior water rights. Junior water rates, those dating after 1905, are curtailed in low water years to ensure senior water right holders get the water they are entitled to. Surface water not appropriated by mid-1905 was claimed by the USBR to support its Yakima Basin agricultural irrigation project. Authorized by Congress, the USBR operates five reservoirs to supply water to irrigators in Kittitas, Yakima and Benton counties. Since 1990, post-1905 water rights within the irrigation project (junior rights holders) have been prorated to less than 50 percent of a full water supply. In addition to the USBR water irrigation project, Congress has mandated that stream flows be maintained to support fisheries available to the Yakama Nation, which holds time immemorial water rights related to these stream flows. See Washington Department of Ecology, "Pending Water Right Applications in Subbasins of the Yakima River Basin," undated; Bob Barwin, "Water Right/Mitigation: Banking in the Yakima Basin," Water Banking/Exchange Workshop, Edmonds, Wash., October 6, 2008; Courtney Flatt, "Water Banks Help Washington Land Developers," KLCC Public Radio (Eugene, Ore.), April 4, 2012; Washington Department of Ecology, "Water Banking and Trust Water Programs: Important Water Management Tools," Publication No. 09-11-035, December 2009.

[92] Clifford, Laundry, and Larsen-Hayden, 2004; Laura Ziemer and Ada Montague, "Can Mitigation Water Banking Play a Role in Montana's Exempt Well Management?" *Trout Unlimited*, September 2011.

(Swauk Basin), the SwiftWater Ranch bank (Teanaway Basin), the Masterson Ranch bank, the Reecer Creek Golf Course bank (Lower Kittitas Basin), the Williams and Amerivest bank (Lower Kittitas Basin) and the Roth-Clennon bank (Upper and Lower Kittitas Basin). The Upper and Lower Kittatas Exchange banks were the first two to become operational. These banks may have either groundwater or surface water that is "deposited" into a bank either temporarily or permanently for either in-stream or out-of-stream use.

The Department of Ecology manages a clearinghouse of available mitigation credits and determines whether the transaction is water-neutral. Mitigation water rights that are offered by sellers are held by the trust (rights are not forfeited and maintain their priority date while in the trust). Each bank operates for a defined area to ensure that the mitigation water is applied within the same area affected by the new demand. The market determines prices.[93] While the Kittitas Exchange makes it easier for small landowners and developers to obtain water rights, the local community initially expressed considerable distrust over the Upper Kittitas Exchange. The suspension of well permits was viewed as anti-growth. Some of the more recent issues identified with water banking in Washington State include:

- USBR limitations in pricing water because of Federal Acquisition Regulations (FAR). USBR leases and procures water in the state and partners with the Department of Ecology on water trading activity. However, the FAR does not allow for water prices to exceed the value of the land combined with the water, which tends to lead to undervaluing the water rights and puts the Bureau at a disadvantage in the market.
- Managing localized economic effects with out-of-basin water rights transfers. While an analysis confirmed that water transfers benefit the overall state economy, more can be done to reduce or alleviate economic effects on communities selling water rights.
- The Department of Ecology is limited to water purchases. In some instances, sellers are reluctant to sell water separately from the land because they are concerned that the remaining land would have limited uses and value. The department would have more options and could benefit multiple environmental concerns if this limitation was lifted.
- The first large transaction with the development company Suncadia distorted prices in the Lamb and Anderson bank located in the Upper Kittatas Exchange pilot.
- A lack of marketing and brokerage assistance such as technical assistance, purchasing expertise and managing and monitoring.[94]

Arkansas River Water Bank Pilot: Banks for Maintaining Water Rights

The first experiment with water banking in Colorado was the Arkansas River Water Bank pilot program. The Arkansas River Basin in southeastern Colorado is Colorado's largest river basin.

---

[93] Prices in the Lamb and Anderson bank located in the Upper Kittitas Exchange pilot were distorted by the first, large transaction with Suncadia, a development company. Suncadia purchased several pre-1905 irrigation water rights at approximately $1,300 per af and has sold more than 30 af of mitigation credits for nearly $40,000 per af. See Washington State Department of Ecology, "Suncadia Water-rights OK'd for Water-banking Program," February 11, 2010.

[94] Peggy Clifford, "2010 Report to the Legislature: Water Banking in Washington State," Water Resource Program, Washington Department of Ecology, June 2010.

The Arkansas River supports recreation, several cities, and a large agricultural community of farmlands and rangelands that use a large web of ditches to transport water. Within the Arkansas River Basin, where approximately 20 percent of the state's population lives, virtually no water is available for new uses. Permanent sales of water to urban centers (Aurora and the Rocky Flats) as well as to investment groups (High Plains A&M[95] and the Fort Lyon Canal Company) affected agriculture in the basin and led to public concern over threats to agriculture, local economic conditions, and permanent access to water rights. The purpose of the bank was to allow users of Arkansas River water more flexibility in meeting water needs while still meeting interstate compact obligations to Kansas.[96] In effect, the water bank was a way to help agricultural senior rights holders in the Arkansas River Basin maintain their water rights and financially benefit from those rights, while providing water to high-growth urban areas that were junior rights holders.

Colorado's legislature established the bank in 2001 and it became active in 2003. The State Engineer provides regulatory oversight, and the Southeastern Colorado Water Conservancy District (SCWCD) administers the bank. The Arkansas River Water Bank is a clearinghouse that facilitates water transfers through an online bulletin board. Water transfers are short-term (one year) bilateral trades between willing demanders (urban users) and willing providers (agricultural users), and market-based negotiations between parties determine prices. Dry-year options, or interruptible supply agreements, based on specific weather conditions and allowing for a schedule of prices also are possible.

Even though the bank conducted no transactions after nine years, the lessons from it are valuable. Community support, administrative procedures, geography, pricing, and timing have been suggested as reasons why the bank has failed to complete transactions. Originally, the bank was allowed to approve out-of-basin transfers without mitigation requirements. As a result of this possibility, the community did not support the bank for the fear of social and economic losses in the present and future. Nor did the community have knowledge of a working water market, which may have contributed to its general distrust of water banks. Additionally, participants in the bank were asking for prices that were higher than the going market price of short-term leases, so demanders were able to acquire water in the lease market at a lower price. The administrative processes were expected to average three months, which is proportionately a significant amount of time to invest for a one-year lease.

---

[95] Southwestern, affiliated with High Plains A&M, LLC, began acquiring water rights in 2001. High Plains purchased more than 18,280 acres of land and more than 21,300 shares (equivalent to more than 60,000 acre feet) of senior 1883 water rights in the Arkansas River Valley. See Chapter Five for more details. Also see Southwestern Investor Group, "High Plains LLC Water Rights Project," undated; and Dan Gordon, "Water Rights: Power Struggle in Southeastern Colorado," *Denver Post*, September 9, 2012.

[96] Kansas has been claiming since 1902 that Colorado takes too much of the river's water. There have been several lawsuits before the U.S. Supreme Court despite the interstate compact agreement between Colorado and Kansas in 1949. The most recent lawsuit, initiated in 1985, largely was resolved by 2009.

Despite these failings, many in Colorado believe that water banks hold promise for dealing with water management issues such as the vulnerability of municipalities to drought, permanent loss of agricultural land to development, and a potential compact call on Colorado River water. In fact, several other water bank proposals have been put forward. For example, the Arkansas Basin and Gunnison Basin roundtables[97] are considering a joint project to see if pre-1922 water rights in the Gunnison River basin could be banked in the Blue Mesa Reservoir as a hedge against a Colorado River compact call. Banking water in the Blue Mesa Reservoir potentially could protect those who rely on post-1922 water rights such as the Colorado–Big Thompson Project water, Denver and other municipalities in the region, Twin Lakes, and the Fryingpan Project should a compact call or shortages occur in the future (after the agreement for water shortage sharing signed by the states in 2007 expires in 2026).[98]

Strengths and Weaknesses of Water Banks

Water banks increasingly are being considered because they are an alternative approach to water management that is flexible (by allowing changes to place of diversion, timing of use, and type of use). They also are open to participation by qualified buyers and sellers, yet still maintain certain standards (on third-party effects, return flow, etc.). They can use market prices to attract and distribute water resources. By simplifying and standardizing the transfer process, reducing transactions costs (such as engineering costs, legal expenses, market information) and speeding up the overall process, water banks may open market transfers to a wider range of participants who can experiment with temporary trades to their advantage. And water banks can increase the availability of water-related information, including its price.

Water banks have broad application in that they can be used for a variety of purposes:

- Firming water supplies for junior rights holders in the face of drought (or other threats to water supply such as an interstate compact call on the Colorado River)
- Reallocating water in areas where water rights are over-appropriated (and demand is high due to growth or where water rights, such as Native American rights, are not well quantified)
- Maintaining groundwater levels or stream flows in ecologically sensitive areas.

Another benefit of water banks is that they are a way of enabling and rewarding water conservation while maintaining water rights. Most water banks also have an objective to

---

[97] The Colorado Water for the 21st Century Act established nine roundtables representing the eight major river basins and the Denver metropolitan area to address water management in a collaborative and inclusive manner and to encourage locally derived solutions. The roundtables include designated members that represent each county, each municipality and each water conservation district within the river basin. Other members represent the various interests such as agricultural, recreational, local water providers, industrial and environmental interests within the river basin, half of which must own water rights.

[98] Chris Woodka, "Blue Mesa Seen as a Water Bank," *Pueblo Chieftain*, February 12, 2012; Colorado Water Conservation District, "Basin Roundtables," undated.

maintain stream flow, either through conserving surface water withdrawals or by using groundwater on balance with surface water.

However, water banks are challenging to establish. They may require changes to state laws and must cover a large enough area to have enough traders while being responsive to hydrological, ecological, and community concerns. As we have seen with some of the examples in this section, many bank pilots have failed because they were not properly structured (leading to high transaction costs or cumbersome processes).As a result, the incentive to participate was low.

Another problem is that potential participants and stakeholders do not trust the bank. Specifically, bank administration or ownership may affect participation and stakeholder acceptance. In Kern County, California, distrust and conflict has arisen over the effects that a quasi-public/private water bank had on groundwater and the pumping costs incurred by the municipal water supplier. Stakeholders in the county also have expressed concern over handing over a public resource to a private firm (technically the bank is administered by a quasi-public agency, but commercial interests have a majority interest) that would reap profits off the deal. Another example of how ownership affects perception of the bank is in Yakima Basin, Wash., where there is distrust of the federal government over water distribution, and the bank was seen as slowing development. A recent blog characterizes this situation well:

> Western water issues are complex and the Yakima River basin is one of the rare places where we see all of the water conflicts concentrated. Here, we have salmon and steelhead runs on the brink of extinction; an economy dependent on irrigated agriculture; recent droughts; a politically active Indian Tribe that both defends salmon and is a major irrigator; and a water supply particularly sensitive to climate change. All of which is a recipe for a fight.

> The most recent round of controversy involved several years of efforts to build the Black Rock project, a pipeline and new reservoir that would pump water from the Columbia River to the Yakima. This $7 billion federal project made no economic or environmental sense and threatened to spread radioactive groundwater from the Hanford Reach Nuclear Reservation to the Columbia River. . . ."[99]

Bank administration must be streamlined and responsive. It also must be relatively easy to sell or buy with a bank, and procedures to verify appropriate water rights must be in place. Otherwise, potential participants will have no incentive to participate. Limited price information may hamper transactions if seller prices are too high or purchase prices are too low. Community support likely will facilitate participation and increase bank activity, so local community concerns regarding the potential lost economic, environmental, or cultural opportunities

---

[99] Michael Garrity, "Big Opportunity (and Big Misperceptions) in the Yakima River Basin," American Rivers River Blog, November 23, 2010. Other sources for these examples are O'Donnell and Colby, 2010; Felicity Barringer, "Storing Water for a Dry Day Leads to Suits," *New York Times*, July 26, 2011; California Water Impact Network, "The Kern Water Bank," undated; Kelly Zito, "Suit to Get Kern Water Bank Returned to State," *San Francisco Chronicle*, July 12, 2010.

associated with the water must be addressed and managed. The bank administrator plays a significant role in this regard.

The Future of Water Banks

The existence of successful water banks such as the California Drought Banks and the Arizona Water Bank Authority suggests that they can be an effective mechanism for allocating water. Growing interest in water banks is evidenced by the development of banks in Kansas and Nebraska, along with additional proposals for banking in Colorado and the expansion of banking in Washington State. In addition, the Western Governors' Association is looking into developing tools and strategies for encouraging innovative water transfers and water conservation, with water banking being one option. Water banks, however, face many challenges in design and implementation, many of which are situation-dependent, as identified in the previous section. In addition, pilots can take years to establish, evaluate, and subsequently restructure (the Arkansas River Bank was a nine-year pilot that had no transactions, while the Central Kansas Water Bank Association had been a pilot for seven years and was being restructured in response to a five-year review). Because of these reasons—complexity, uniqueness, unfamiliarity to buyers and sellers—they are unlikely to expand rapidly.[100]

Block Pricing

Block pricing is not a method for transferring water, but an approach to charging for water use. Unlike the previous water market mechanisms such as water banks and water rights sales and leases described in previous sections, block pricing does not transfer water rights temporarily or permanently. We include it as a market mechanism here because as usage rises, price is used to signal to consumers the scarcity of supply and the costs of maintaining additional infrastructure for peak demand. In other words, a market tool is used to change consumer behavior.

Water suppliers (utilities) largely use block pricing to determine fees charged to water users (customers). A water supplier that uses block pricing charges a different rate for each increment of water use. Under increasing block pricing, the price paid for water increases with the quantity purchased. Decreasing block pricing, which often used for large or industrial customers, is when the rate *declines* with more water use. Some utilities use both increasing and decreasing block pricing for different types of customers. For instance, Norfolk, Virginia, uses an increasing block price for residential customers and a decreasing block price for its only very large industrial user. Block pricing also is known as tiered water rates, while increasing block pricing also is known as inclining block pricing and decreasing block pricing also is called declining block pricing.

---

[100] Tom Iseman, "Innovative Water Transfers: West-Wide Practice and Potential," American Water Resources Association, April 27, 2011; Tom Iseman, "Banking on Colorado Water," *PERC Reports*, Vol. 28, No. 1, Spring 2010; and Washington State Department of Ecology, "Current Status of Water Banking," undated.

Water utility rate structures must ensure that *costs* for providing the water are recovered. These costs, which include infrastructure, system maintenance and operations expenses, and the costs to develop future supply, are recovered through the water-rate structure, tap fees, and other surcharges.[101] Water utilities charge for water in any number of ways. These rate structures, graphically displayed in Figure 4.6, can be flat fees (a fee is charged with no relation to the quantity consumed); uniform fees (the same fee is charged per gallon of water consumed); seasonal rates (price is higher during seasons of low supply/high demand); and block pricing (increasing where the price charged rises with the quantity used and decreasing where the prices declines with increasing water use). Flat fees actually inhibit conservation, since no incentive exists to spend money to reduce water usage, while uniform rates do not distinguish between essential and non-essential uses.

Block Pricing Can Encourage Conservation

Increasing block pricing is one pricing scheme that may use price at various thresholds to encourage conservation while maintaining revenues for the utility. Seasonal rates also can provide an incentive to reduce discretionary water use (swimming pools, irrigation) during peak demand periods and encourage conservation. Thus, increasing block pricing and seasonal pricing are two approaches that use price to reduce water usage. Additionally, because block pricing is not a strict cost-recovery pricing scheme, utilities should be able to sustain stable revenues with conservation.

Figure 4.6. Water Rate Structures

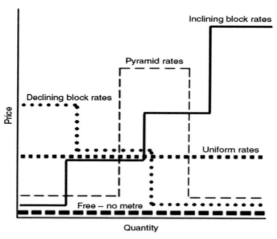

SOURCE: Young-Doo Wang, William James Smith, and John Byrne, "Water Conservation-Oriented Rates: Strategies to Extend Supply, Promote Equity, and Meet Minimum Flow Levels," American Water Works Association, 2005.

---

[101] Recall that the *value* of water also includes the social and environmental opportunity costs, such as recreational opportunities and watershed health.

Increasing block pricing is a useful management approach when utilities are nearing system capacity because they can decrease demand during peak periods and encourage conservation, thereby extending existing water supplies. Increasing block pricing also is of interest because it has been found to promote conservation more effectively than other, prescriptive approaches. For example, a study of western utilities found that increasing block pricing led to a 10 percent to 30 percent reduction in water use.[102] And a 2007 study found that Phoenix, which used flat rate pricing, had an annual per-capita water consumption that was 75 percent greater than per-capita water consumption in Tucson, which used increasing block rate.[103] And Spalding County, Georgia, switched from decreasing to increasing block rate in 1991 and achieved a more modest conservation effect. No other conservation measures were used, and daily water used declined from 243 gallons to 231 gallons per capita, a 5 percent decrease.[104] A systematic analysis of increasing block pricing in Santa Cruz, California, found that *high-use* households responded to marginal price increases, although these effects were modest—a 12 percent decrease in response to a doubling of the marginal price.[105]

To encourage conservation effectively while maintaining utility revenues, the "blocks" must be structured correctly.[106] This is not a trivial task, and developing this price structure will require some expertise.[107] In particular, blocks must not be larger than typical usage. Because of

---

[102] Western Resource Advocates, "Water Rate Structures Structuring Water Rates to Promote Conservation," undated.

[103] This comparison, however, does not control for differences in the cities such as consumers conservation ethic, experience with drought, etc. which may also account for differences in consumers responses to water price. See Glennon, 2009.

[104] Megan Baroni, "Pricing Strategies to Manage Water Demand," in *Whose Drop Is It Anyway? Legal Issues Surrounding Our Nation's Water Resources*, American Bar Association, April 2011, pp. 135–157.

[105] Shanthi Nataraj and W. Michael Hanemann, "Does Marginal Price Matter? A Regression Discontinuity Approach to Estimating Water Demand," *Journal of Environmental Economics and Management*, Vol. 61, No. 2, 2011, pp. 198–212.

[106] The sensitivity of demand to price (price elasticity) is a key economic measure used to determine the effectiveness of pricing for conservation. Studies have shown urban residential demand elasticity is relatively inelastic (at current price levels and in the short run at least). For example, a 10 percent increase in the marginal price of water can be expected to diminish demand by about 3–4 percent in the short run (other studies over different time frames have found elasticities ranging from –0.31 to –0.64, Nataraj found high-end users' in Santa Cruz, California, water price elasticity to be inelastic, at –0.12 (although since these were high-end users the results are not likely applicable to other water users). However, in the long run conservation may be greater as people replace equipment and invest in water conservation technologies. Additionally, there is a broad range to price elasticity which can vary with place and time; be affected by income levels or information provided to water users; the overall price level and the application of block pricing. In particular, should water prices rise to a rate that more closely reflects its true value, price elasticity would be expected to increase. See Sheila M. Olmstead, and Robert N. Stavins, "Comparing Price and Non-price Approaches to Urban Water Conservation," Cambridge, Mass.: National Bureau Of Economic Research Working Paper 14147, June 2008.

[107] When considering alternative rate structures, it must be noted that increasing block pricing is complicated to implement and may lead to unintended consequences when considering fairness to the poor, to the utility's ability to recover costs, and the effectiveness of the conservation signal. Consequently, depending on the local circumstances other rate structures may more effectively accomplish the same objectives.

this, block pricing must distinguish among certain classes of customers, as water use is not the same for residential customers and industrial customers (an industrial customer could be using water efficiently[108] but because it is using so much water in total, it would be pay a high price if blocks were based on residential use). The second element of encouraging conservation with block pricing is that price increases must be noticeable to consumers. The consumers' actual price may include a fixed and variable cost. Usually the fixed cost includes the cost of meter reading, billing, and in some cases debt service, while the variable cost is based on water usage. For block pricing, the elements of price that will determine the conservation effect are the combination of fixed cost, the number and size of the blocks (the point at which price increases), and the relative price increase at each block. These factors determine the price signal given to consumers. If the fixed-cost portion is too high, and the relative price increases for the blocks are low, the conservation signal to consumers will be weak (the average price curve becomes flat and for customers the price of each additional unit of water will appear to be constant). To implement increasing block pricing effectively, utilities will have to have the capability to predict demand responses to different rates, calculate the cost of service for each block, and to determine the appropriate break-points (or the amount of water consumption) within each block for each customer class within its market.

Figure 4.7 presents a graph of block pricing for a group of Colorado cities to illustrate the relative merits of various utility pricing schemes. Three cities still use a uniform rate structure (Loveland, Fort Morgan, and Broomfield) that provides no incentive for conservation and distributes costs evenly across customers, regardless of consumption.

---

[108] *Efficiently* is used here in the sense of the physical processes using as little water as technically feasible (for example, a given quantity of output is produced with fewer inputs), and in not the economic sense in which the price paid for water equals the value derived from its use.

Figure 4.7. Marginal Price Curves for Thirteen Colorado Cities

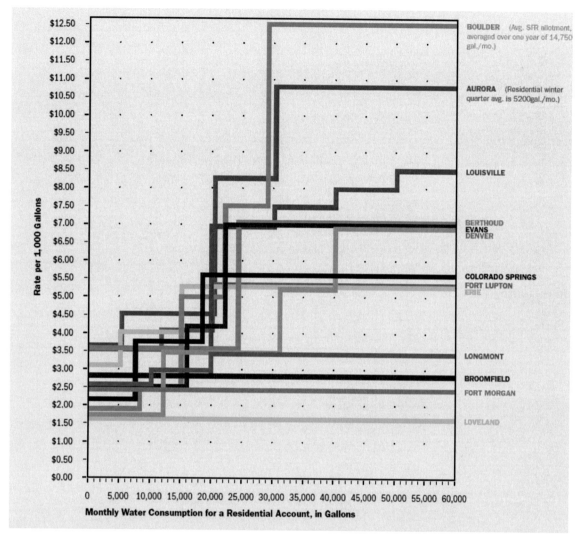

SOURCE: Western Resource Advocates, "Front Range Water Meter: Water Conservation Ratings and Recommendations for 13 Colorado Communities," Boulder, Colo., November 2007.

Boulder and Aurora send the strongest signal to consumers with price structures that have small increments between blocks (more blocks or tiers) and rapid price increases. Both cities have a second block with a relatively small price increase, but subsequent blocks have substantial increases that send a strong price signal to consumers to conserve. Aurora's first block is offered at a relatively low price to cover primary needs for cooking, cleaning, and bathing. The second block, or tier, typically covers water use for large families or efficient outdoor uses. Fort Lupton's price structure shows how having a high fixed cost (at about $3.50 per 1,000 gallons) relative to the block increases (which top out at less than $5.50 per 1,000 gallons) can send mixed signals to consumers. With this combination, the average price actually becomes flat (visualize a curve above the blocked lines in Figure 4.7), and so the conservation signal is weak.[109]

In terms of encouraging conservation, increasing block pricing is considered for several reasons to be more effective than other policy instruments such as requiring specified technology (low-flow toilets or shower heads), lawn watering limits, caps on usage, education or voluntary reductions). First, because increasing block pricing allows consumer flexibility, conservation is achieved at lower cost. Under a block-pricing paradigm, consumers have the flexibility to make the investments that make economic sense for them and therefore are allowed to respond differently. A few studies comparing cost savings from increasing block pricing with prescriptive approaches have been performed. One study compared the cost of using mandatory outdoor watering restrictions (allowed two days per week) with drought pricing in eleven urban areas. The researchers found that for the same level of water usage reduction, the drought price approach would have resulted in an $81 increase in overall welfare, which was about one-quarter of the average household water bill for these eleven cities. In effect, overall consumers benefited from making their own choices to reduce water usage.[110] Second, enforcement is less necessary in a pricing scheme, whereas with prescriptive measures monitoring or inspections are necessary, such as with limits on irrigation or building codes. Therefore, for the most part, this approach obviates the need for expenditures associated with enforcement and eliminates any reduction in attaining conservation goals because of undetected violations. A study of 85 municipal water utilities found that during a prolonged drought in California, more than half of their customers did not comply with quantity-of-use restrictions.[111] Third, with increasing block pricing utilities can maintain their economic viability, which is a challenge with cost-based structures. If structured properly, increasing block pricing can increase revenues enough to cover the income lost from the reduction in demand. With a traditional cost-based structure, revenues will decline as soon as consumption declines, putting the utility at risk for being able to cover expenses.

> This occurred in 1991 in southern California. During a prolonged drought, Los Angeles water consumers responded to the Department of Water and Power's request for voluntary water use reductions. Total use and total revenues fell by more than 20 percent. As a result, the Department requested a rate increase to cover its growing losses.[112]

---

[109] This discussion is a general overview of the relative merits of these cities' pricing schemes. The specific fees, block structures, demand curves, etc. will be the ultimate determinate of the conservation signal sent by these pricing schemes. For example, the overall conservation signal sent by the city of Fort Lupton's pricing scheme ultimately will depend largely on the relationship of the fixed fee to the value of the water in addition to the small marginal increases due to the blocks.

[110] Olmstead and Stavins, 2008.

[111] Sheila M. Olmstead, "Managing Water Demand: Price vs. Non-Price Conservation Programs," Pioneer Institute, July 2007.

[112] Olmstead and Stavins, 2008.

For other policy concerns, such as equity and predictability of outcomes, increasing block pricing and prescriptive approaches have their own advantages and disadvantages. It can be politically difficult to raise prices for water, so this has to be considered when proposing increasing block pricing.[113]

Establishing a block pricing structure requires in-depth knowledge of consumers and sufficient analytical capability to be able to establish the rates and sizes of each block based on assessments of consumer response to price as well as the cost of service. Therefore, under block pricing, projected revenue streams will be subject to greater uncertainty, because now revenues will vary with the number of customers as well as with water usage across several price points. Sufficient analytical capability and close monitoring will be required to ensure that revenues are within an acceptable range—not so low that costs are not covered, but not so high as to violate allowable rates of returns determined by regulators. Financial management approaches such as reserve funds or revenue stabilization funds will help mitigate revenue uncertainty. However, utilities must have this capability and also must have the administrative capability to manage these pricing structures.[114]

The Future of Block Pricing

Block pricing is commonly used, but the question is how much it will be used to encourage conservation through increasing block pricing. There is limited readily available data on the trends regarding the use of increasing block pricing. For example, a U.S. EPA survey in 2000 of 1,200 water systems found that 20 percent used flat rates, more than 50 percent had uniform rates, 19 percent had decreasing block rates, and 20 percent had increasing block rates (which were more common among larger suppliers (27.5 percent) versus smaller ones (7 percent). Meanwhile, a 2008 survey of 280 utilities found 28 percent used decreasing block rates, 40 percent used increasing block rates and 11 percent used seasonal rates.[115] The American Water Works Association cites a 2002 survey of utilities servicing residential customers that found 40 percent of the utilities surveyed used uniform rates, while 30 percent used declining block rates, and 30 percent used inclining block rates (as opposed to 1996, when the distribution was 32 percent, 36 percent, and 32 percent, respectively). They note that decreasing block pricing is used more frequently in the Midwest and North Central states while increasing block pricing is more common in California and the Sun Belt states.[116]

While the trends are unclear, interest in increasing block pricing may rise as utilities seek to maintain revenues while encouraging water conservation to extend existing supplies or to

---

[113] Olmstead and Stavins, 2008.

[114] American Water Works Association, *Principles of Water Rates, Fees, and Charges*, Sixth Edition, Denver, Colo.: American Water Works Association, 2012, p. 111; Wang, Smith, and Byrne, 2005, pp. 31–38.

[115] Baroni, 2011, pp. 135–157.

[116] American Water Works Association, "Tiered Water Rates," 2012.

mitigate supply variability. As experience with managing revenue uncertainties grows either through the development of analytical capabilities or financial management approaches, this may help extend interest in increasing block pricing as well.

Water Quality Trading

Water market mechanisms also have been used to seek improvements to water quality in several areas of the country, as pollution and urban growth continue to stress waterways. Water quality affects the amount of usable water, since water quality must match the intended purpose. Therefore, water quality directly and indirectly affects the economic opportunities and environmental health that waterways provide. In this section, we discuss water quality trading.

Water quality trading is a voluntary, market-based approach that both point- and nonpoint polluters within a watershed can use to ensure the watershed meets the Clean Water Act standards.[117] Water quality credits, accrued by investing either in equipment that lowers discharges below permitting requirements (point sources), or by following best-management practices for specific pollutants for those without permits (nonpoint sources) can be traded with other sources within a watershed to reduce the cost of compliance overall (because of cost differentials among polluters for reducing discharges). Therefore, the market creates flexibility to meet Clean Water Act standards by providing alternative means to achieve the standard potentially improving water quality at a faster pace or higher level by incorporating the unregulated, nonpoint polluters.

Driven by the Challenges in Addressing Water Quality Problems

According to U.S. EPA, more than 20,000 water bodies (rivers, lakes, or estuaries) within the United States are considered impaired. An impaired water body is one that cannot fully support its aquatic life or conform to fishable and swimmable water-quality standards set by states, territories, or authorized tribes. The leading contributors to water impairment are excess sediment, nutrients such as nitrogen and phosphorus, and pathogenic microorganisms.[118] Point sources of pollution, such as industrial facilities, concentrated animal farms, active mines, and wastewater treatment facilities, can more readily measure their effects on water quality. Nonpoint sources, such as development, agricultural operations, silviculture,[119] abandoned mines, and military training lands, also introduce contaminants into waterways depending on the

---

[117] Point sources of pollution are those that discharge into the air or waterway from a specific location such as pipes, conveyance channels, ditches, etc. In contrast, nonpoint sources of pollution are those that are discharged over a wide area of land, possibly from multiple sources, and carried to lakes and streams by surface runoff. As this runoff moves across the land surface, it picks up soil particles and pollutants, such as nutrients, oils, and pesticides.

[118] Allison Shipp and Gail E. Cordy, "The USGS Role in TMDL Assessments," U.S. Geological Survey Fact Sheet FS 130-01, undated.

[119] *Silviculture* means controlling the establishment, growth, composition, health, and quality of forests and woodlands.

specific land use approaches used. Contaminants such as sediment, nitrogen, phosphorous, mercury, automotive oils, and pesticide residue enter into the waterways when erosion (due to vegetation removal) occurs, or through direct transport of excess contaminants from runoff into waterways.

The classic example of water quality trading is the Vittel Bottling Company story. In the 1980s, members of the company realized that the cow dung and fertilizers from upstream farms were entering the aquifer that supplied water to their mineral water plant after farmers had replaced filtering grasslands with corn. Eventually, Nestlé (which acquired Vittel) purchased 600 acres of sensitive habitat upstream and signed long-term conservation contracts with farmers to better manage their animal waste and reforest riparian zones, improving the water quality within the aquifer that supplied their mineral water.[120]

The Basics of Water Quality Trading

First applied to air pollution, market principles now are being used to meet water pollution control targets. Water quality trading allows polluters who are reaching regulatory limits to buy credits from others who have excess credits. Water quality trading occurs when sources within a watershed have different costs to control the same pollutant. This provides polluters with the flexibility to meet regulatory standards, since those who need credit have more opportunities to identify a lower-cost option. These options also may be advantageous in terms of timing. Water quality trading programs originated in the 1980s, with point source-to-point source trading on the Fox River in Wisconsin and point- to non-point trading on the Dillon Reservoir in Colorado.

Water quality trading gained interest after the success of air-pollution trading with the Acid Rain Program (in which sulfur dioxide emissions permits were traded). However, water quality trading has some differences compared to air quality programs. First, as opposed to air pollutants, water pollution remains within the confines of the watershed; trading opportunities are limited to a specific watershed. Additionally, stream flows, and the nature of water pollutant transport, make water quality trade more prone to hotspots (areas of high pollutant concentrations). Because of this tendency, some programs limit trading with upstream sources, in which buyers can only purchase credits from upstream sources (the downside is that trading opportunities are further reduced within a specific watershed). In general, water quality trading has cost and technical challenges related to transaction costs, environmental equivalence, hot spots, nonpoint source reduction quantification and verification, and program enforcement.[121] The most challenging aspect is with measuring and monitoring nonpoint sources. While trades do occur among point sources for cost savings, greater gains are expected from including nonpoint sources within market trading.

---

[120] Ecosystem Marketplace Team, "Water Trading: The Basics," April 16, 2008.

[121] Mark Kieser and Andrew Feng Fang, "U.S. Water Trading: The Infrastructure," Ecosystem Marketplace, undated.

The Clean Water Act (CWA) does not have any formal requirement for trading programs, as do the Clean Air Act Amendments (CAAA), but the U.S. EPA developed a water quality trading policy in 2003 (described further below) that outlines how the agency sees water quality trading as consistent with the Clean Water Act. In addition, nine states have established statewide regulatory authority for trading. Even though the CWA does not explicitly require water quality trading, the driving factor in water quality trading markets is regulation.[122] The increasing number of Total Maximum Daily Loads (TMDLs)[123] that the Clean Water Act and local standards require drives the need for new investments to meet water quality standards. According to U.S. EPA statistics, states and tribes have issued 45,740 TMDLs since 1995, with more than 9,200 issued in the peak year of 2008.[124] Urban growth and development also will put additional pressure on discharges to waterways.

For example,

> With TMDLs specifying pollutant caps in watersheds, one area of opportunity for further application of trading in the U.S. is for offsetting growth. Growth stresses the capacity of municipal wastewater treatment plants. In addition, federal storm-water regulations have begun to treat sources of storm-water as they do point sources: requiring NPDES permits for stormwater discharges from urbanized areas and most construction sites.
>
> Stormwater and wastewater treatment often involve expensive capital investments and high operation and maintenance costs, resulting in high marginal costs of pollutant load reduction. This presents an opportunity for such sources to trade with other nonpoint sources, especially agriculture, where marginal costs of load reduction remain low. Again, such trading markets will be limited in specific watersheds where growth rate is high and agricultural operations account for a substantial portion of the total pollutant load.[125]

Indeed, in 2008 the Township of Fairview, Pennsylvania, purchased credits from local farmers through an aggregator, the Red Barn Trading Company, and was able to avoid spending $6.4 million to upgrade its sewage treatment system. The credits purchase was 75 percent less costly than the upgrades would have been, and kept the town in compliance with water quality

---

[122] Willamette Partnership, Pinchot Institute for Conservation, and World Resources Institute, *In it Together: A How-to Reference for Building Point-Nonpoint Water Quality Trading Programs*, July 2012, p.7.

[123] TMDLs are the analytic basis for allowable discharges into a water body. They are determined based on state-established standards for the specific uses of the water body (drinking water, recreation, fishery, etc.) and are the total pollutant load that a specific water body can assimilate without exceeding these standards. Once the TMDL is established, dischargers are allocated limits so that in effect they act as a limit or cap to any water quality trading program. Point sources receive a waste-load allocation, while nonpoint sources receive a load allocation. Point sources include all sources subject to regulation under the National Pollutant Discharge Elimination System (NPDES) program, such as wastewater treatment facilities, some stormwater discharges and concentrated animal feeding operations. Nonpoint sources include all remaining sources (such as farms, housing developments, etc.) as well as anthropogenic and natural background sources. TMDLs also must take seasonal variations in water quality into account and include a margin of safety.

[124] Willamette Partnership et al., 2012, p. 9.

[125] Kieser and Feng Fang, undated.

regulations. Red Barn Trading Company accumulated and guaranteed the credits, deriving them from chicken farmers in the watershed.[126] Fairview was the first municipality in the Chesapeake Bay Watershed to meet its water quality improvement requirements entirely through water quality trading.[127]

In the case of water quality trading, programs may be structured to allow exchanges among point sources, among nonpoint sources and between point and nonpoint sources. Point sources are regulated under the Clean Water Act through the National Pollutant Discharge Elimination System (NPDES). NPDES permits will identify how much of a particular pollutant a point source is allowed to discharge into the waterway. Nonpoint sources (such as agricultural operations, silviculture operations, and stormwater runoff from rural areas) are not regulated, yet they account for more than 80 percent of the nitrogen and phosphorous in waterways.[128] Because of this, policymakers are very interested in using water quality trading programs to encourage trades between point and nonpoint sources, although some challenges have resulted because point sources are regulated and nonpoint sources are not.[129]

Most water quality trading programs include non-regulated, nonpoint sources of water pollution but have had limited success in attracting them. First, it is difficult to measure, monitor, and verify the contribution of a nonpoint source to pollutant loads along a stream bank where several such sources are present. Since discharges are not continually monitored, as they are with regulated entities, water quality trading programs typically have methods for calculating baseline discharges, reductions achieved through best-management practices (BMP), and methods for ensuring that BMPs are working.[130] If these credits were to be traded with regulated point sources, some level of certainty is required—not only in ensuring that anticipated reductions in pollutants discharges are met, but also that these reductions will continue. This adds uncertainty and risk to the transaction for point sources that have regulatory requirements that must be met,

---

[126] Red Barn Trading Company actually purchased manure from henhouses on 22 farms and took it to strip-mined land outside the Chesapeake Bay watershed. The company then sold the credits for phosphorous and nitrogen to wastewater treatment plants. As an aggregator, buying credits off farmers and selling to waste treatment plants, Red Barn Trading Company assumes the liability for ensuring that these credits are secure (that the farmers' discharges will be at the levels expected). See Alice Kenny, "Pennsylvania Water Deal: Blip or Boom?" *Ecosystem Marketplace*, June 3, 2008.

[127] Mindy Selman, Suzie Greenhalgh, Cy Jones, Evan Branosky, and Jenny Guiling, "U.S. Water Quality Trading: Growing Pains and Evolving Drivers: The State of the Market Today," *Ecosystem Marketplace*, May 14, 2008.

[128] Ecosystem Marketplace Team, "Water Trading: The Basics," *Ecosystem Marketplace,* April 16, 2008; Willamette Partnership, 2012, p. 2.

[129] Ecosystem Marketplace Team, 2008.

[130] BMPs for agriculture may include bank stabilization, crop tillage rotation, wetland restoration, conservation buffers along the edges of croplands, grazing management practices, alternative pest control methods, and other erosion and sediment controls. The actual discharges off of land employing BMPs will vary with site-specific conditions such as precipitation, the effect of soils, cover and slope on pollutant transport to waters, variations in land management practices due to site operations and time lag between implementation of practices and improvement.

or they may be at risk for a non-compliance violation. Secondly, nonpoint sources, such as farms, may not want to get involved in regulatory processes and face additional scrutiny (since transparency is preferable within a regulatory process and businesses tend to prefer privacy). Most trades thus far have been from point-to-point sources (as dominated by the Connecticut Long Island Sound program).[131]

U.S. EPA's Water Quality Trading Policy

In 1996, U.S. EPA formally supported water quality trading programs, and several additional programs were begun. In 2003 U.S. EPA issued a formal water quality trading policy.[132] It outlines the key elements of a successful trading program:

- **Legal Authority and Mechanisms.** The CWA provides authority for U.S. EPA, states (and tribes) to develop trading programs through the NPDES permitting process for point sources. Since states (and tribes) implement this program, they provide the legal authority for trading through legislation, rule making, and NPDES permitting of TMDLs or watershed plans.
- **Units of Trade.** A unit of trade may be pollutant-specific credits, measured to be consistent with NPDES permits.
- **Creation and Duration of Credits.** Credits should be generated either before or during the compliance period covered by the NPDES permit and will apply as long as the pollution controls or management practices are working.
- **Quantifying Credits and Addressing Uncertainty.** States must develop procedures to incorporate credits in NPDES permits to include accepted protocols to quantify pollutant loads, load reductions, and credits as well as procedures to monitor discharges to verify and assess compliance. Suggestions for these procedures include: monitoring; the use of greater than 1:1 trading ratios between nonpoint and point sources; using either demonstrated performance values, or conservative estimates of the effects of management practices; using site- or trade-specific discount factors; and retiring a percentage of nonpoint source reductions for each transaction or amount of credits. Another option is for states to establish a reserve pool of credits that would be available for use if planned for credits do not perform as expected. Among active programs, a handful use a set of standard BMP efficiency rates, nearly half of them use site-specific indicators and models, and about half of them use custom calculations to estimate pollutant removal.
- **Compliance and Enforcement Provisions.** Programs must have procedures for establishing and ensuring compliance. The NPDES permittee ultimately is accountable for discharge limits, whether or not it uses credits from other sources.
- **Public Access to Information and Participation.** Transparency, or stakeholder access to information on the program's activities, will build acceptance, reduce transaction costs, and facilitate trades and enhance program credibility.

---

[131] Kieser and Feng Fang, undated; Paul Quinlan, "Chesapeake Bay: Cap and Trade for Water Pollution—Trendy, Hip, Glitzy and Controversial," *Energy and Environment News*, May 8, 2012.

[132] Kieser and Feng Fang, undated; Quinlan, 2012.

- **Program Evaluation.** Programs should be assessed routinely to verify water quality conditions, the effectiveness of nonpoint mitigation activities, and the economic effectiveness of the program.
- **U.S. EPA's Oversight Role.** The U.S. EPA has an oversight role to ensure that trading programs are consistent with the CWA.[133]

Different Examples of Water Quality Trading Programs

As of 2012, around 26 water quality trading programs in the United States have had at least one transaction, and a few others were in development. Programs may run statewide or be focused on a specific watershed. Figure 4.8 shows the states that have programs: three were in development (Florida, Maryland, Minnesota); three were watershed-specific programs (Connecticut, Delaware, Virginia) and seven were statewide programs (Colorado, Idaho, Michigan, Ohio, Oregon, Pennsylvania, Vermont). Pollutant-specific trading has occurred in more than a dozen states (Arizona, California, Colorado, Connecticut, Delaware, Georgia, Massachusetts, Minnesota, North Carolina, New Jersey, New York, New Mexico, Nevada, Ohio, Oregon, Vermont, and Wisconsin).[134] Water quality markets exist for phosphorus, nitrogen and temperature; some also in sediment and ammonia. More than 85 percent of the programs allow point-to-nonpoint trades. The active programs use standard BMP efficiency rates, site-specific models, and custom calculations to determine the pollutant discharge reductions into the waterways from nonpoint sources.[135]

---

[133] Modified from U.S. Environmental Protection Agency, "Final Water Quality Trading Policy," Washington, D.C., January 13, 2003.

[134] Note that statewide refers to the fact that the program is operated at the state level. Markets will be centered on a specific watershed or region that forms the basis for the TMDL determinations.

[135] U.S. Environmental Protection Agency, "State and Individual Trading Programs," undated.

Figure 4.8. States with Water Quality Trading Programs

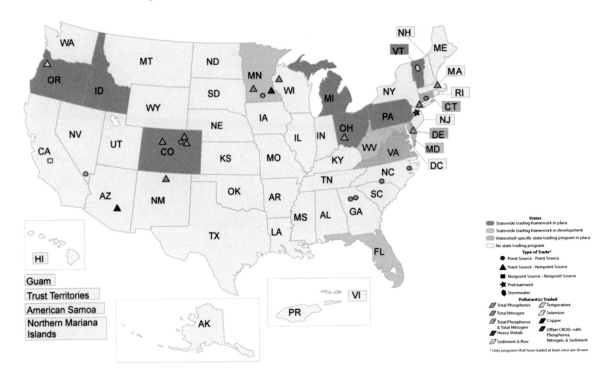

SOURCE: U.S. Environmental Protection Agency, "State and Individual Trading Programs," undated.

The market structure of these programs can vary. The programs' five different market structures are described in Table 4.4. An analysis of programs found that of the active programs, 67 percent use bilateral trades, 46 percent use sole source offsets, 21 percent use an auction platform, and 17 percent use an exchange.[136]

Some of the more active programs are Southern Minnesota Beet Sugar Cooperative (point-nonpoint trades for phosphorous), Minnesota River (point to point for phosphorous), Neuse River in North Carolina (point-to-point nitrogen), and the Connecticut Nitrogen Credit Exchange Program on Long Island Sound (point-to-point trades for nitrogen). For example, the Connecticut Nitrogen Credit Exchange Program on Long Island Sound involves point source-to-point source trades for nitrogen. In 2001 a TMDL was issued for dissolved oxygen in Long Island Sound. Point sources were given caps for nitrogen discharges in proportion to their discharge volumes, while new sources would have to trade to offset 100 percent of their discharges. In response to the new standard, the Connecticut Legislature established a nitrogen credit exchange, which is run by an advisory board. The board determines the price of a nitrogen credit each year in consultation with the Connecticut Department of Environmental Protection (CTDEP). The credit price is based on the cost of the projects to remove nitrogen and the pounds of nitrogen removed. A facility can sell credits if its discharges are below its permitted limits. Within this statewide program, 79 sewage-treatment plants have traded credits; more than one-third of them had

---

[136] Willamette Partnership, 2012, p. 14.

lowered nitrogen discharges below permit limits within three years. This program accounts for a large majority of water quality trades made in the country. From 2002 to 2005, an average of 43 municipalities purchased credits.[137]

The Southern Minnesota Beet Sugar Cooperative (SMBSC) is engaged in a phosphorous trading program. It is a farmer-owned cooperative that runs a beet processing facility on one square mile of land in south-central Minnesota. Near the facility are settling ponds, water-holding lagoons, receiving sites and roughly 1,100 acres of tillable land, part of which is planted with alfalfa and grasses that are irrigated with wastewater from the ponds and a wastewater treatment facility. Discharges from these operations enter the Minnesota River upstream from Lake Pepin and the Gulf of Mexico.[138]

In 1999, the SMBSC sought to expand operations and build a wastewater treatment facility (WWTF) to accommodate the capacity increase as well as odor problems resulting from the lagoons and irrigation. The Minnesota Pollution Control Agency issued a NPDES permit for the WWTF with the condition that the SMBSC completely offset the phosphorous discharges from the plant with nonpoint source reductions and required the SMBSC to establish a $300,000 trust fund to finance nonpoint reductions. Given the uncertainties with nonpoint sources, a ratio of 2.6 to 1 is used for nonpoint-to-point credits to ensure that discharges are within TMDL limits. The SMBSC is responsible for ensuring nonpoint source phosphorous reductions are achieved, and an independent auditor must certify these reductions. Best-management practices, which the Minnesota Pollution Control Agency must approve, are used to achieve the reductions and site-specific factors are used to calculate credits. Best-management practices include cattle exclusions, buffer strips, constructed wetlands, set-asides, alternative surface tile inlets, and cover cropping.[139]

According to its website,

> SMBSC has facilitated the reduction of 150,000 pounds of phosphorus into the Minnesota River Watershed by creating incentives for farmers and cattle ranchers. Each year SMBSC funds projects that protect and improve water quality, like phosphorous retaining cover crops, drain tile screens and fencing to keep animal waste and activity from streams. The Minnesota Pollution Control Authority reviews and certifies the projects to determine the total amount of phosphorous pollution that has been averted.[140]

---

[137] Rob Luke, "Voluntary Water Markets: The Demand Dilemma," *Ecosystem Marketplace*, May 26, 2008; U.S. Environmental Protection Agency, "List of State and Individual Trading Programs: Excel Spreadsheet," undated; Selman, Greenhalgh, Branosky, Jones, and Guiling, 2009; U.S. Environmental Protection Agency, "Water Quality Trading Program Fact Sheets," undated, pp. A-13–A-23.

[138] Southern Minnesota Beet Sugar Cooperative, "SMBSC Facts," undated. Southern Minnesota Beet Sugar Cooperative, "Southern Minnesota Beet Sugar Cooperative Far Exceeds Annual Phosphorous Offset Requirements," undated.

[139] U.S. Environmental Protection Agency, "Water Quality Trading Program Fact Sheets," undated, p. A-47.

[140] Southern Minnesota Beet Sugar Cooperative, undated.

As of 2006, the SMBSC had significantly exceeded its offset requirements by achieving a reduction of 15,767.5 phosphorous pounds per year (2,500 pounds of phosphorous per year is the point source permitted limit; therefore, at a 2.6-to-1 ratio, 6,500 pounds of phosphorous offsets is required). The cost to SMBSC for each credit was $18.65 per pound in 2000–2001, which is comparable to the treatment costs for a WWTF to meet a 1 mg/L phosphorous limit. Given that the Southern Minnesota Beet Sugar Cooperative wastewater treatment facility NPDES permit required completely offsetting discharges, the SMBSC felt the approach saved money, with the added cost savings from protecting young beet crops with the cover crop BMPs.[141]

In the Chesapeake Bay, meanwhile, water quality trading programs also have been tried in the past. They recently have been revisited as a means of achieving nitrogen and phosphorous goals when population growth (about 3 million people by 2030) and development in the watershed is expected. The Chesapeake Bay Watershed encompasses more than 64,000 square miles and parts of six states. It is the United States' largest estuary, providing cultural, recreational, and economic opportunities. Nonetheless, most of the bay is impaired due to the overabundance of nitrogen, phosphorus, and sediment. Both point and nonpoint sources contribute pollutants to the bay, primarily wastewater treatment plants and agricultural and urban/suburban runoff, respectively.[142] Agriculture contributes perhaps 30 percent of the Chesapeake Bay's nutrient load; it contributes just 0.5 percent to the area's economy.[143] Around 2002, Virginia, Pennsylvania, and Maryland all developed trading and offset programs for phosphorous and nitrogen. Virginia also included stormwater pollution loads in its program. Each allowed point-to-point as well as point-to-nonpoint trading. However, trading in each of the programs was low, and the final blow came when Maryland initiated a tax on urban sewer users and rural well users to fund new discharge technologies and agricultural mitigation practices. The tax nearly eliminated demand for water quality credits and contributed to the water quality trading program's demise.[144] Nevertheless, water quality trading is being revisited in the watershed. A recent study estimated that the cost of compliance for point sources potentially could be cut by at least 20 percent if trading is only allowed within the same drainage basin and from point-to-point source, and up to 49 percent if water quality trading occurred on a large scale (across all six states and point-to-nonpoint sources).[145]

[141] U.S. Environmental Protection Agency, "Water Quality Trading Program Fact Sheets," p. A-53.

[142] Willamette Partnership, 2012, p. 32.

[143] Luke, 2008.

[144] Luke, 2008; Quinlan, 2012.

[145] Quinlan, 2012; Luke, 2008.

Table 4.4. Typical Market Structures of Water Quality Programs

Bilateral negotiations: Trades characterized by one-to-one negotiations where price and many other factors are typically arrived at through a process of bargaining rather than simply observing an existing price on the market. This market structure generally has high transaction costs. Examples: Dillon Reservoir, Colorado, and Minnesota River Basin Trading Program, Minnesota.
Sole-source offsets: Structures in which both point and nonpoint sources are allowed to increase pollutant loads at one point if they reduce or "offset" pollution elsewhere (either onsite or off-site). Transaction costs tend to be low. Examples: Southern Minnesota Beet Sugar Cooperative, Minnesota, and Delaware Inland Bays, Delaware.
Brokered trades: Structures in which the link between sellers of credits and buyers of credits is brokered by an intermediary (sometimes called an aggregator or a clearinghouse). Intermediaries convert a product with a variable price and quality into a uniform product. For example, an aggregator might pay several farmers to install BMPs and then offer pollutant reduction credits to buyers at a fixed price. Transaction costs tend to be incurred by the intermediary and are higher than for buyers and sellers.
Auction platforms: This structure relies on brokered trades, but provides an independent platform for buyers and sellers to bid on and/or offer credits. The auction platform is a means of setting prices, and can be designed in several ways. Examples: Maryland Chesapeake Bay Plan, Ohio River Basin Trading Project, Ohio.
Exchange market: Structure that matches buyers and sellers anonymously. Often exchanges use auction pricing, but participants may provide bids and offers online, and the exchange system matches the trade. An exchange is characterized by its open information structure and fluid transactions between buyers and sellers. Transaction costs are typically low. Examples: Long Island Sound Nitrogen Credit Exchange Program, Connecticut, Neuse River Basin Total Nitrogen Trading Program, North Carolina.

SOURCE: Willamette Partnership, 2012, p. 11.

The Future of Water Quality Trading

Water quality trading program can potentially improve the cost and timeliness of complying with the CWA and create new revenue streams for nonpoint sources (such as farmers). Implementation of these programs also will improve the information and tools available for tracking discharges from nonpoint sources and create additional environmental and other benefits beyond water quality through improved land management practices. The challenges with water quality trading programs are to verify and monitor nonpoint source pollution reductions credits over time and to ensure that localized pollution hotspots are not created. Attracting nonpoint sources also may be difficult, since they may be required to share more information with the public because of regulatory requirements on those with whom they trade. The World Resources Institute has been reviewing water quality trading programs within the United States and world-wide, and its assessment is that despite low trading volumes in general, these kinds of programs will grow as they become further refined and developed and as better information is gained through experience. Furthermore, water quality issues are not diminishing. (The review notes that within 12 years the number of globally identified hypoxic zones has increased by a factor of four.) As a result, approaches will be needed to manage these water quality issues. Since U.S. trading credits are generated on water bodies that are regulated for TMDLs, trends in those standards will continue to drive any potential expansion in the use of water quality trading.[146] As a water market mechanism, water quality trading programs are likely to continue, evolve, and increase in use.

Water Market Mechanism Conclusions

Water market mechanisms still are fairly new and evolving. Most of these mechanisms are not fully developed, and the market opportunities are fragmented. There have been many pilots and limited examples of implementation. Water market approaches will improve based on such experiences. We are starting to see a handful of investment firms and service firms that are investing in or supporting water markets, a sign of a developing marketplace.

Water market mechanisms are complicated and require specialized expertise for each location. The actual implementation of the mechanism, whether it is a water auction, bank, or other forms of transferring water utilizing price, must account for the local conveyance opportunities in addition to legal, hydrological, and environmental circumstances and considerations. The heterogeneity of water rights and the costing complexities of water and applying the water market approach in a specific location are another complicating factor.

However, despite these challenges, interest in and use of water market mechanisms is likely to continue for a variety of reasons:

- The pressures on water supply will continue, since water is a finite resource that more people and interests are competing for. Increasing uncertainty over changing patterns of water supply and demand resulting from climate change and persisting water quality concerns will also help encourage greater use of the different water market mechanisms.
- The traditional approaches of developing more water transfer and storage infrastructure, such as constructing dams, reservoirs, and conveyance canals, are nearing their limits. Most water sources are fully allocated and often over-allocated, leading to legal battles over water rights.
- Water market mechanisms are attractive because they can be voluntary, economically efficient, and flexible. They can help more accurately reflect the full cost of water, which can help promote wiser use of water and conservation.
- The trade press' promotion of water firms as good investments shows that market opportunities are growing.
- Increasing numbers of organizations and groups of organizations, such as the WGA, are helping to facilitate innovative water transfers and water market approaches.

Given all these reasons, experimentation, innovation, and growth in different water market approaches is likely to continue and evolve.

Implications for Army Installations of the Different Water Market Mechanisms

Water market mechanisms are an opportunity for Army installations, especially in the future as the markets evolve and grow. Basic water leasing and selling, water auctions, and water banks

---

[146] Selman, Greenhalgh, Jones, Branosky and Guiling, 2008.

are potential opportunities for installations to lease or sell excess water, including effluent, to fund installation water efficiency and infrastructure investments. They also are an opportunity to procure water if existing installation water supplies are not sufficient. Obviously, these opportunities will vary by location and are still limited for much of the country, but are likely to grow.

Water banks, especially those used to balance available water supply during seasonal fluctuations or periods of drought, can help installations in another important way. Such water banks, like the California Drought Water Banks, potentially could be used as a hedge against drought or other water emergency. Obviously, challenges and risks are associated with participating in a regional drought water bank, such as concerns about ceding control over some installation water to store it in a common water bank. But the benefits, including working with key local agencies regarding drought issues, could be worth considering such issues.

Block pricing can affect the price that an installation pays for water. Installations that buy their water from a public utility, as Fort Carson does from Colorado Springs Utilities, may be subjected to increasing or decreasing block pricing. Increasing block rates potentially can help promote water conservation on Army installations and may make it easier for those installations to find funding for water efficiency projects. Namely, since the water is more expensive with greater use, some water efficiency projects now can show greater cost savings because it helps the installation move to a lower cost per unit used in the block pricing system. On the reverse side, if an installation is a large user, akin to an industrial water customer, it can potentially negotiate to have a decreasing block price and save money on its water utility bills. However, even though this would save some money in the short term for that installation, it would not help that installation to invest in and meet its water conservation goals. For installations that charge tenants for their water use based on the amount of water they use, installation personnel potentially could use increasing block rates to help motivate tenants to conserve water.

Water quality trading also could provide some future potential opportunities for some installations. If an installation that is located in an area with a water-quality trading program, owns its own wastewater treatment plant and faces the need to upgrade or rebuild the plant, it potentially could buy some water quality credits from a local farmer to avoid the need to implement such costly improvements to the plant. That is, the farmer would implement BMPs to decrease its pollutant load so that the installation does not have to put in as costly measures to decrease its pollutant load. The cost of the credits would need to be less than the investments needed to upgrade the installation's wastewater treatment facility. Similarly, if this same installation instead could reduce its water pollution load by upgrading its wastewater treatment plant or implementing BMPs to reduce erosion and sedimentation pollution effects, then it potentially could sell some water credits and earn funding for investment in its water infrastructure or efficiency.

In summary, water market mechanisms have the potential to help installations have sufficient water for their missions, help with other aspects of water security, save money, or invest in water efficiency and infrastructure projects.

5. Colorado and Fort Carson Water Case Study

As was discussed earlier, the state and local circumstances with respect to water help determine what an installation's present as well as future partnership and water markets opportunities. This is because of several factors:

- Water availability varies by location. Some locations have limited water resources relative to the needs, which makes water more scarce and valuable and can provide more market opportunities. Population growth, pollution, and droughts have created more pressures on water sources, even in the East; those pressures can create more water market and partnership opportunities. Water sources often cross state boundaries, complicating water's allocation among competing needs.
- Individual state water rights laws affect how water markets can work. Legal complexities govern water use and ownership. In addition, regional, state, and local agreements and treaties often prescribe how water is allocated, stored, and used.
- Water distribution infrastructure affects the costs associated with moving water, which is an important aspect of water markets.
- Regional, state, and local water demands by different sectors, including agriculture, industry, mining, hydropower, and municipalities, impact water markets. Understanding the historical development patterns of water demand and, in many places, the new water pressures (such as those associated with recent urban and suburban growth) helps to understand both present and future water demands.
- The influence of federal, state, and local environmental laws, constraints, and requirements for water in a specific location can affect water markets.

In this chapter, we present an in-depth case study to understand the challenges and opportunities for installation partnerships and water markets in a specific location, namely, in Colorado and at Fort Carson, located south of Colorado Springs. We set the context by explaining the water situation in Colorado, including the complexity of water demand, supply, historical agreements, rights, and distribution. We present examples that illustrate issues related to water markets as well as water sources that supply water to Fort Carson. Currently, Fort Carson relies on Colorado Springs Utilities (CSU) to supply 95 percent of its potable water, so it is important to understand where CSU gets its water, which is discussed next. Then we explain Fort Carson's water situation, including the installation's water management program, strategic goals, sources, and conservation activities. This overview helps to understand what is going on at Fort Carson, how water partnerships and markets currently work, and where opportunities may be in the future.

Colorado Water

Colorado generally is an arid Western state where water is carefully managed. Growing urban areas over the last 30 to 40 years have increased demands for water. Colorado's main source of water is the snowpack in the Rocky Mountains; runoff from this snow feeds rivers and other surface water sources. The state has complex inter- and intra-state agreements for managing water sources and rights. Colorado water rights are based on the prior appropriation system for both surface and groundwater. Since the late 1800s, Colorado has developed extensive infrastructure to transfer water, because often the water has not been where the people want and need it. Past droughts, ongoing population growth, and different groups' competing water demands have increased concerns about current and future water shortages. Given such pressures, water markets have started to develop as a way of reallocating existing supplies.

In this section, we explain this situation in Colorado. Understanding the history and how things have evolved over time is important to understanding the present circumstances. First, we discuss water supply and demand. Then we summarize some of the state's inter- and intra-state water projects and agreements to help understand water allocation and large-scale infrastructure. Existing infrastructure is important for water markets, because of the ability to transfer water readily from one party to another party. We then briefly present an overview of state water rights and regulations. Finally, we discuss some of the state's evolving water markets.

Colorado Water Supply and Demand

Colorado, like many western states, is feeling the pressure of water being a finite resource with multiple competing and increasing demands placed on it. Patterns of water use have changed in response to these pressures. And with supply limited, as demand increases it creates opportunity for water markets to develop to further redistribute this finite supply. However, despite these measures, a future supply gap is predicted. Thus, we begin this section by briefly explaining the history and changes in the state's water supply-and-demand story.

Evolving Water Demand

Colorado is the eighth largest state by land area and has a population of more than 5.3 million.[1] In 2005, Colorado used approximately 13,500 million gallons per day (MGD) of water,[2] 86 percent for agriculture and the remaining 14 percent for municipalities, industry, energy generation, and non-consumptive uses such as environmental protection and recreation. Demand for water is growing. Colorado's population increased 30 percent from 3.3 million in 1990 to 4.3 million in 2000.[3] It is projected to grow up to 8.5 to 10 million people by 2050.[4]

---

[1] U.S. Census Bureau, "State & County QuickFacts: Colorado," March 31, 2015.

[2] This use is for withdrawals. Tamara Ivahnenko and Jennifer L. Flynn, "Estimated Withdrawals and Use of Water in Colorado 2005," U.S. Geological Survey, 2010, p. 46.

[3] U.S. Census Bureau, "Resident Population Data," undated.

Moreover, much of this growth is occurring in low-density areas, leading to significant urban and rural sprawl.[5] Sprawl is associated with higher water use, principally for outdoor use such as lawn maintenance.[6] Therefore, population growth is another area of water demand. Other growing demands for water include energy development (both renewable and non-renewable), as well as non-consumptive recreational and environmental purposes. In addition to an expected increase in water demand from growth in urban and suburban areas, Colorado, like many states, has seen an increased demand for water for recreation and environmental purposes over the last few decades. For instance, along the upper Arkansas River in Colorado, urban water interests, agricultural irrigators, anglers, and whitewater boaters started fighting over the flow and control of the river in the early 1990s after an increase in recreation on the river.[7] A state energy study also projected that oil shale development alone could require up to 170,000 AF/year of additional water resources by 2050.[8] Colorado also needs water for other industries such as snow-making and breweries. While per-capita water demands have fallen, this will not be sufficient to stem an aggregate increase in demand. Even with conservation, the Colorado Water Conservation Board (CWCB) projected that Colorado would need an additional 538,000 to 812,000 AF of additional water to meet municipal and industrial demands in 2050.[9]

A study comparing water withdrawal data from 1985 to 2005 suggests that the use of water in the state has shifted significantly in past decades in response to population growth and changing needs, even though overall withdrawals have not changed. Table 5.1 shows that during that period, publicly supplied water and self-supplied domestic and self-supplied industrial water increased substantially,[10] while water for mining and livestock decreased. Overall water use for

---

[4] Colorado Water Conservation Board, "Colorado's Water Supply Future: State of Colorado 2050 Municipal and Industrial Water Use Projections," July 2010, p. ES-3.

[5] Rural sprawl is low-density residential development and commercial strip development along roads scattered outside suburbs and cities. Rural sprawl is rural residential development at exurban densities with 1.7 to 20 acres per housing unit. See David M. Theobald, *Defining and Mapping Rural Sprawl: Examples from the Northwest U.S.*," Fort Collins, Colo.: Colorado State University, Natural Resource Ecology Lab and Department of Recreation and Tourism, September 16, 2003.

[6] Western Resource Advocates, "Smart Water: A Comparative Study of Urban Water Use Across the Southwest," December 2003, p. 96.

[7] For more details about this conflict, see Robert Benjamin Naeser and Mark Griffin Smith, Playing with Borrowed Water: Conflicts Over Instream Flows on the Upper Arkansas River," *Natural Resources Journal*, Vol. 35, Winter 1995.

[8] Projections were made for three development scenarios. Of the energy supplies assessed, oil shale development by far had the greatest variation among scenarios and accounted for the largest increase in water demand due to energy production. See Colorado Water Conservation Board, "Colorado's Water Needs," undated ; and URS Corporation, "Energy Development Water Needs Assessment," September 2008.

[9] Colorado Water Conservation Board, "The Municipal & Industrial Water Supply and Demand Gap," undated.

[10] Public supply refers to a publicly or privately owned water system for public distribution. Self-supplied domestic water refers to water withdrawn by those not served by a public system; in Colorado, self-supplied domestic water comes from groundwater wells. Self-supplied industrial water is water used primarily for manufacturing and also is not supplied by a public system.

irrigation, which constitutes 90 percent of the water withdrawals examined in this study, remained largely unchanged (although there are clearly shifts in water use practices).[11]

Table 5.1. Comparison of Colorado Total Withdrawal Estimates, by Select Category, 1985 and 2005

Category	1985	2005	Percent increase or decrease, 1985–2005
Water use (Mgal/d)			
Total withdrawals	13,549.67	13,581.22	0.2
Irrigation	12,413.70	12,362.49	-.4
Irrigation—groundwater	2,128.28	2,357.82	10.8
Irrigation—surface water	10,285.42	10,004.67	-2.7
Public supply	737.08	864.17	17.2
Public supply groundwater	86.00	101.86	18.4
Public supply surface water	651.08	762.31	17.1
Self-supplied domestic	16.70	34.43	106.2
Self-supplied industrial	120.35	142.44	18.4
Livestock	60.74	33.06	-45.6
Mining	91.32	21.42	-76.5
Thermoelectric	109.78	123.21	12.2
Population (thousands)			
Served by public supply—groundwater	446.96	667.07	49.3
Served by public supply—surface water	2,562.51	3,689.50	44.0
Self-supplied domestic	221.73	298.61	34.7
Irrigated acres (thousands)			
Total irrigated acres	3,353.07	3,023.25	-9.8
Irrigated acres flood	2,678.57	1,875.24	-30.0
Irrigated acres sprinkler	674.50	1,147.57	70.1
Irrigated acres microirrigation	NA	3.16	NA

SOURCE Ivahnenko and Flynn, 2010, p. 46.

---

[11] It is important to note that this table shows how the water intensity of agriculture has changed only slightly, but statewide public supply intensity (per capita) has gone down significantly.

This trend has been echoed on a local scale in recent years. Denver Water, the utility for the Denver area, reports that system demand changed little from 2004 to 2010 (Figure 5.1) despite a sharp increase in the number of customers. However, this trend is not expected to continue. Denver Water projects that future water demands will increase as the number of customers increases in the coming decades (Figure 5.2).

Figure 5.1. Historical Trends in Demand Among Denver Water Customers

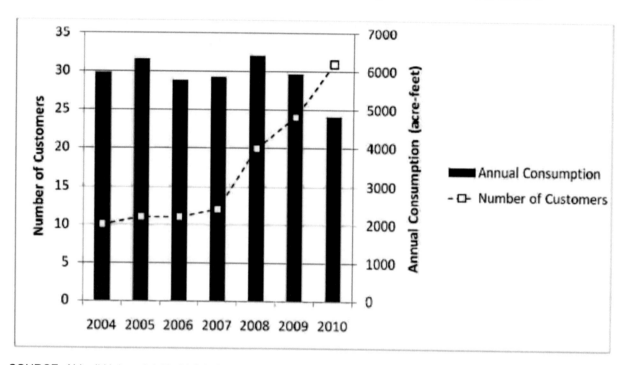

SOURCE: Abigail Holmquist, Todd Cristiano, Marc Waage, Steve Price, and Mary Stahl, "Denver Water's Plan for a Sustainable Reuse System," presentation at the 2011 WateReuse Symposium, September 12–14, 2011, p. 8.

Figure 5.2. Projected Trends in Demand Among Denver Water Customers

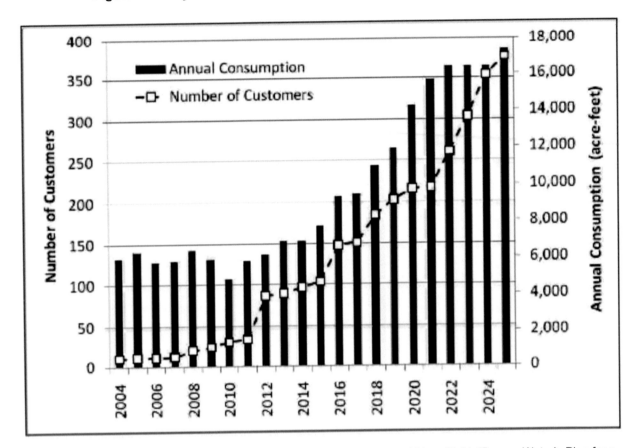

SOURCE: Abigail Holmquist, Todd Cristiano, Marc Waage, Steve Price, and Mary Stahl, "Denver Water's Plan for a Sustainable Reuse System," presentation at the 2011 WateReuse Symposium, September 12–14, 2011, p. 8.

Infrastructure, socio-economic, and institutional structures such as water rights legal regimes, water conservation and conservancy districts, and markets affect how readily water can be directed to emerging demands. As illustrated in Figure 5.3 a CWCB analysis suggests that even if water providers are successful in implementing projects planned to extend current water supplies (such as conservation, reuse, new supply, and agricultural transfers), a 20 percent statewide shortfall in meeting demands will result. Gaps in specific regions may be more or less severe). Moreover, should some of these projects be less successful than anticipated, the gap will be larger (or occur earlier).[12] Overall statewide demand for water (which has been fully allocated) is growing and the likelihood that water will be scarce in areas of the state, or at a given time period, are high.

---

[12] Colorado Water Conservation Board, "The Municipal & Industrial Water Supply and Demand Gap," undated.

Figure 5.3. Colorado's Projected Water Supply Gap

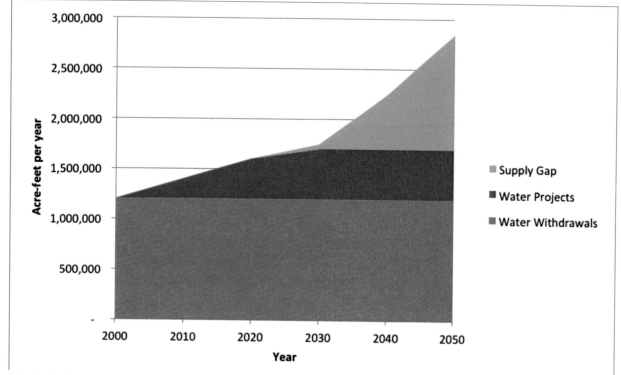

SOURCE: Colorado Water Conservation Board, "The Municipal & Industrial Water Supply and Demand Gap," undated.

In addition to an expected increase in water demand from growth in urban and suburban areas, Colorado, like many states, has seen an increased demand for water for recreation and environmental purposes over the past few decades. For example, along the upper Arkansas River in Colorado, urban water interests, agricultural irrigators, anglers, and whitewater boaters started fighting over the flow and control of the river in the early 1990s after an increase in recreation on the river.[13]

Colorado Water Supply

Colorado also is one of the driest states, with an average annual precipitation of only 15.47 inches.[14] The state relies primarily on two sources to meet its water demands. Surface water from lakes, rivers, and streams accounts for 81 percent (11,000 MGD in 2005) while groundwater accounts for 19 percent (2,500 MGD in 2005).[15] Snowmelt primarily drives these sources: Most areas in Colorado receive less than 20 inches of rainfall a year and, although precipitation in high

---

[13] For more details about this conflict, see Naeser and Smith, 1995.

[14] U.S. Department of the Interior and U.S. Geological Survey, "Colorado," National Atlas of the United States of America, undated.

[15] Ivahnenko and Flynn, 2010, p. 1.

mountain regions can exceed 20 inches, it primarily falls as snow.[16] From this runoff, Colorado's river systems generate 16 MAF each year. The headwaters of four major river basins begin in the Colorado Rockies: the Arkansas, Colorado, Rio Grande, and South Platte. Three of these rivers flow away from most of the state's population centers: the Colorado River flows mostly west and the Arkansas and Rio Grande rivers flow east and south. Only the South Platte River flows east and north through Denver and up into Nebraska.

Precipitation in Colorado does not fall where it is needed. As indicated in Figure 5.4, 80 percent falls on the West Slope, high in the Rocky Mountains. However, today, 80 percent of Colorado's population lives in the Front Range—an area just west of the Great Plains and east of the Rockies (see Figure 5.5). The Front Range (also known as the East Slope) runs north to south and includes the cities of Fort Collins, Boulder, Golden, Denver, Colorado Springs, and Pueblo. Thus, from its early days, much of Colorado's water infrastructure has aimed at moving water from the West Slope to the Front Range. Because many rivers start in the Rocky Mountains and flow into other states, numerous inter-state compacts and agreements require two-thirds of Colorado's water to leave the state for other areas.

Figure 5.4. Average Annual Precipitation in Colorado, 1971–2000

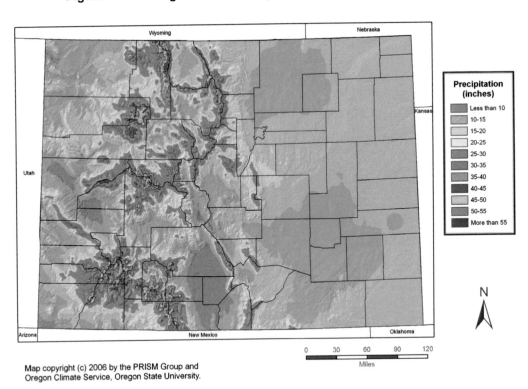

Map copyright (c) 2006 by the PRISM Group and Oregon Climate Service, Oregon State University.

SOURCE: PRISM Group and Oregon Climate Service, "Average Annual Precipitation 1971–2000 Colorado," 2006.

---

[16] Colorado Water Conservation Board, "Water Supply Planning," undated.

Figure 5.5. 2010 Population in Colorado, by County

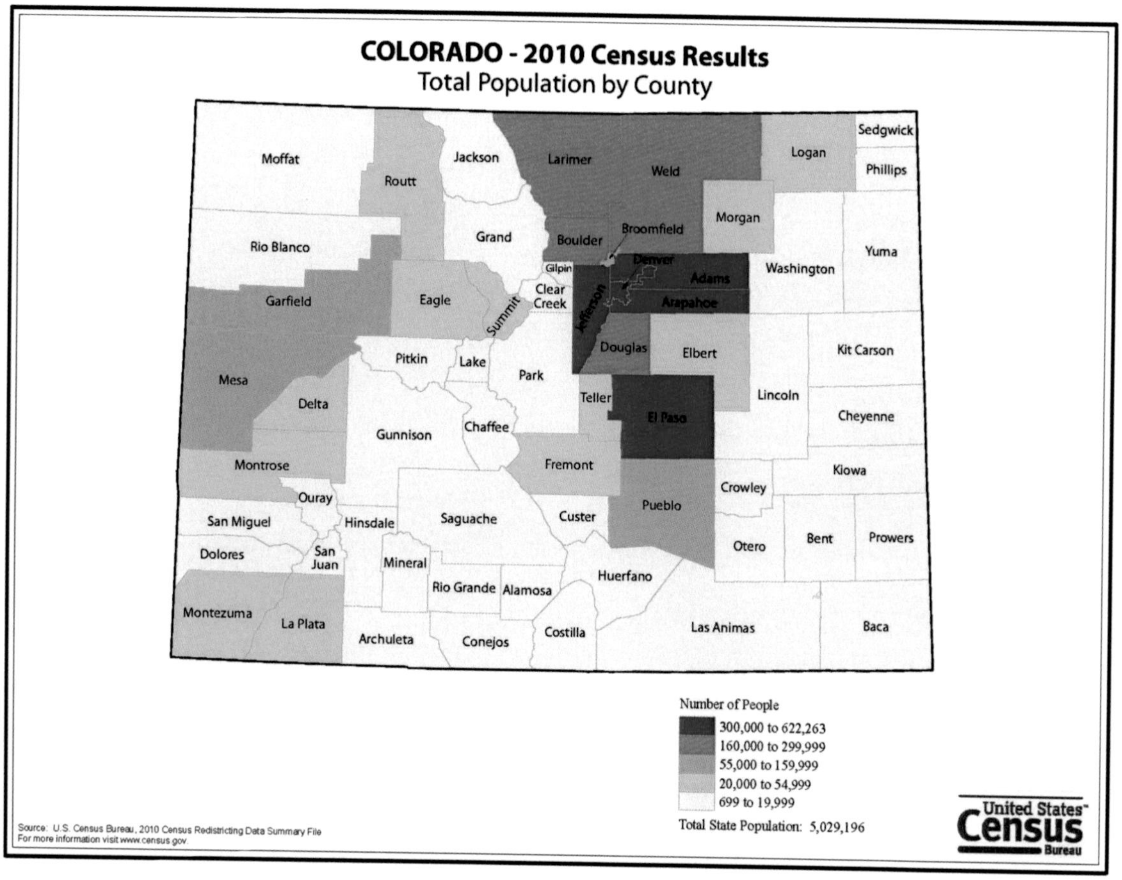

SOURCE: U.S. Census Bureau, "Colorado 2010 Census Results, 2012."

Colorado has a long history of moving water to meet development needs. Settlers in the 1800s sought ways to divert water for mining and agriculture. In 1880, the state's first trans-basin diversion, the Ewing Ditch, diverted water from the Eagle River in the Colorado River Basin in the west to the Arkansas River watershed in the southeast for mining.[17] As was the case elsewhere, the mining rights shaped water rights: The philosophy of "first in time, first in right" governed claims to mining rights, which formed the basis for prior appropriation water laws in the state.

As with other states in the West, Colorado is facing increasing uncertainty about its future water supplies. Increasing population, extended periods of drought, long-term climate change, and over-allocation of water resources suggest that it may become harder to meet water needs in the future.

---

[17] Peter D. Nichols, Megan K. Murphy, and Douglas S. Kenney, "Water and Growth in Colorado: A Review of Legal and Policy Issues," Boulder, Colo.: Natural Resources Law Center, University of Colorado School of Law, 2001, p. 9.

The price of municipal water around the country is increasing for numerous reasons. These reasons include increasing operating costs, aging and degrading infrastructure that has to be replaced, and recent drought periods that have reduced the available amount of water from usual sources. One study compared water rates from 2010 to 2011 and 2012 for a number of cities around the United States. This study found an increase of about 13 percent in Denver Water's water rates in 12 months.[18] One explanation for this increase is that "Denver—which gets half its water from the Colorado River watershed—has junior water rights, meaning that it is first to be cut off when water is short, according to the prior appropriation system used by most western states."[19] Denver Water notes that rehabilitation of water infrastructure is behind these rate increases.[20]

Colorado Springs Utilities (CSU) also is increasing rates, by 12 percent from 2010 to 2016 to pay for new infrastructure—the Southern Delivery System described later in this chapter—to provide more water to its growing service area.[21]

Interstate Water Agreements

Rivers and watersheds cross state boundaries, and many of these water sources have competing demands. This competition has led to disputes and controversies among stakeholders since the 1800s. Today, Colorado is party to a complex array of interstate compacts and agreements that have emerged from decades of efforts to resolve disputes and establish water rights. Colorado is directly involved in two international treaties with Mexico regarding the Rio Grande, Tijuana, and Colorado Rivers; two U.S. Supreme Court decrees that govern water in the North Platte and Laramie Rivers with Nebraska; and nine interstate compacts that also affect water allocation.[22] These nine interstate compacts are as follows:

1. Colorado River Compact
2. La Plata River Compact
3. South Platte River Compact
4. Rio Grande River Compact
5. Republican River Compact
6. Costilla Creek Compact
7. Upper Colorado River Compact
8. Arkansas River Compact

---

[18] Denver was the only city in Colorado included in the study. See Brett Walton, "The Price of Water 2011: Prices Rise an Average of 9 Percent in Major U.S. Cities," *Circle of Blue*, May 5, 2011.

[19] Walton, 2011.

[20] Denver Water, "2012 Billing Rates," undated.

[21] This rate increase is lower than originally anticipated to pay for the new infrastructure, due to lower-than-expected financing rates and efficient contract management. See Daniel Chacon, "SDS Water Rate Hikes May be Lower than Planned," *The Gazette*, April 18, 2012.

[22] Colorado Division of Water Resources, "A Summary of Compacts and Litigation Governing Colorado's Use of Interstate Streams," 2006.

9. Animas-La Plata Project Compact.

In this report, we focus on two compacts that are relevant for water markets, Fort Carson, and the Colorado Springs area. These are the Colorado River Compact and the Arkansas River Compact. These two examples also illustrate the complexities of water allocation, which can make it more difficult for water markets to develop.

Colorado River Basin Compact

The 1,450-mile Colorado River flows from the Rocky Mountains, across Colorado and Utah, and along the borders of Arizona, California, and Nevada, into Mexico to the Gulf of California. The Colorado River's tributaries make up the Colorado River Basin, which also covers the states of Wyoming and New Mexico. Figure 5.6 shows a map of the Colorado River Basin and the seven states that share it.

Figure 5.6. Map of the Colorado River Basin

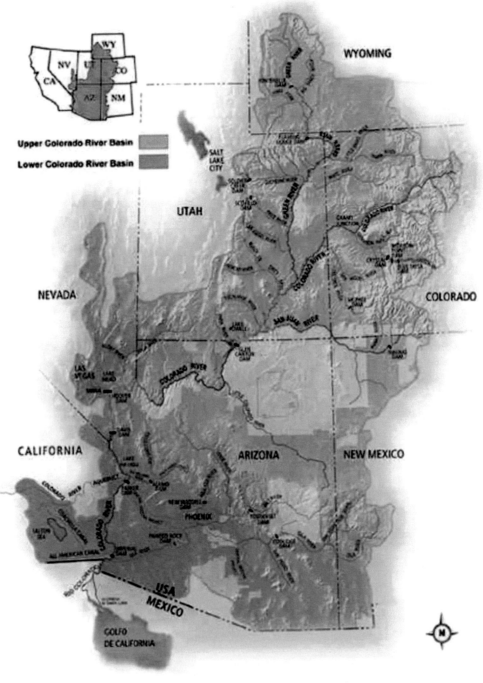

SOURCE: Glen Canyon Dam Adaptive Management Plan, undated.

The shared use of the Colorado River Basin among these seven states—Arizona, California, Colorado, New Mexico, Nevada, Utah, and Wyoming—required negotiation and has been controversial over the last century. The "Law of the River" collectively describes numerous

200

federal laws, court decisions, interstate compacts, and regulations designed to allocate resources and resolve disputes.

Central to the "Law of the River" is the Colorado River Basin Compact, an agreement among the seven states to divide water from the Colorado River equitably, establish the relative importance of different uses of water, ensure development of the basin, and reduce existing and future controversies.[23]

The compact requires the Upper Basin states (Arizona, Colorado, New Mexico, Utah, and Wyoming) to deliver 7.5 MAF of water per year on a 10-year rolling average to the Lower Basin states (Arizona, California, and Nevada). If the 10-year rolling average falls below this level, the Lower Basin states may make a "call" on the compact, forcing the Upper Basin states to reduce water consumption.[24] In 1928, Congress divided the Lower Basin allocation, giving California 4.4 MAF, Arizona 2.8 MAF, and Nevada 0.3 MAF each year.[25]

By late 2013, such a call had never occurred. However, much controversy surrounds the Colorado River Basin Compact, and the implications of a call are unclear. The original delivery requirements of 7.5 MAF were based on wetter-than-average data, and there is concern about meeting release requirements given recent severe droughts. In particular, Front Range cities, which are junior water rights holders relative to agriculture-based rights holders, would be required to curtail their water use. Climate change, demographic changes, and other factors threaten the sustainability of the Colorado River Basin. As a consequence of all of these factors, the USBR undertook the Colorado River Basin Study to assess and manage risks in the coming decades.

Arkansas River Compact

The Arkansas River Compact is another key interstate compact, especially since this river supplies water to southeastern Colorado near Fort Carson. As shown in Figure 5.7, the Arkansas River flows from the Collegiate Peaks in the Colorado Rocky Mountains northwest of Colorado Springs down by Pueblo and into Kansas, then on to Oklahoma and Arkansas, where it joins the Mississippi River. As early as 1902, Colorado and Kansas have disputed and litigated over the apportionment of Arkansas River waters. The states negotiated the Arkansas River Compact in 1948, seeking to settle existing disputes, forestall future disputes, and to divide the water equitably between the states.

---

[23] Watson and Scarborough, 2009; Pitzer, Eden, and Gelt, 2007; Colorado Division of Water Resources, 2006.

[24] Watson and Scarborough, 2009, p. 5.

[25] Pitzer, Eden, and Gelt, 2007, p. 12.

Figure 5.7. Map of the Arkansas River Basin

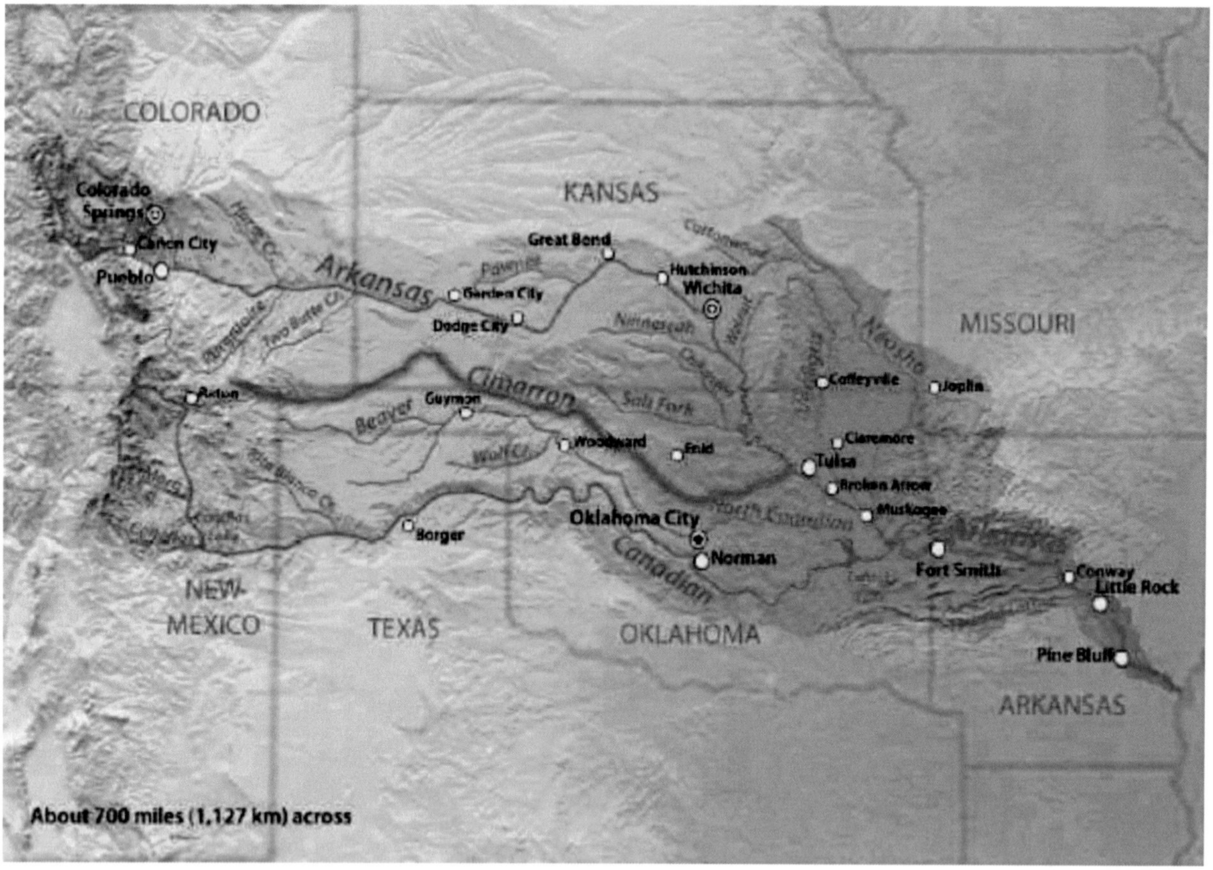

SOURCE: "File:Arkansasrivermap.jpg," Wikipedia Commons.

The Arkansas River Compact was intended to protect the status quo of water-supply use in 1949 and to allocate remaining water supply between the two states. The Compact stipulates that future development and use in Colorado should not materially deplete flows to Kansas, but does not specify a particular allocation. Instead, it allows each state to call for up to a maximum rate of water to be released from the John Martin Reservoir in Bent County, Colorado, regardless of whether the other state makes a similar call. The absence of a clear allocation led to a "race to the reservoir"—both states sought to use any stored water as quickly as possible, before the other state could do the same.

In the late 1970s, the states sought an alternative, more efficient way of allocating water from the river. In 1980, the states negotiated an operating plan that allocated 60 percent of the water stored in the John Martin Reservoir to Colorado, and 40 percent to Kansas. But this allocation did not prevent disputes. The Kansas Department of Agriculture documents one such dispute:

> After the compact … Colorado allowed high-capacity irrigation wells to be developed in the Arkansas River valley. The well pumping reduced river flow and materially depleted water that would have been available to Kansas. Kansas filed *Kansas v. Colorado*, No. 105, Original, in 1985 to enforce the terms of the Arkansas River Compact… In 1995… the court found that Colorado's post-

compact well pumping violated the compact. ...In April 2005, Colorado paid
Kansas more than $34 million in damages for Colorado's compact violations
from 1950 through 1999 and more than $1 million in legal costs in June 2006.[26]

This compact is important to Fort Carson because CSU gets some of its water from the
Arkansas River. If there is a "call" on this river, some parties in Colorado are senior rights
holders; however, as this quotation just illustrated, not all of them are, and some could be
affected during time of drought. It is unclear what would happen to CSU's water supply in such a
situation. Colorado Springs Utilities realizes these pressures and demands on its water sources
and has been taking steps to try to ensure water supply for its customers now and in the future. It
has an aggressive water conservation program, uses increasing block rate structure for water
pricing, has drought plans in place, and has diversified the water sources it draws water from.
However, if there is a significant drought throughout the state, it would likely effect CSU's
diverse water sources, so it is unclear what the impact of a "call" on this one source would be.
This example illustrates some of the complexities and uncertainties associated with water sources
in Colorado that could affect water supply (and costs) in the Colorado Springs area.

Intra-State Water Projects

The state has numerous water projects in each of its river basins to ensure its supplies can meet
diverse demands. In 2005, for example, 43 total inter-basin transfers took place that diverted
nearly 1 million acre-feet in total between different hydrologic sub-regions.[27] In this report, we
focus on agreements and projects that are particularly relevant to water markets and Fort Carson.

Colorado–Big Thompson Project

The Colorado–Big Thompson (C-BT) Project has helped facilitate water markets in Colorado by
providing the key infrastructure to enable water to get from point A to point B, as was explained
in Chapter Four. The Colorado–Big Thompson Project, the largest trans-mountain diversion in
the state, diverts approximately 260,000 af of water from the headwaters of the Colorado River
to the Front Range. The Western Slope collection system traps runoff and conveys water to the
Alva B. Adams tunnel that spans the Continental Divide and serves agriculture, municipalities,
industry, and energy projects in the Eastern Slope.[28]

The uses of water from this project have evolved and echo the overall growth of the state:

> When the C-BT Project was completed in 1957, 98 percent of its water deliveries
> were for agricultural purposes. Today, nearly half of the C-BT water delivered

---

[26] Kansas Department of Agriculture, "Kansas-Colorado Arkansas River Compact Fact Sheet," August 2009.

[27] Ivahnenko and Flynn, 2010, pp. 7–8.

[28] U.S. Bureau of Reclamation, "Colorado–Big Thompson Project," last updated July 18, 2013.

annually goes to towns and cities and rural domestic suppliers, which provide water to smaller towns and rural residents.[29]

Key municipalities served by the Colorado–Big Thompson Project's East Slope distribution system include Fort Collins, Boulder, Greeley, and Loveland. Figure 5.8 shows some of the municipalities and infrastructure of the Colorado–Big Thompson Project. Within this figure, the two top left maps show the locations of the project within the state, the map on the lower left the West Slope collection system and the map on the right shows the East Slope distribution system. The East and West Slope maps have significantly different scales.

Figure 5.8. Maps of Colorado–Big Thompson Project

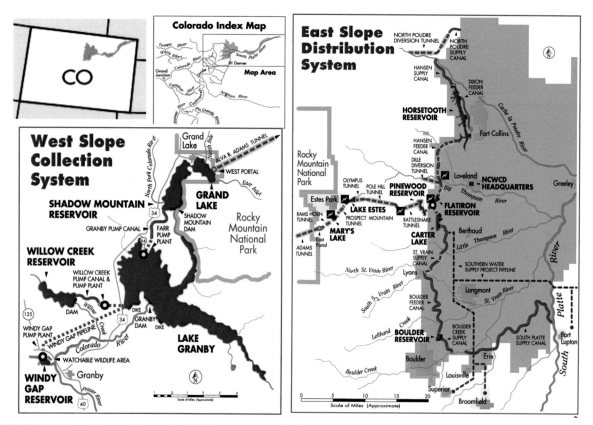

SOURCE: Northern Water, "Colorado–Big Thompson Project," undated.

Figure 5.9 shows what percentage of the Colorado–Big Thompson Project water has gone to municipal use over time and the total water delivery each year.

---

[29] Northern Colorado Water Conservancy District, "The Colorado Big Thompson Project: Historical, Logistical, and Political Aspects of This Pioneering Water Delivery System," undated, p. 22.

Figure 5.9. Percentage of Colorado–Big Thompson Project Water Supplied for Municipal Use

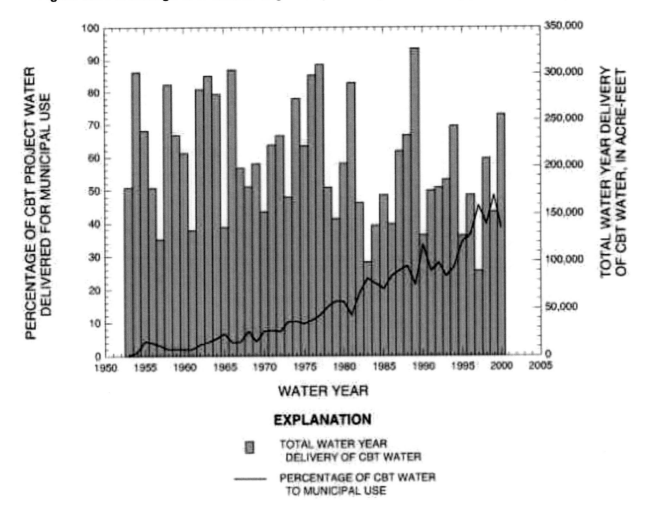

SOURCE: Michael Stevens, "Water Quality and Trend Analysis of Colorado–Big Thompson System Reservoirs and Related Conveyances, 1969 Through 2000," Reston, Va.: U.S. Geological Survey, Water-Resources Investigations Report 03–4044, 2003, p. 5.

As was discussed in Chapter Four, different Colorado water users have used the Colorado–Big Thompson Project infrastructure to lease and sell water because it provides a readily available way to transfer the water physically.

Fryingpan-Arkansas Project

The Fryingpan-Arkansas Project is a trans-mountain diversion of 48,500 acre-feet of surplus water from the Fryingpan River on the West Slope to the Arkansas River in the east. CSU gets some of its water from this project. Figure 5.10 provides overview of this project's infrastructure. It shows how Ruedi Dam and Reservoir provides storage of water for the Western Slope users for irrigation, municipal benefits, recreation, and fish and wildlife enhancement. The North and South Side collection systems on the Western Slope collect runoff and water from the Fryingpan and Roaring Fork River Basins flow into the Charles H. Boustead Tunnel, which conveys all the water from the two collection systems through the Continental Divide to Turquoise Lake. On the

Eastern Slope, conduits carry water from Turquoise Lake to Twin Lakes, from which water is released to Lake Creek and the Arkansas River for delivery to users upstream of Pueblo Reservoir or for storage in Pueblo Reservoir. Finally, water is released from Pueblo Reservoir to the Arkansas River for irrigation and municipal purposes, to the Fountain Valley Conduit for municipal purposes, to the Bessemer Ditch for irrigation, and to the Pueblo Fish Hatchery for the fishery.[30]

Figure 5.10. Map of the Fryingpan-Arkansas Project

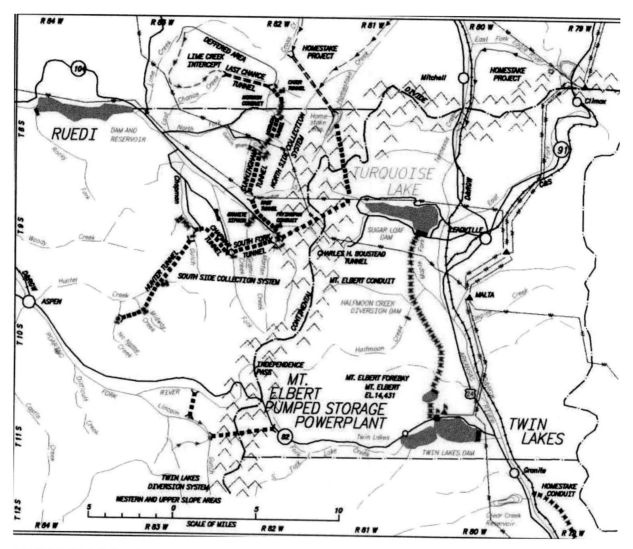

SOURCE: "Twin Lakes Diversion System, Western and Upper Slope Areas," undated.

Water rights allow the project to divert 120,000 AF but are not to exceed 2.3 MAF in any consecutive 34-year period. This diverted water, coupled with water from the Arkansas River

---

[30] U.S. Bureau of Reclamation, "Fryingpan-Arkansas Project," last updated April 4, 2013.

Basin, provide 80,000 AF annually for municipal and domestic use and for irrigation of 280,000 acres in the Arkansas Valley.[31]

The Southern Delivery System

Colorado Springs Utilities is undertaking a key new project—the Southern Delivery System. CSU has water rights to water in Arkansas River Pueblo Reservoir, but access is insufficient, particularly given growing demands. A second delivery system, the Southern Delivery System, is currently under construction to enable deliveries from Pueblo Reservoir to Colorado Springs, Fountain, Security, and Pueblo West to meet growing demands in the region. Phase 1 of the project involves pipeline construction, and deliveries are expected in 2016. Phase 2 will add two reservoirs to extend raw-water delivery capacities. Construction is expected to start in 2020 and be complete in 2025. Phase 2 also will expand the water treatment plant and pump stations built during Phase 1 to better meet peak demand of more than 100 million gallons of treated drinking water per day.[32]

Potential for Future Water Projects

As water demands grow, the state may need to carry out additional water projects or find other ways to meet growing demands for water. There are three broad, non-exclusive options for meeting demands. The first is to increase water supplies through new projects or rehabilitating or expanding existing projects. A second approach is to purchase and transfer water rights, principally from the agricultural sector, which has more senior water rights in comparison to urban users. The third approach is to reduce demand through conservation and energy projects. Each of these options has varying socioeconomic and environmental implications.

A recent study found that the cost of these options varies significantly. Comparing the costs of different types of water projects, the study estimated that new water projects cost approximately $16,200/AF, while agriculture-urban transfers cost $14,000/AF. Some conservation projects, such as turf replacement rebates, leak reduction, efficient toilet and washer rebates, and conservation-oriented pricing schemes, can cost as little as $5,200/AF and yield up to 300,000 AF annually.[33]

Transferring water from agriculture to urban users can do much to meet increasing urban water demands. However, there can be negative consequences for reducing agriculture as well,

---

[31] U.S. Bureau of Reclamation, last updated July 23, 2012.

[32] Paula Moore, "Southern Delivery System Pipeline Working its Way toward Colorado Springs," *ENR Mountain States*, April 23, 2012.

[33] Douglas S. Kenney, Michael Mazzone, and Jacob Bedingfield, "Relative Costs of New Water Supply Options for Front Range Cities," *Colorado Water*, September/October 2010, p. 5.

including reduced food production, potential loss of habitat and ecosystems,[34] and economic decline of rural communities.

Water transfers are a significant concern in the lower Arkansas Valley of Colorado. Since 1950, 80,000 acres (15 percent of historically irrigated land) have been left fallow because of urban water transfers. A similar area of land is projected to be lost to "buy and dry" programs in the next 20 years. However, temporary water leases are seen as more viable, as will be discussed more in the markets sub-section.

Water Sources for Colorado Springs Utilities

Since most of Fort Carson's water supply comes from CSU, next we briefly discuss the water sources and projects that supply its water. CSU obtains its water from mountain streams, some local streams, local groundwater, and purchases from Fountain Valley Authority's Fryingpan-Arkansas Project.

The CSU system is largely a gravity system stretching 11 miles west to east and 16.5 miles north to south. The system collects water from the West Slope to Homestake Reservoir, where it is piped under the Continental Divide to Turquoise Lake and Twin Lakes, where power is generated for the USBR. A pumping station pumps water 750 feet up over the Mosquito Range, where a pipeline transports water to Colorado Springs. As was discussed earlier, CSU is building the Southern Delivery System, which will be an additional source of water starting in 2016.

CSU has strategically planned to ensure it has access to water to meet future demands. However, given the state's water demands just discussed and the uncertainty from factors such as drought, population growth, and climate change, it is uncertain whether enough water will be available for future demands. If CSU did have a shortage of water, what would the effect be on Fort Carson?

Colorado Water Rights and Regulations

As mentioned, water rights in Colorado are governed by prior appropriation, which applies to both surface water and groundwater tributaries to surface water. When water is scarce, a senior rights holder may invoke a "call" on a source of water to obtain the full supply to which it is entitled. Junior water rights holders may lose their access to this water source.

Water rights shape how the state meets growing demands. For instance, Colorado water rights and laws have historically restricted rainwater capture. Section 36-20-103 C.R.S. states that the state has rights to precipitation as a public resource, except where it is under-

---

[34] Some farms and ranches can help maintain habitat for certain species and ecosystems compared to sprawling urban or suburban development.

appropriated or where one has priority to collect water. However, most Colorado streams are over-appropriated, effectively eliminating rainwater capture.[35]

However, a Colorado Water Resources Board study found that only about 3 percent of rainwater reaches streams; the rest is consumed by native vegetation.[36] The state legislature has enacted two laws to allow limited rainwater harvesting. The first, Senate Bill 09-080, allows limited collection and use of precipitation for Colorado landowners of residential properties that are supplied by a well or could qualify for a well permit. An issue brief from the Colorado State Council Staff describes this law:

> [T]he new law allows precipitation to be collected from up to 3,000 square feet of a roof of a home. However, only homes that are not connected to a domestic water system, such as the Denver Water Department, may collect the water. Eligible persons must also have an exempt well permit or qualify for such a permit. The collected water may only be used for ordinary household purposes, fire protection, watering of animals and livestock, and irrigation of up to one acre of gardens and lawns. However, persons with exempt well permits are limited to using the collected water under the same restrictions as their well permit. For example, an owner of a well permit that is restricted to in-house use only would not be allowed to use harvested water to irrigate a lawn or garden.[37]

Another law, House Bill 09-1129, allows developers to apply for approval to be one of ten statewide pilot projects that harvest rainwater and put it to beneficial, but non-essential, use in a subdivision.

Water rights also affect reclaimed water, which is "domestic wastewater that has received secondary treatment by a domestic wastewater treatment works and such additional treatment as to enable the wastewater to meet the standards for approved uses."[38] Thus it is distinct from rainwater capture and greywater.[39]

There are two main kinds of water reuse: direct and indirect reuse. In direct reuse, "return flows from reusable supplies are physically reclaimed for potable and non-potable purposes. For example, a water utility captures reusable treated water leaving its wastewater treatment plant (WWTP) and uses this water again for urban, agricultural, recreational, environmental, or industrial purposes." In indirect reuse, "return flows can be reused under substitution or exchange arrangements. An example of indirect reuse is when a water utility lets reusable water

---

[35] David Beaujon, "Rainwater Harvesting in Colorado," Issue Brief 09-02, Colorado Legislative Council Staff, August 2009, p. 2.

[36] Leonard Rice Engineers, Inc. et al., "Holistic Approach to Sustainable Water Management in Northwest Douglas County," January 2007, p. 1.

[37] Beaujon, 2009, p. 2.

[38] Colorado Department Of Public Health and Environment, "Regulation No. 84 Reclaimed Water Control Regulation," 5 CCR 1002-84, August 2010, p. 2.

[39] As was explained in Table 2.1, greywater means water captured from sinks, baths, showers, and residential laundries that can be treated and reused. It does not include water from kitchen sinks or dishwashers.

leaving its WWTP flow downstream for diversion by an irrigator, and the utility diverts an equivalent amount of water into its system upstream."[40]

Colorado law regarding reclaimed water is complex but generally supportive of reclamation projects. A review of Colorado's regulatory framework regarding reuse suggests that "reuse has been well-accepted and has enjoyed long-term, political support in Colorado for a number of reasons, including the state's arid climate and relatively long history with the practice, as well as the leadership provided by early municipal practitioners."[41]

A key regulation, "Regulation No. 84: Reclaimed Water Control Regulation," was designed "to further promote reuse of reclaimed domestic wastewater by providing a comprehensive framework which, when followed, will assure responsible management of operations and a product of quality compatible with the state's goals of protecting the public health and the environment."[42] It approves reclaimed water for industrial use, landscape irrigation, commercial use, and fire protection. Potential developers of water-reclamation systems must submit a letter of intent to the Water Quality Division of the Colorado Department of Public Health and Environment. If approved, potential users must submit letters of intent indicating their interest in receiving reclaimed water.

A key element of the letter of intent is that it must affirm that the project will not "materially injure water rights."[43] In Colorado, a water right only allows one use of water, unless it is diverted water, and water uses must maintain historical return flows to streams. In an effort to protect water rights, reuse is limited to water from certain kinds of supplies:

- Non-tributary groundwater
- Trans-basin diversions, which in the Front Range are mainly water imported to the South Platte River Basin from other rivers
- Changing the historically consumed portion of water rights from one use to another, such as from agricultural to municipal
- Water diverted under a water right that allows reuse.[44]

The traditional source of reclaimed water is effluent.[45]

CSU has one of the oldest direct-reuse systems in the state. In the 1960s, it began its non-potable reuse system for irrigation purposes. Today, approximately 26 percent of CSU's

---

[40] Western Resource Advocates, "Meeting Future Urban Water Needs in the Arkansas River Basin," undated, pp. 25–26.

[41] Nathan Bracken and the Western States Water Council, "Water Reuse in the West: State Programs and Institutional Issues," July 2011, p. 18.

[42] Colorado Department of Public Health and Environment, "Regulation No. 84: Reclaimed Water Control Regulation," Denver, Colorado: Water Quality Control Commission, 5 CCR 1002-84, August 2010, p. 18.

[43] Colorado Department of Public Health and Environment, August 2010, p. 4.

[44] Western Resource Advocates, "The Reuse Strategy," undated, p. 25.

[45] Kelly DiNatale, "Purple Mountain Majesties—Water Reuse in the Rockies," presentation at the 2009 Water Reuse Workshop, August 13, 2009.

demands are met with reused water, either directly or through exchanges and augmentation.[46] Reuse is becoming a major part of meeting anticipated future water gaps. The Pikes Peak Area Council of Governments covers El Paso County and the City of Colorado Springs. Its Draft Regional Sustainability Plan seeks to use 100 percent of the region's reusable water supplies to satisfy 2030 demands with existing water supplies.[47]

Colorado Water Market Activities

The stresses on water, and the water demands described above, have spurred the development of some water markets. A number of factors contribute to Colorado's activity in developing water market mechanisms. First, Colorado has a long history of developing and relying on water conveyance infrastructure. Second, historically agricultural, municipal, and commercial enterprises comprised the bulk of water use, but in the last few decades, in-stream uses as hydropower, fish and wildlife preservation, and recreation have augmented these earlier uses. Third, during the last few decades Colorado has also seen significant population growth (on the Front Range), and greater interest in uses that rely on in-stream flows such as recreation. Moreover, as mentioned previously, growing population centers on the East Slope are left with junior water rights whereas those rights holders on the Western Slope, primarily agricultural, are senior rights holders, putting much of the population at risk during low water-supply periods. Fourth, Colorado's waters are fully appropriated. All of these factors have created a demand for water rights and provide the contextual factors that give Colorado the potential for water rights market transfers. Water markets have existed in Colorado for a long time, facilitated by the extensive infrastructure for conveyance, most notably the Big-Thompson project but also the numerous ditch companies.

In Colorado, a water right may be owned by individuals or by water-distribution organizations (such as irrigation ditch companies, conservancy districts, municipalities, or firms). As in other states, a state authority controls water rights transfers, in part to prevent and mitigate any third-party effects. In Colorado, water transfers generally are administered through the water courts (other western states utilize the office of the state engineer or other water agencies). Oversight by the Colorado water courts considers third-party effects only on other water users and not on public values such as water quality, ecosystems health, and community values. Note that courts can modify water rights and water shares as a condition of transfer approval.[48]

---

[46] Western Resource Advocates, "Filling the Gap: Meeting Future Urban Water Needs in the Arkansas Basin," March 2012, p. 29.

[47] Pikes Peak Area Council of Governments, "Pikes Peak Region Sustainability Project, Regional Sustainability Stretch Goals for 2030," December 6, 2010, p. 3.

[48] Charles W. Howe and Christopher Goemans, "Water Transfers and Their Impacts: Lessons from Three Colorado Water Markets," *Journal of The American Water Resources Association*, Vol. 39, No. 5, October 2003, pp. 1055–1065.

Market mechanisms for water rights transfers in Colorado have taken the form of leases and sales. There have also been plans for, and pilots of, water banks. Many Colorado cities use price as an incentive to conserve water by implementing block pricing for water users.

Water Leases and Sales in Colorado

Colorado has a significantly larger *number* of water rights transfers than any other western state, based on the most commonly used dataset of water transactions as reported in the *Water Strategist* for the period 1987–2005.[49] An analysis of all water rights transfers (short- and long-term leases and sales) in Colorado based on annual flow in the time period 1987–2005 found them to be dominated by agriculture-to-urban trades (51 percent), followed by agricultural-to-agricultural (29 percent), and urban-to-urban (20 percent) trades.[50]

Most of these transactions occur within the Colorado–Big Thompson (C-BT) Project. Unlike other areas of Colorado, the Northern Colorado Water Conservation District administers the water rights created by the project, as well as transfers of these rights. The market transactions include short-term leases and permanent sales of water. Consisting of mostly bilateral, transfers are of shares of the C-BT project (water rights to the C-BT are defined in terms of shares of available water since water availability varies from year-to-year) are facilitated and reviewed by a board of the Northern Colorado Water Conservation District (NCWCD). The district reviews transfers to ensure that water is put to beneficial use and prevent speculation (since water is non-native and transfers are intra-basin, third party effects are considered minimal). The market determines the price.[51]

Markets thrive in the C-BT Project for several reasons. First, the water rights are homogeneous—in other words, they are defined in the same units of water and have the same priority, so they are entirely comparable when trading. Second, unlike other parts of the state, the NCWCD board administers water rights transfers, and because the transfers do not have to pass through the water courts the process is smooth and efficient. Lastly, there is no return flow responsibility for these water rights (water rights cannot be transferred outside the NCWCD boundaries).[52] These factors facilitate the transfer of water rights within the NCWCD and the C-BT Project. In addition to these transfers, the Northern Water Conservancy District holds periodic auctions of unallocated Colorado River water available through the C-BT Project.

---

[49] According to the *Water Strategist*. However, the *Water Strategist* stopped publication in 2009. See also Brewer, Glennon, Ker, and Libecap, 2008, pp. 91–112.

[50] Brewer, Glennon, Ker, and Libecap, 2008, pp. 91–112.

[51] David S. Brookshire, Bonnie Colby, Mary Ewers and Philip T. Ganderton, "Market Prices for Water in the Semi-Arid West of the United States," *Water Resources Research*, Vol. 40, 2004; and Janis M. Carey and David L. Sunding, "Emerging Markets in Water: A Comparative Institutional Analysis of the Central Valley and the Colorado–Big Thompson Projects, *Natural Resources Journal*, Vol. 41, 2001, pp. 283–328.

[52] Howe and Goemans, 2003.

Trading may not be so easy in other regions and conservation districts, however, since the water court and not the conservation board handles sales.

A study of water-rights transfers specifically in the South Platte and Arkansas River basins found that within the South Platte basin, water rights transfers are primarily sales shares of ditch company water rights, but also include C-BT water (which provides about 30 percent of the region's water). In the South Platte, most of the trading that occurred between 1970 and 1995 was from agricultural uses to urban uses within the same basin for both the native water and the C-BT water. Several large sales of native water rights held by agriculture to major cities (Thornton and Fort Collins) occurred during this time period. In comparison to native waters, the transfers of the C-BT water in the region tended to include more agricultural-to-agricultural trades, shorter-term, and smaller transfers. This suggests, again, that the structure of the water rights, the infrastructure, and the relatively simple administrative process make trading easier and allowing water users, including farmers, to adapt more quickly to changing water needs. In contrast to the South Platte, water transfers in the Arkansas Valley were almost exclusively agricultural-to-urban and interbasin. There were fewer transfers of larger amounts than in the South Platte, including two large sales of Rocky Ford water and most of the water in the Colorado Canal to the city of Aurora in the mid-1980s. Given that there were fewer and more infrequent water transfers, and most of these sales were interbasin, it suggests that the transaction cost associated with each transfer is fairly high, discouraging smaller transfers.[53]

Other water transfers have become controversial as oil and gas companies seek water for their hydraulic fracturing operations, an example in which financial benefits (the increased value added from water use) are increased, but social and environmental concerns are present. These companies have purchased excess water and wastewater from cities and the NCWCD as well as water rights from ditch companies, farmers, and others. As mentioned earlier, the Northern Water Conservancy District in Colorado routinely auctions off surplus Colorado River water and in a recent auction sold water to an oil and gas company for hydraulic fracturing, setting off concerns about maintaining the hydrological cycle in the region. And the city of Aurora, Colorado, recently leased $9.5 million worth of treated wastewater (2.4 million gallons) over five years to the Anadarko Petroleum Company for use in oil and gas drilling.[54] While the city benefited financially from this transaction, among the concerns is that the water sold to these companies will be used out-of-basin and will not be available for other downstream users. In addition, while the quantities are fairly small relative to the amount of water used by farmers, oil and gas companies can afford to pay significantly more for water, thousands of dollars per acre foot as opposed to tens of thousands.[55] Oil and gas companies estimate that they will use about

[53] Howe and Goemans, 2003.

[54] Justin Dove, "Three Easy Ways to Invest in Water," *Investment U*, No. 1742, April 2, 2012; "Across the USA: Colorado," 2012, p. 9A.

[55] Jack Healy, "For Farms in the West, Oil Wells are Thirsty Rivals," *New York Times*, September 5, 2012.

6.5 billion gallons of water in 2013, representing 0.1 percent of overall water use in Colorado, or less than the amount used for golf courses or snow making. However, others have suggested this figure is an underestimate, and the true water use for hydraulic fracturing is anywhere from 7.2 billion to 13 billion gallons. (Note that while the total amount used statewide is small, the effects on a particular watershed may be significant.)

Water Banks in Colorado

The first experiment with water banking in the state of Colorado was the Arkansas River Water Bank pilot program. The Arkansas River Basin in southeastern Colorado is Colorado's largest river basin, draining nearly 25,000 square miles. The Arkansas River supports recreation, several cities, and a large agricultural community of farmlands and rangelands that use a large web of ditches to transport water. Within the Arkansas River Basin, where approximately 20 percent of the state's population lives, virtually no water is available for new uses. Permanent sales of water to urban centers (Aurora and the Rocky Flats) as well as to investment groups (High Plains A&M[56] and the Fort Lyon Canal Company) affected agriculture in the basin and prompted public concern over threats to agriculture, local economic conditions, and permanent access to water rights. The purpose of the bank was to allow users of Arkansas River water more flexibility in meeting water needs, while still meeting interstate compact obligations to Kansas.[57] In effect, the water bank was a way to help agricultural senior rights holders in the Arkansas River Basin maintain their water rights and financially benefit from their water rights, while providing water to high-growth urban areas that were junior rights holders. A map of the Arkansas River Basin within Colorado is shown in Figure 4.7.

---

[56] Southwestern, affiliated with High Plains A&M, LLC, began acquiring water rights in 2001. High Plains purchased more than 18,280 acres of land and more than 21,300 shares (equivalent to more than 60,000 acre feet) of senior 1883 water rights in the Arkansas River Valley. These water shares are delivered through the Fort Lyon Canal, the longest canal in the valley. The canal is 100 miles long and maintained by the Fort Lyon Canal Company. While the land is currently in production (farming and grazing), in the future the company intends to sell some of this water to Front Range communities. According to the company website, these water sales will be "sensitive to the interests and needs of the agricultural community on the Arkansas River," perhaps because of the strong reaction by the community to the original purchase. In 2006, High Plains sold water and land interests to Pure Cycle Corporation (after state courts determined the purchase of Fort Lyon water was speculative, failing to meet laws requiring a "beneficial" use, and therefore was in violation of Colorado water law). Pure Cycle, a water and wastewater company, intends to build a conveyance project that could bring Arkansas River water to Denver-area buyers. Many competing interests for water in the Arkansas River Basin—agricultural, urban, recreational and environmental—at times have had a contentious history. Pure Cycle Corporation is working with natural resource, governmental, and agricultural stakeholders to seek ways of maintaining the agricultural community (by finding ways to use water more efficiently) in the basin while allowing some of the water to be sent to urban centers on the Front Range. See Southwestern Investor Group, "High Plains LLC Water Rights Project," undated; Dan Gordon, "Water Rights: Power Struggle in Southeastern Colorado," *Denver Post*, September 9, 2012.

[57] Kansas has been claiming that Colorado takes too much of the river's water since 1902. There have been several lawsuits before the U.S. Supreme Court despite the interstate compact agreement made between Colorado and Kansas in 1949. The most recent lawsuit, initiated in 1985 was largely resolved by 2009.

The bank became active in 2003. It had been established in 2001 by Colorado House Bills 1354 (allowing the State Engineer to promulgate banking rules) and in 2003 by House Bill 1381 (which made the bank permanent and restricted out-of-basin transfers). The State Engineer provides regulatory oversight and the Southeastern Colorado Water Conservancy District (SCWCD) administers the bank. The Arkansas River Water Bank is a clearinghouse that facilitates water transfers through an online bulletin board. Water transfers are short-term (one year) bilateral trades between willing demanders (urban users) and willing providers (agricultural users) and prices are determined by market-based negotiations between parties. Dry-year options, or interruptible supply agreements, based on specific weather conditions and allowing for a schedule of prices also are possible.

In its first nine years there were no transactions, yet the lessons from this bank are valuable. Several reasons having to do with community support, administrative procedures, geography, pricing and timing have been suggested as to why the bank failed to complete any transactions. Originally, the bank was allowed to approve out-of-basin transfers without mitigation requirements. As a result of this possibility, the community did not support the bank for the fear of economic loss in the present and future, as well as significantly altering the community socially. Nor did the community have knowledge of a working water market to the extent that agriculture within the C-BT project area had that knowledge. This may have contributed to general distrust of water banks. Additionally, participants in the bank were asking for prices that were higher than the going market price of short-term leases, so demanders were able to acquire water in the lease market at a lower price. The administrative processes were expected to average three months, which is a significant amount of time to invest for a one-year lease.

Despite these failings, many in Colorado believe that water banks hold promise for dealing with water management issues such as the vulnerability of municipalities to drought, permanent loss of agricultural land to development and a potential compact call on Colorado River Water. In fact, several other water bank proposals have been put forward. For example, the Arkansas Basin and Gunnison Basin roundtables[58] are considering a joint project to see if pre-1922 water rights in the Gunnison River basin could be banked in the Blue Mesa Reservoir as a hedge against a Colorado River Compact call. While there has never been a call on the Colorado River, recent droughts and possible consequences of climate change suggest the upstream states (CO, NM, UT and WY) could be vulnerable to a compact call from downstream states (AZ, CA and NV). Banking water in the Blue Mesa Reservoir could potentially protect those who rely on post-1922 water rights, such as the Colorado–Big Thompson Project water, Denver and other

---

[58] The Colorado Water for the 21st Century Act established nine roundtables representing the eight major river basins and the Denver metropolitan area in order to address water management in a collaborative and inclusive manner and to encourage locally derived solutions. The roundtables include designated members that represent each county, each municipality and each water conservation district within the river basin. Other members represent the various interests such as agricultural, recreational, local water providers, industrial environmental interest within the river basin; half of whom must own water rights.

municipalities in the region, Twin Lakes, and the Fryingpan Project, should a compact call or shortages occur in the future (after the agreement for water shortage sharing signed by the states in 2007 expires in 2026).[59]

The last market mechanism that is used in Colorado is block pricing of water by utility providers.

Block Pricing in Colorado

Many of the Front Range cities use an increasing tiered, or block pricing, structure. Discussed in greater detail in Chapter Four, increasing block pricing can provide an incentive to conserve water if structured properly by sending a price signal to consumers to promote water conservation (and thereby stretch existing water supply and infrastructure). A study of major cities in Colorado found that nine of the 13 cities studied employ a block-pricing scheme, although the structure of these pricing schemes varied in terms of the strength of the conservation signal. Three components of block pricing influence the degree to which price sends a conservation signal: a fixed charge, the amount of water usage included in each block, and the price increase for moving from block to block. The larger the fixed charge, in general, the lower the conservation signal that is sent. In addition, block sizes can be based on standard usage by customer class or historical usage for the given customer, to represent typical indoor use requirements, basic outdoor use, and more extensive outdoor use. Regardless of how these block sizes are established, they only will be effective if they accurately reflect patterns of water consumption. Third, the price increase needs to be substantial enough to change consumer behavior. Slight increases will likely have little effect.

When plotting the three components of block pricing, some cities in Colorado have had fairly flat curves, meaning either that the price rise for increasing water usage is low or the blocks are relatively large, so that no real conservation signal is sent. Other cities have high price increases in which the cost of water rises sharply and quickly with increasing water use, offering a strong conservation signal. Still others are somewhere in between these two schemes. Figure 4.9 in Chapter Four presents the price curves for the 13 Colorado cities studied.

The three largest water utilities in Colorado are Denver Water, Aurora, and CSU. At the time that the 13 cities were studied, Boulder and Aurora sent the strongest signal to consumers with price structures that have small increments between blocks (more blocks or tiers) and rapid price increases. Both cities' blocks are based on past use. The first block intends to cover essential indoor uses such as cooking, cleaning, and bathing. The second block intends to cover water use for large families, or minimal outdoor use. In these cities, there is a relatively small price increase. However, subsequent blocks have substantial increases that send a strong price signal to consumers to conserve. The increasing block rate structures in Denver and Evans send a more

---

[59] Chris Woodka, "Blue Mesa Seen as a Water Bank," *Pueblo Chieftain*, February 12, 2012; Colorado Water Conservation District, "Basin Roundtables," undated.

modest conservation signal, targeting those moderate- to high-volume users with a price increase up to four times the price paid by lower-volume users. In contrast, Colorado Springs and Louisville target the low- and moderate-volume users by implementing a sharp price increase after the first block, which is easily exceeded during the summer irrigation season.[60] In fact, because of continued drought, which led to lower reservoir levels, CSU has issued water restrictions and instituted a larger price increase for exceeding the first block's threshold. This action met with complaints from residents. Recently CSU increased the first block's threshold and reduced the price increase.[61] Other cities studied—Berthoud, Longmont, and Erie—have slight price increases, so the conservation signal is modest; in the case of Fort Lupton, the price signal is negative because of the high fixed fee. Still others, including Loveland, Fort Margan and Broomfield, have uniform rates, so usage has no effect on the price paid.

Fort Carson Water Case Study

Fort Carson is a 138,303-acre major Army training site located directly south of Colorado Springs in El Paso County, Colorado. The post purchases most of its potable water from CSU, but owns and maintains the water lines and other water infrastructure on post. Its average annual water bill in 2011 was over $2.1 million. Fort Carson also owns and operates its own wastewater treatment facility.

Fort Carson has over a decade of experience in implementing water efficiency and conservation measures. In fact, water consumption at the fort was reduced more than 45 percent between 2002 and 2011.[62] This reduction occurred while the post population increased substantially. Over the years, Fort Carson has implemented many water-efficiency projects because of a strong sustainability program, leadership support, drought problems in Colorado, and federal and Army requirements, such as Leadership in Energy and Environmental Design (LEED) implementation and Executive Order 13514 water-use reduction requirements. The Colorado drought problems caused the installation to take additional measures to conserve water, which the post has kept in place. In 2011, Fort Carson was selected by the ASA(IE&E) as one of two integrated Net Zero pilot installations with the goal of reaching net zero energy, water and waste by 2020. Its Net Zero Water Goals also have helped the installation invest more in water conservation. Fort Carson water conserving activities have included the following:

- Irrigating the golf course with treated wastewater
- Enviro-transpiration—a smart water irrigation system
- Landscaping and turf management changes

---

[60] Western Resource Advocates, "Front Range Water Meter: Water Conservation Ratings and Recommendations for 13 Colorado Communities," Boulder, Colorado, November 2007.

[61] Colorado Springs Utilities, "Water Shortage Rate Changes Effective August 1, 2013."

[62] Andrea Sutherland, "Fort Carson Builds toward Energy, Water, Waste Goals," *Public Works Digest*, Vol. XXIII, No. 5, September/October 2011.

- Requiring military units to use the Central Vehicle Wash Facility, which recycles its water
- Using low-flow fixtures in new facilities, such as pint urinals
- Water line and tank replacements.[63]

Fort Carson has developed a strategy for becoming a Net Zero Water Installation by 2020. It consists of two main steps. First, the installation plans to continue investing in water-efficiency activities to conserve water. Second, the installation plans to find and use alternative sources of water so it can reduce its potable water use. Its most significant alternative source of water comes from a plan to expand its use of reclaimed water by expanding its wastewater treatment facility. The post also hopes to use stormwater to help with some on-post irrigation. As just discussed, state law presents some legal challenges to doing this, but given current activities in the state to address these, this is likely to become a more viable option in the future. Fort Carson also may explore the possibility of using greywater collection system within a building to help with toilet flushing. Lastly, the post has some limited surface and groundwater rights that are also potential sources for water.

Reducing potable water usage provides a range of benefits for Fort Carson. First, it can help the post meet its Net Zero Water Goals. Second, it will save them money on their CSU utility bill. Third, it can help during droughts because the post will not be needing or using as much potable water. Lastly, it can help enhance the post's water security, since it does not need as much water and has some alternative sources for water, which is especially important if water shortages are a future problem within Colorado.

Understanding more about such activities at Fort Carson helps to understand the installation's experiences and opportunities with partnerships and potential future water markets.

Overview of Fort Carson and Water Goals

We begin this section by providing a brief overview of Fort Carson and some of its major activities, such as its sustainability program, that drive its water goals.

Background on Fort Carson

Fort Carson stretches south along Interstate 25 into Pueblo and Fremont Counties. The installation measures about two to 15 miles wide (east to west) and 24 miles long (north to south). The cantonment area of Fort Carson lies in the northern part of the installation, with most of the major training areas to the south and east on the post, such as three major gunnery ranges located near the southern end.

As of mid-2010, the following major units were stationed at Fort Carson: the 4th Infantry Division Headquarters, the four Brigade Combat Teams of the 4th Infantry Division, the 43rd Sustainment Brigade, 10th Special Forces Group (Airborne), the 71st Explosive Ordnance

---

[63] More details on these and other water efficiency projects are discussed later in this chapter.

Group, and the 10th Combat Support Hospital. Training at the installation involves tanks, other tracked and wheeled vehicles, and rotary wing aircraft. Some joint use and training with other Services, such as the Air Force, Marine Corps, and other agencies, also is conducted at Fort Carson. Fort Carson also manages the 235,368-acre Piñon Canyon Maneuver Site (PCMS) in southeast Colorado.

Fort Carson has seen a lot of growth due to Base Realignment and Closure (BRAC) and the expanding Army. Post population almost doubled between 2005 and 2012 (from about 15,000 in 2005 to 27,000 in 2012). In 2006, Fort Carson had more than 18,000 Soldiers stationed there. By 2009, Fort Carson had more than 19,600 active-duty Soldiers and more than 25,300 family members. In 2008, Fort Carson had about 750 buildings with 12.865 million square feet of heated space. In 2012, about 27,000 Soldiers were stationed at Fort Carson. Since 40 percent of the Soldiers are single, most of them are required to live in the barracks (with a few exceptions, such as more senior Soldiers, and some single Soldiers with children). About 60 percent of the Soldiers are married, and 26 percent of these live on-post.

Fort Carson Sustainability Program

Fort Carson has had one of the most active and long-running installation sustainability programs in the Army, which has included an emphasis on water-conservation issues. Fort Carson's water program has benefited from this strong sustainability program and vice versa.

In 2002, Fort Carson started developing a sustainability program. It held a sustainability workshop to help develop installation sustainability goals and cross-functional teams for sustainability. Before the sustainability workshop, Fort Carson developed a detailed sustainability baseline report covering descriptive statistics, key issues and challenges, and existing sustainability activities in nine subject areas: Water, air quality, energy, transportation, lands, materials, wildlife, noise, and cultural resources. Workshop participants included representatives from Fort Carson, Headquarters Department of the Army (HQDA) and subordinate headquarters, other local military installations, civic leaders, regulatory agencies, and the community. They organized into six teams of 25 to 35 people each (community well-being, energy and transportation, facilities and installation, materials, training lands and ranges, and sustainability management system) to develop 25-year sustainability goals. The teams produced 12 goals (see Table 5.2, which shows each along with the major proponent for the goal). For each goal, the teams defined the problem being addressed, the desired end state, metrics, and intermediate objectives.[64]

---

[64] Fort Carson, "Fort Carson Sustainability Baseline," Fort Carson, Colorado, August 2002.

Table 5.2. Fort Carson's 25-Year Sustainability Goals

Goal Number	Goal Name	Description	Proponent
1.	Sustainable Water and Energy	Sustain all facility and mobility systems from renewable sources, and reduce the total water purchased from outside sources by 75 percent.	DPW
2.	Sustainable Transportation	Reduce automobile dependency and provide balanced land use and transportation systems.	DOL
3.	Improve Communication	Improve communication to foster understanding and attain a "Community of One."	PAO
4.	Sustainability Partnerships	Enhance partnering to collaboratively develop, integrate, and implement regional sustainability.	DECAM
5.	Hazardous Air Pollutants	The total weight of hazardous air pollutant emissions is reduced to zero.	DECAM
6.	Master Plan	Further integrate sustainability principles into the Fort Carson land use planning, Real Property Master Planning, and Military Construction, Army (MCA) programming processes.	DPW
7.	Platinum Buildings	All applicable facilities at Fort Carson will be high-performance buildings that meet or surpass the Platinum Standard of SPiRiT or LEED by 2027.	DPW
8.	Sustainability Training	Key stakeholder groups are trained, compliant and motivated toward sustainability principles.	SPPO
9.	Sustainable Procurement	All DOD and Fort Carson procurement actions support sustainability by 2027.	DOC
10.	Zero Waste	The total weight of solid and hazardous waste disposed of is reduced to zero by 2027.	DPW
11.	Training Lands	Training Ranges (land and associated air space used for live fire ranges, maneuver, testing and urban development designated for Military Operations in Urban Terrain [MOUT] training) capable of supporting current and future military training to standard.	DPTM
12.	Sustainability Management System (SMS)	Advance a sustainable mission and Fort Carson by adopting an SMS/EMS and by imparting (passing on) a personal commitment and enthusiasm for sustainability.	DECAM

SOURCE: Fort Carson, "Sustainability Program Home Page," Fort Carson, Colo.: U.S. Army Environmental Compliance Division, last modified on February 28, 2008.

Such sustainability goals have helped provide support for the post's water program, especially in its pursuit of water conservation investments and education. For example, in 2011, Fort Carson issued a sustainability guide to help educate Soldiers, family and staff about ways to conserve water. This guide encouraged units to use the Central Vehicle Wash Facility for washing tactical vehicles, and individuals to Xeriscape landscapes with native plants, install low-flow shower heads, and to check toilets for leaks.[65]

---

[65] For more information, see Fort Carson, "Fort Carson Sustainability Guide 2011–2012: Knowing Your Piece of the Sustainability Puzzle," 2011, p. 11.

Because of Fort Carson's long history in sustainability, the water program and its staff coordinate and interact with the sustainability program staff regularly, and there is overlap in duties and interest. Also, the sustainability program has given additional visibility to water efforts. On the other hand, the water program also has its own responsibilities and activities that do not directly fall under sustainability.

Fort Carson Water Conservation Requirements and Progress

Fort Carson, like other Army installations, must meet federal, DoD, and Army water-conservation requirements. Signed in 2007, Executive Order 13423 directs each agency beginning in FY 2008 to reduce water consumption intensity, relative to the baseline of the agency's water consumption in FY 2007, through life-cycle cost-effective measures by 2 percent annually through the end of the fiscal year 2015 or 16 percent by the end of FY 2015.[66] The 2009 Army Energy Security Implementation Strategy also specifically identified water conservation as a related priority.[67] The 2010 Department of Defense *Strategic Sustainability Performance Plan: FY2010* outlined annual targets for meeting the following goals:[68]

- Reducing facilities' potable water consumption intensity by 26 percent of FY 2007 levels by FY 2020
- Reducing industrial and irrigation water consumption by 20 percent of FY 2010 Levels by FY 2020
- Maintaining, to the maximum extent technically feasible, pre-development hydrology for all development and redevelopment projects of 5,000 square feet or more

Executive Order (EO) 13514 requires all federal agencies to set a baseline for potable water use at FY 2007 and reduce potable water use intensity (WUI) by 2 percent per year based on the FY 2007 baseline through FY 2020, which is a total reduction of 26 percent.[69] Fort Carson has been very successful at reducing its water use. "The post reduced water consumption by more than 45 percent since 2002 through landscaping and turf management changes, using low-flow fixtures in new facilities, irrigating the golf course with treated wastewater and requiring military units to use the Central Vehicle Wash Facility."[70] This success and strong sustainability program and management support for water conservation helped the installation be chosen as a Net Zero installation pilot in 2011.

---

[66] See Executive Office of the President, 2007.

[67] U.S. Department of the Army, 2009, p. 12.

[68] U.S. Department of Defense, *Department of Defense Strategic Sustainability Performance Plan,* 2010, pp. II-14-15.

[69] Even though such water-conservation goals are Army-wide requirements, each installation is expected to strive toward meeting them at the individual installation-level. Namely, the Army has spread these goals equally to all installations rather than requiring more water conservation from some installations compared to others.

[70] Sutherland, 2011.

Fort Carson Net Zero Installation

In 2011, the Army launched the Net Zero Pilot Program to include six Net Zero Energy pilot installations, six Net Zero Water pilot installations, six Net Zero Waste pilot installations, and two integrated Net Zero pilot installations pursuing Net Zero Energy, Water and Waste by 2020. The integrated Net Zero pilot program was designed to conserve natural resources and to address the interrelationships between such resources. In April 2011, the ASA-IE&E selected Fort Carson as one of the integrated Net Zero pilot installations. The Army defined a Net Zero Water Installation as one "that reduces overall water use, regardless of the source, increases use of technology that uses water more efficiently; recycles and reuses water, shifting from the use of potable water to non-potable sources as much as possible; and minimizing interbasin transfers of any type of water, potable or non-potable, so that a Net Zero water installation recharges as much water back into the aquifer as it withdraws."[71]

All the net zero pilots that involve water, including Fort Carson, now have an accelerated WUI reduction goal compared to EO 13514. The Army WUI reduction goal for net zero water installations is 26 percent by FY 2015 and 50 percent by FY 2020.

Fort Carson's strategy for net zero water is first to continue to invest in water efficiency to conserve water and then to find and use alternative sources of water to reduce the potable water that it uses. A key way to reduce potable water use is to expand the use of reclaimed water on post by creating a zero-discharge sewage treatment plant and reusing all of its water. Fort Carson water personnel also would like to use stormwater and some of the limited surface water and groundwater sources to help with irrigation to reduce potable water use. However, this approach raises legal and other challenges, which is discussed later in this chapter.

Reducing potable water use through activities such as water conservation and reclaimed water use has other advantages apart from contributing to achieving net zero goals. It helps to save money and maintain activities during drought. It also enhances water security, especially if Colorado experiences a water-shortage problem.

Fort Carson Water Supply and Infrastructure

This section provides an overview of Fort Carson's water infrastructure and water supply.

Fort Carson Water Infrastructure

Fort Carson purchases more than 95 percent of its water from CSU through two supply mains. The post owns and maintains its own water supply distribution system. Fort Carson also owns and operates its own onsite WWTF.

The wastewater treatment facility discharges to Fountain Creek. Fort Carson's WWTF is a traditional biological treatment plant with primary and secondary clarifiers and aerobic digesters.

---

[71] U.S. Department of the Army, "Army Directive 2014-02 (Net Zero Installations Policy)," Washington, D.C.: Secretary of the Army, January 28, 2014.

This facility produces reclaimed water, which is delivered and stored at the golf course retention pond for supplemental irrigation, mostly during the summer months. The reclaimed water is created by treating a percentage of the discharge with ultraviolet light to remove additional biological contaminants before the reclaimed water is used in irrigation. "Due to Colorado water right laws, Fort Carson pays a fee (called an 'augmentation fee') to CSU for the right to use this water. Fort Carson's reclaimed water permit has been grandfathered in with the State of Colorado because the system was in use before formal reclaimed system permit standards were developed."[72]

Fort Carson Sources for Water

For the potable water it purchases from CSU, Fort Carson pays a flat fee for water use that varies seasonally. CSU charges Fort Carson a seasonal, military rate—the summer rate, charged for the months of May through October, is nearly 85 percent more than the winter rate.[73] Other fees, such as tap fees and development fees, may be charged if Fort Carson requests new connections to the water supply infrastructure or exceeds its capacity limit (a maximum amount, based on historical usage), so water use is monitored closely.

CSU collects the water it provides to Fort Carson from both sides of the Continental Divide in Colorado. CSU source waters on the Western Slope are from Homestake Creek, Roaring Fork River, and Ivanhoe Creek. The water from the Western Slope is piped to the regional watershed. The source waters in the regional watershed are the Arkansas River and Fountain Creek, a tributary of the Arkansas River. The other 5 percent of Fort Carson's water comes from some limited surface water and groundwater sources on post, which will be discussed in more detail below.

Relationship with Colorado Springs Utilities

Fort Carson has a good working relationship with CSU. It helps that CSU is a progressive utility that has similar water-conservation goals. In fact, CSU has an aggressive conservation program to assist its customers in saving water. Colorado Springs Utilities developed it first water conservation plan in 1998.[74] Colorado Springs Utilities' conservation goals for the most recent plan, the *2008–2012 Water Conservation Plan*, include maintaining low residential use per capita and gaining a better understanding of how commercial customers use water in order to inform conservation and efficiency actions. The plan focuses on conservation programs and does not address long-term supply and distribution issues or drought planning.[75]

---

[72] Kate McMordie Stoughton, "Fort Carson Net Zero Water Balance," *PNNL*, April 2012, pp. 1.1–1.2.

[73] The military commodity charges are: summer rate is $0.0416 per cf and the winter rate is $0.0226 per cf. See Colorado Springs Utilities, "Water Rate Schedule," March 26, 2013.

[74] Because of the water concerns discussed earlier in this chapter, the State of Colorado requires relatively larger water providers to develop and implement water plans that encourage customers to use water more efficiently.

[75] Colorado Springs Utilities, "2008–2012 Water Conservation Plan," December 31, 2007.

Colorado Springs Utilities has many programs to encourage water and energy conservation in both the residential and commercial sectors, although we focus on water conservation in this discussion. It encourages conservation through a public education program, Xeriscaping classes and a demonstration garden, rebates for water saving equipment, and increasing block pricing.

Education and outreach is accomplished through the Colorado Springs Utilities Conservation and Environmental Center, which has information, experts available for consultation, and limited classes on energy and water efficiency as well as a Xeriscape demonstration garden and displays of low-flow and efficient equipment. Figure 5.11 provides a photograph of one of the Center's exhibits that shows water and dollar savings for a more water-efficient faucet. Colorado Springs Utilities also provides speakers on various water-related topics such as water sources, water quality and treatment, watershed management, water conservation, Xeriscaping, or water law. In addition, CSU has partnered with the Center for ReSource Conservation to provide indoor and outdoor water audits.

Figure 5.11. Faucet Water Saving Exhibit at CSU Conservation and Environmental Center

SOURCE: Photo by Beth Lachman.

Equipment rebates are offered to residential customers by CSU for WaterSense®-labeled toilets and irrigation equipment such as rain sensor shut-off devices, sprinkler heads with check valves, rotating matched precipitation spray nozzles, and weather-based irrigation controllers. Showerheads can be exchanged for WaterSense® heads (as long as supplies are available). Similar rebates are offered to commercial customers for high-efficiency toilets and urinals, smart irrigation controllers, high-efficiency matched precipitation nozzles, and pressure-regulating sprinkler heads with check valves; there also is a toilet-recycling program.[76]

In addition, the Colorado Springs Utilities block pricing was found to be one of the strongest throughout Colorado in terms of sending a price signal to consumers to motivate water conservation.[77] It was estimated that this rate structure alone saved 750 AF of water (or approximately 2,100 gallons per residential customer) between 2001 and 2007.[78]

Fort Carson Water Demand

Fort Carson's average annual water demand is about 853 million gallons. This water is supplied to the entire installation, including family housing, which is entirely privatized.[79] Because of Army requirements regarding privatized housing, Fort Carson does not report the water used in the privatized family housing in the Army Energy and Water Reporting System (AEWRS). This exclusion is because the post no longer has any control over water use in the housing areas, given the way in which the privatization contract was set up. Thereby, privatized family housing technically is not considered in the installation's water use intensity or Net Zero Water Goals. Similarly, since the county runs the schools on post, they are not included in Net Zero goals.

Table 5.3 shows Fort Carson's four highest water-use categories: irrigation on-post (excludes privatized family housing irrigation), the plumbing fixtures in family housing, family housing irrigation, and the plumbing fixtures in barracks (including laundry and kitchen uses). These four uses make up 82 percent of Fort Carson's total water consumption.[80]

---

[76] Colorado Springs Utilities, "Residential," August 29, 2013; Colorado Springs Utilities, "Business," August 29, 2013.

[77] Western Resource Advocates, "Water Meter: Water Conservation Ratings and Recommendations for 13 Colorado Communities," Boulder, Colorado, November 2007.

[78] Western Resource Advocates, "Conservation Measures that Make Cents," Boulder, Colorado, 2008.

[79] This statistic is based on the typical use between FY 2008 and FY 2011. See also Stoughton, 2012, p. 1.2.

[80] Stoughton, 2012.

Table 5.3. Fort Carson Water Use, by Major End-Use Category

Water Use Category	Average Water Use (1,000 gallons)	Percent of Total Use
On-Post Irrigation (excluding Family Housing irrigation)	294,993	36%
Family Housing Domestic Plumbing	193,700	23%
Family Housing Irrigation	106,700	13%
Barracks Domestic Plumbing	83,300	10%
Distribution System Losses	51,700	6%
Military Daytime Domestic Plumbing	50,100	6%
Hospital	27,200	3%
Civilian/Contractors Domestic Plumbing	12,900	1%
Dining	9,500	1%
Central Chiller Plant	4,200	<1%
Central Vehicle Wash Facility	3,500	<1%
Motor Pools	2,700	<1%
Central Heating Plant	1,000	<1%
Morale, Welfare, and Recreation Swimming Pools	800	<1%

SOURCE: Kate McMordie Stoughton, "Fort Carson Net Zero Water Balance," *PNNL*, April 2012, p. iii.

Fort Carson has large summer peak water demand loads, which are mainly due to grass and other landscape irrigation (see Figure 5.12). Summer peak water use on-post ranges between 91 and 134 Mgal per month. Wastewater discharge at Fort Carson, averages about 33 Mgal per month. "According to the water use data, winter use and wastewater discharge are closely correlated. This suggests that relatively small losses in the system are likely and that the supply meter and wastewater discharge meter are fairly well calibrated."[81]

---

[81] Stoughton, 2012, p. 1.2.

Figure 5.12. Fort Carson FY 2007–FY 2011 Total Potable Water Supply and Wastewater Discharge

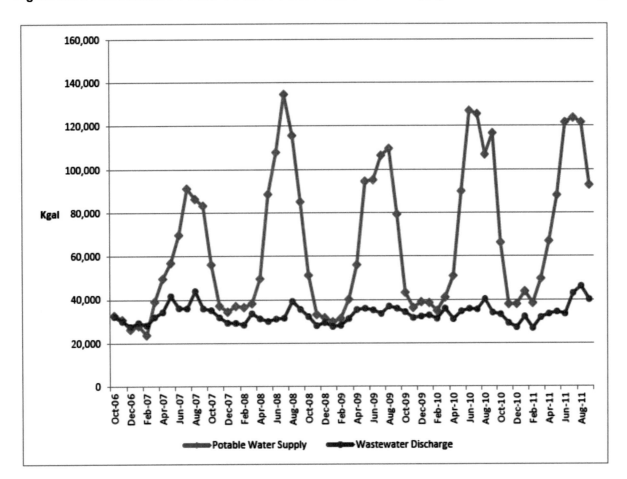

SOURCE: Kate McMordie Stoughton, "Fort Carson Net Zero Water Balance," *PNNL*, April 2012, p. 1.2.

Fort Carson's population almost doubled between 2005 and 2012. Therefore, overall water use had increased, but water intensity had decreased because of all of the water conservation activities on post.

Fort Carson Water Program

Next, we briefly provide an overview some of Fort Carson's main water activities to better understand partnership and water rights opportunities. First, we briefly describe water staffing responsibilities and some water efficiency projects. Then we discuss potential alternative water sources, some of which could potential lead to market opportunities in the future. Lastly, we discuss funding issues because this shows some use of traditional partnerships and what may drive Fort Carson's long-range interest in future water partnerships and markets.

Water Staff

Fort Carson has several different staff members who deal with water issues. The utility program manager oversees the water program and manages water as well as energy utilities. Another staff member deals with water issues regarding irrigation and landscaping. DPW engineers have responsibility for maintaining and running the potable water infrastructure and WWTF. Another staff member deals with water-rights issues with the help of an installation attorney.

Water Efficiency Projects

Over the years, Fort Carson has implemented a range of water efficiency projects. Many of its water-conservation activities started because of a two-year drought that began in 2001 and because of the post's sustainability water goals. At that time, installation personnel changed their behavior by doing simple things, such as not watering certain areas, limiting the allowable times for irrigation, not allowing water to run down the street, and education programs. Also at that time, the post population was down as a result of deployments, so that helped keep water use down during the drought period. Fort Carson continued some of these practices even after the drought was over to keep water use down. Over the years, Fort Carson has implemented many water efficiency projects because of the combination of the drought, sustainability, and Net Zero Water goals, and Executive Order 13514 and other federal and Army requirements, such as LEED implementation.

Examples of water efficiency projects have included:

- **Army LEED standards for new construction** have included a range of more water-efficient technologies and appliances being installed in new buildings. LEED buildings, for example, have included more water-efficient appliances and low-flow water fixtures. Growth on post would have led to much more water use had these standards not been implemented, and the increase demand for water would have led to capacity increases to the water infrastructure as well as the wastewater treatment plant. Unfortunately, there is no way to take credit for the costs avoided by implementing LEED standards.
- **Installing an Enviro-Transpiration system**, a smart-water irrigation system. This system was installed in May/June 2012 at a cost of $600,000. It uses weather and rain sensors to determine the irrigation needs, and can be described as the brains of the irrigation system. Payback is estimated to be $200,000/year, a three-year simple payback.
- **Investing $200,000 to improve the efficiency of irrigation systems.** The Enviro-Transpiration system is the irrigation system brains, but one needs the system infrastructure to apply water evenly and efficiently, so the post is investing in more efficient and reliable irrigation heads and placing the heads in more effective locations.
- **Requiring military units to use the Central Vehicle Wash Facility** (CVWF), which recycles and reuses the wastewater after cleaning tanks and other tactical vehicles. This water-recycling system includes a water storage capacity of 10 million gallons and a treatment scheme that includes grit chamber, sand filters, oil skimmers, and aeration basins. Fort Carson estimates that its CVWF saves 150 million to 200 million gallons per year in potable water.[82]

---

[82] This successful design has been copied at more than 25 Army installations. For more information see Scholze, 2011, p. 23.

- **Pint water urinals.** Fort Carson originally had tried to use waterless urinals, but they had some maintenance problems. They are high maintenance, so it switched to pint urinals, which are more water efficient than the new Army standard, which is for 1-gallon urinals.
- **Water conservation policy.** Fort Carson developed and implemented a water policy, water restrictions, and water education activities that emphasize water conservation. Many of these came out of the drought requirements, which were incorporated into installation policy.
- **Seasonal water rates to their customers**, which helps motivate them to reduce water use, especially irrigation.
- **Building water efficiency projects** through an ESPC between Fort Carson and Johnson Controls Inc. (discussed more later).
- **Installing water-efficient washing machines**, which are part of the Energy Star program.
- **Stormwater management projects**, such as bioswells that collect runoff water.
- **Water line and tank replacement.** Fort Carson has made repairs to leaking water lines, and a leaking water tank was replaced. The tank was leaking so much that the hill it was located in was sliding. DPW personnel found leaks on about 30 percent of the water lines and plotted them on a map to help identify and manage the problem areas. A limited number of posts have systems in place to identify leaks and may have continuous monitoring systems. Fort Carson has a SCADA (supervisory control and data acquisition) water system that monitors water use. Fort Carson now loses only 2 percent of its total water from leaks, and water personnel would know if they had a major leak because the water tanks would not refill, levels would drop, and an alarm would sound.
- **Landscape and turf management changes that reduce water use.** In many areas, Fort Carson has been able to reduce green grass areas around some buildings by planting some trees and native bushes with rocks underneath them instead of having the entire area covered with grass (see an example in the photograph shown in Figure 5.13). The rocks used in landscaped areas instead of grass cost less to maintain, as they need no irrigation or mowing, but they are not maintenance-free. They still have to be sprayed for weeds. The property needs to have a landscape master plan to properly manage the landscaped areas, including those with rocks.

Figure 5.13. Fort Carson Landscaping Reducing Grass Use

SOURCE: Photo by Beth Lachman.

The post has faced some challenges in implementing water conservation projects. For example, LEED implementation has not been very good at conserving water in appropriate landscaping. It has resulted in "zero-scaping" instead of Xeriscaping to meet LEED requirements. Often this is because many in the Army still want green turf instead of Xeriscaping. They typically lay down seed for native grasses and use temporary irrigation systems (the installation loses LEED points for permanent irrigation) to get the seed started. However, often the contractor does not do it right and after some use and disturbance, the grasses get trampled or die off, resulting in unsightly weeds—hence the term "zero-scaping."

Maintaining grass for fields continues to be a high-water use as well. Fort Carson has many fields that require irrigation. One post water expert suggested that for on-post training and ball fields, Fort Carson might get better results by using a smaller number of fields more intensively and perhaps using artificial turf in the high-use areas. Artificial turf is expensive. However, in a one-to-one comparison of artificial turf to grass, turf will lose, but when considering turf for high-use areas, such as physical training (PT) areas and playgrounds, it might make more economic sense.

Taking Advantage of Alternative Sources for Water

Besides buying potable water, Fort Carson has three other alternative water sources that it potentially can take advantage of:

- Reclaimed water
- Stormwater runoff
- Water rights to other surface water and groundwater sources.

We discuss each of these below.

Reclaimed Water

As was mentioned earlier, Fort Carson uses reclaimed water from its WWTF to irrigate its golf course. In fact, reclaimed wastewater has been used on the golf course since the 1970s. In 2012, this was the only use of reclaimed water on post.

Fort Carson water personnel would like to expand the use of reclaimed water, ultimately developing a zero discharge WWTF. The goal is to eventually use all the WWTF wastewater output to avoid any discharges to Fountain Creek, which means they will be using all their reclaimed wastewater. They have conducted a non-potable water expansion study to assess how the WWTF, reclaimed water piping, and storage system would need to be upgraded and expanded to create such a system. They have developed a plan to do this non-potable water expansion. It will cost $8 million and they have started securing funding to implement this plan. This non-potable water expansion project is key for the post achieving its net zero water goals. CSU is very supportive of this reclaimed water project.

Fort Carson has several things it would like to do with an expanded non-potable water system. Post water personnel would like to use the additional reclaimed water for tree and other landscape irrigation. Another potential use is to have a non-potable water fill point to meet the water needs for building construction and training units. The post also considered using reclaimed water at the Central Vehicle Wash Facility (CVWF). However, the CVWF already uses recycles water and is not a major water user, so there is not much benefit to using reclaimed water at the CVWF, and this idea was rejected.

Another possible project would be to do a pilot within a new building of a separate greywater system for toilet water using the WWTF. The WWTF would do the work for this greywater collection system since it is easier to run wastewater through the existing plant on a large scale and then use the existing infrastructure either to discharge into the creek or use as reclaimed water as an alternative to a distributed wastewater treatment system. However, one lawyer advising the post does not think Fort Carson should take the liability risks in doing such a pilot.

The golf course always will have first rights on the wastewater; in other words, if they have to augment the pond with potable water because other uses of the reclaimed water extend usage beyond that which is available in the pond, the golf course will continue to be charged only the augmentation fee.

There are some challenges and technical issues associated with reclaimed water that would need to be addressed. Reclaimed water tends to be high in salts and acid. It is possible to treat the water for acids. However, not much can be done about the salt intensity, which can be a problem in irrigation. For example, after 30 years, the extra salts likely would kill ponderosa pines. Another challenge is a tradeoff between reducing water usage and using reclaimed water. If Fort Carson conserves a lot of potable water while expanding the use of reclaimed water, it might not have enough reclaimed water for its purposes.

Fort Carson pays an augmentation fee to CSU for the use of reclaimed wastewater because of Colorado state law. This is because its contract is only for a single use of the water. CSU has return flow requirements into Fountain Creek for some of the Fort Carson wastewater. The augmentation fee allows CSU to exchange water for the return flow requirements. This fee is 80 percent less than the potable water fee Fort Carson pays CSU. Namely, during the summer, potable water costs about $5 per 1,000 gallons, while reclaimed water is $1 per 1,000 gallons. The reclaimed water price also is more stable since price increases would be hard to justify because no infrastructure costs need to be wrapped up into it.

Stormwater Runoff

Fort Carson would also like to do more with stormwater runoff, for example, to help with irrigation needs. However, as discussed earlier, Colorado state law limits some potential uses of stormwater. In Colorado, property owners, including Fort Carson, are not authorized to "retain" water because of the effects on downstream water users; but they can "detain" it for less than 72 hours. The implications are that they can build such things as swells, rainwater gardens, and limited detention ponds to prevent flooding problems, but they cannot use rain barrels or create permanent ponds. Fort Carson has, for example, installed permeable pavers and built bioswells at the end of parking lots to help manage stormwater runoff. The issue is that water must be sent downstream within 72 hours.

To retain water, it is necessary to do an extensive pre-development hydrological study to prove it does not affect downstream water users. There are some efforts within the state, including some demonstration projects, to change this policy. A demonstration project at Sterling Park is doing the analysis in harvesting rainwater to prove it is not taking this water away from downstream users.

Fort Carson also has a requirement to show that any new development does not affect pre-development hydrological conditions. They must do this for EISA 2007, Section 438.

Water Rights to Other Surface Water and Groundwater Sources

Fort Carson has some very limited water rights, some of which were acquired with land purchases. These rights are to some reservoirs, wells, and spring water sources. For example, at the Turkey Creek Ranch, a remote recreation area, there are some limited water rights. It has a surface water aquifer used for drinking water at this facility. An expensive RO system is used to purify the water. Historically, windmill pumps have been used to pump some water from wells. In the 1990s, the installation used solar power to pump water for wildlife in some remote areas. Fort Carson has 10 groundwater wells that are tracked for beneficial uses. However, Fort Carson was at risk of losing some of its water rights because of the state of Colorado's legal action. Because of personnel turnover, the fact that the rights were very small and not used for much, and the complex legal issues involved, the installation had not kept up in the past with some of the state reporting requirements regarding its water rights. It addressed this issue and was able to maintain these water rights.

Because the City of Fountain and Fort Carson both have water rights to Keeton Reservoir on post, they have a partnership to share this resource. Fort Carson uses some of their water rights to fill this reservoir. City of Fountain has a pipeline that runs across Fort Carson from this reservoir that the city has historically maintained. The city originally got potable water from Keeton Reservoir, but now uses it as exchange water that goes in Fountain Creek. Fort Carson maintains a pipeline from it that feeds some of Fort Carson's other reservoirs. Fort Carson water personnel would like to use some of this water to irrigate fields.

Another location where Fort Carson would like to use some of the post's own water is at the Wilderness Road Complex, a remote location where a battalion from the 47th Brigade Combat Team Brigade has been located. Fort Carson has been doing $700 million worth of construction at this location for this battalion, including LEED buildings to save water and energy. In fact, the new battalion headquarters has received LEED-Platinum certification.[83] The Wilderness Road Complex has 5,000 Soldiers stationed there, but is too far from the main cantonment area to make it worth extending the reclaimed water infrastructure to it. All uses of water at this location have been potable water purchased from CSU. So the challenge is, how does Fort Carson incorporate this community into Net Zero Water activities in addition to water-conservation activities? Fort Carson water personnel would like to use some of their own water resources or even stormwater runoff instead of potable water to at least help with irrigation.

Funding for the Water Program

In this section, we discuss how Fort Carson funds different aspects of its water program, including smaller water conservation and efficiency projects and potable water and sewage treatment infrastructure projects. We begin by discussing how they pay for the post potable water system.

Potable Water System

For potable water, Fort Carson has to pay a consumption rate to CSU as well as pay for the infrastructure that they maintain on post. The on-post water distribution system has infrastructure maintenance as well as replacement and upgrade costs. Installation water personnel try to seek

---

[83] For more information, see ENR Mountain States, "Fort Carson Brigade Battalion Headquarters Earns LEED Platinum," July 5, 2012.

reimbursements for water use from non-Army garrison water users on post. These on-post customers include non-appropriated funds (NAF)-category C operations; such morale, welfare, and recreation (MWR) activities as bowling, golf clubhouses, outdoor recreation, and sales; Army and Air Force Exchange Service (AAFES); the Commissary; Burger King; etc. Fort Carson water personnel started billing on-post customers for water about three to four years ago. They are billed a usage rate to account for the commodity cost paid to CSU plus some of the water infrastructure costs. However, they are not charged the full cost of providing the water. In particular, funding from non-installation sources for water infrastructure projects is not included in this rate. For example, in the past, Fort Carson has received some stimulus money and other congressional money spent on water infrastructure, which was not included in these rates. Fort Carson DPW personnel pay the CSU water bill and then bills on-post users for their water usage.

One of these reimbursable customers challenged the rate: In particular, this customer only wanted to pay marginal rates for water, namely, what Fort Carson pays CSU for water. The customer did not want to pay for the water infrastructure costs. However, Fort Carson water staff did not change the rate, especially since the customer did not have to pay the full cost of water.

Military family housing is the biggest water user, and Fort Carson water personnel make sure housing pays its fair share. Policies differ at other installations as to how family housing is billed for utilities, so other installations do not take the same approach to billing housing for water use compared with Fort Carson. Currently, housing is billed like other on-post customers for water use. Initially, the privatized housing contractors on Fort Carson had not metered irrigation water, so they were not paying for it. Meters have since been installed, and they are now paying for what they use. Since housing is privatized, the contractors now have an incentive to conserve water in the housing areas because they can save costs.

New construction projects on post also must pay for the water they use. Construction sites use large amounts of water—millions of gallons—for site soil compaction. A contractor can easily use $10,000 or more worth of water in a month. Some construction projects spend $100,000 on water from start to finish. Fort Carson water personnel bills construction projects for this water use.

Construction contractors get water from a metered fire hydrant or an overhead fill point. Their trucks fill up at one of these two sites and then take the water to the construction site. Contracts require contractors to be responsible for utilities until the post has "beneficial occupancy" of the new buildings, at which point the charges stop.

Non-Potable Water Expansion Project

Fort Carson has acquired part of the funding for the non-potable water system expansion project. The total project cost would be $8 million. In 2012, the installation secured $4 million in SRM funds. Given that they are SRM funds, they are limited to a maximum of $750,000 for new capacity/expansion, and the rest will be spent on repairing existing infrastructure. The simple payback for this project is less than 10 years. Fort Carson water also received ECIP funding for the remaining $4 million.

Funding for Water Investments

Fort Carson funds its investment in water infrastructure and efficiency projects in a number of different ways, including some traditional partnerships, such as an ESPC. Funding sources for water investments have included SRM, ESPC, ECIP, and special congressional funds. ECIP funds are appropriate for system expansion, whereas SRM funding is limited to $750,000 for new capacity that is complete and usable, so SRM projects only can be used for incremental improvements. For example, SRM funds have been used to replace and upgrade in-building water systems with more water-efficient technologies, such as low-flow toilets. Whenever possible, Fort Carson water personnel also take advantage of special external funding, especially for large-scale water infrastructure investments. For example, Fort Carson has received some past stimulus money to replace water lines and special congressionally appropriated funds for water utility upgrades. Water personnel have been able to find funding for most of the projects they have wanted to do. The sustainability and Net Zero Water programs help ensure senior leadership support. They also have good DPW support from the director.

However, water infrastructure investments are getting more difficult to fund as federal and Army budgets get tighter. Infrastructure upgrades need to be programmed into the system so that if SRM funds become available, or other funding sources for that matter, the installation is ready.

Fort Carson has also gotten some water funding through the Army's Utility Modernization Program. The IMCOM-funded Utility Modernization Program has three priorities: energy and water projects with good paybacks; energy and water projects with longer paybacks; and infrastructure projects.

Fort Carson also leveraged some water infrastructure replacement funds by partnering with the Colorado Department of Corrections. The city of Trinidad supplies potable water through a 30-mile line to PCMS, but Fort Carson owns the water line. A state prison also uses this water line for its potable water. When this water line was leaking and needed replacing, the prison and Fort Carson shared the replacement cost.

When Fort Carson evaluates water conservation projects cost savings, it can be difficult to factor in some important savings that may accrue, especially when it saved a large amount of water during its population and building growth period. For example, the savings calculation does not account for fact that Fort Carson did not have to build additional water capacity, did not have to expand the sewage treatment plant, does not have to buy more water, and does not have to upgrade the sewer collection system. That is, Fort Carson does not get the true cost savings credit for the actual cost of saving water usage.

Fort Carson has an ESPC with Johnson Controls. Johnson Controls is pursuing a range of building water efficiency on task order 1 of its ESPC contract, which includes many water and energy efficiency projects. The paybacks for water at Fort Carson are good, unlike many other places in the United States. Johnson Controls will initiate water projects in which it can make the payback values work. Sometimes it may bundle a longer payback project with a shorter payback one. If a project proposal has a simple payback of 10 years or less, Fort Carson will invest in it. Sometimes Fort Carson will invest in a project with a 15-year payback because of other synergistic benefits to the project, such as quality of life or others that are difficult to quantify (this happens more often with energy projects than water projects). The payback analysis includes cost savings for water and energy; while there also are sewer cost savings, these are not usually included beyond the energy savings.

Summary

To conclude this chapter, we present a summary that provides an overview of Colorado water issues that are most relevant for Fort Carson and then briefly summarize water management at Fort Carson. Given the water challenges within this state and Fort Carson's progressive water conservation and management program, Fort Carson can serve as a useful military installation model for water management.

The Colorado Water Context

Colorado is an arid Western state where water is carefully managed, especially because of a growing population over the last 50 years that places additions stresses on water supplies. Snowpack-fed surface water supplies 81 percent of the state's water, while groundwater accounts for the rest. Since the late 1800s, extensive infrastructure has been built to transport water to mining, agricultural and population centers that are located on the Eastern Slope of the Rocky Mountains, away from the major sources of water on the Western Slope. This massive infrastructure is needed because 80 percent of the state's precipitation falls on the Western Slope, while 80 percent of Colorado's population lives in the Front Range. The Front Range (also known as the Eastern Slope) runs north to south and includes the cities of Fort Collins, Boulder, Golden, Denver, Colorado Springs, and Pueblo.

The state has complex water rights laws and is party to many interstate agreements. Both surface and groundwater water rights are based on the prior appropriation system. Colorado's water systems have been either fully- or over-appropriated. Colorado is party to a complex array of interstate compacts and agreements that have emerged from decades of efforts to resolve disputes and establish water rights. Notably, two river compacts affect Fort Carson since these rivers are water sources for part of Fort Carson's water supply: the Colorado River Compact and the Arkansas River Compact. These compacts and agreements place requirements and limits on water supplies available to meet the state's needs. In the event of low-flow periods they will dictate how water is to be shared. Given the state's prior appropriation system when water is scarce, a senior rights holder may invoke a "call" on a source of water to obtain the full supply to which it is entitled. Junior water-rights holders may lose their access to this water source. The urban areas tend to be the junior water-rights holders.

Demand for water has increased in the state due to growing urban areas and new demands for recreational and environmental purposes, while shifting away from mining and livestock. Water use for irrigation, which accounts for the vast majority of Colorado's water use, has remained largely unchanged. Although per-capita water demands are falling, it may not be sufficient to stem an aggregate increase in these demands. As with other states in the West, Colorado is facing increasing uncertainty about its future water supplies due to long-term climate change, suggesting that it may become harder to meet water needs in the future.

Such stresses on water and water demands in Colorado have spurred the development of some water markets. These markets are facilitated by Colorado's extensive conveyance infrastructure, most notably the Big-Thompson project but also the numerous ditch companies. In Colorado, water transfers generally are administered through the water courts (other Western states utilize the office of the state engineer or other water agencies), which consider third-party effects on other water users. Market mechanisms for water rights transfers in Colorado have taken the form of leases and sales and water banks. Many Colorado cities also use price as an incentive to conserve water by implementing block pricing for water users.

Fort Carson Water Management

Fort Carson is a major Army training site located directly south of Colorado Springs in El Paso County, Colorado. At more than 138,000 acres, the installation purchases 95 percent of its potable water from Colorado Springs Utilities (CSU). The other 5 percent of Fort Carson's water comes from some limited surface water and groundwater sources on post. Fort Carson owns and maintains the water lines and it's wastewater treatment facility on post. CSU collects the water it provides to Fort Carson from diversified water sources and both sides of the Continental Divide, including from the Colorado and the Arkansas River Basins.

Fort Carson's average annual water bill in 2011 was over $2.1 million, comprised of a flat fee charged at a seasonal military rate. Major water-use activities in order of volume used are: irrigation (excluding privatized family housing irrigation), family housing plumbing, family housing irrigation, and barracks plumbing (including laundry and kitchen). These uses comprise over 80% of the total water consumed, about 853 million gallons per year, and peak in the summer months due to landscape needs.

Fort Carson has a long history in implementing water efficiency and conservation measures; it reduced water consumption by more than 45 percent between 2002 and 2011 despite experiencing nearly a doubling of its population. These projects were implemented because of a strong sustainability program (which gave water use greater visibility and some additional staff resources), as well as leadership support, regional drought problems, and federal and Army requirements for water conservation. In 2011, Fort Carson became a pilot for Net Zero Water, Waste and Energy installation by 2020, which provides additional justification for water-saving investments.

Fort Carson water conservation activities are diverse and range from policy statements and education to incorporating new construction codes, water saving operational practices, conservation oriented internal water pricing rates, and technology. These activities have included: using treated wastewater for golf course irrigation, installing an enviro-transpiration smart irrigation system, modifying landscaping and turf management practices, requiring military units to utilize the central vehicle wash facility which recycles its water, installing low-flow fixtures, and repairing leaking water lines and tanks. To meet its Net Zero Water Installation goals for 2020, water managers at Fort Carson plan to 1) continue investing in water efficiency activities to conserve water, and 2) identify and use alternative sources of water in order to reduce potable water use. Fort Carson water provider, CSU, supports the post's efforts to expand such alternative water sources, as well as its conservation activities.

The most significant alternative water source is treated wastewater from the wastewater treatment facility. Possible uses in addition to the golf course include for landscape irrigation, construction, training, and vehicle wash. The post plans to expand its wastewater treatment plant to have a zero discharge sewage treatment plant and more reclaimed water available for such uses. A second alternative water source is stormwater, which may be used for irrigation. Technical and legal issues around water quality and rights must be assessed and addressed for using these alternative sources. A third alternative source, greywater, is being explored, as are additional opportunities for using the limited surface and groundwater sources on post. There are tradeoffs among some of these approaches (i.e., conservation will reduce the amount of reclaimed water available) and costs of servicing remote areas with necessary infrastructure that must also be addressed. However, by reducing potable water usage Fort Carson will save money on their CSU water bill, achieve its Net Zero Water Goals, more easily manage during drought periods, and enhance its' water security.

Installation leadership, previous sustainability efforts, water prices, and overall water pressures in Colorado have created a supportive environment and rationale for water projects. Fort Carson water management personnel have taken advantage of a range of funding sources for water projects through both traditional and non-traditional sources, including SRM, ECIP, ESPC, and special congressional authorizations. However, water infrastructure investments are becoming more difficult to fund as federal and Army budgets tighten. The Fort also has leveraged some water infrastructure investments through public-to-public partnerships. For example, some reservoir water and pipelines are shared in partnership with nearby communities, who share in the costs of maintenance and upgrades. In addition, on the installation, non-garrison organizations such as privatized housing, NAF programs, AAFES, and other organizations are charged for water use. The charge includes the commodity cost for water supplied by CSU and a portion of the infrastructure cost (excluding the cost of special projects).

Given these water conservation and management successes despite the financial, technical, and Colorado legal and water supply challenges, Fort Carson can serve as a model for other military installation's water programs.

6. Arizona and Fort Huachuca Case Study

This chapter presents an in-depth case study for Arizona and Fort Huachuca to show the variations in challenges and opportunities for installation water partnerships and water markets in different locations. Again, we set the context by explaining the water situation in Arizona, including the complexity of water supply, demand, historical agreements, rights, and distribution. We present examples to illustrate some of the differences and similarities in Arizona compared to Colorado and how such circumstances help drive water market and management opportunities and challenges within Arizona. For instance, groundwater use and management concerns are significant within the state because of problems with the over-drafting of groundwater and how groundwater use affects surface water systems, ecosystem health, and endangered species. This situation has had a significant effect on Fort Huachuca. Next, we present an in-depth case study about water at Fort Huachuca. This case study includes background information about the installation, an overview of Fort Huachuca's water-related legal challenges, water conservation requirements, its water supply and infrastructure, its water demand, and its water program. Fort Huachuca's water program discussion includes an overview of how Fort Huachuca, as many progressive communities have done, is taking advantage of alternative water sources, such as reclaimed water, and its challenges and creative approaches to finding funding for water efficiency and infrastructure investments.

Understanding the Arizona Water Situation

Arizona is the sixth-largest state by area, covering 113,909 square miles, and today it is the 16th largest state by population, with nearly 6.6 million residents.[1] Similar to other states in the region, Arizona has grown significantly during the past few decades. For example, its population was 3,665,228 in 1990, and it grew by almost 1.5 million people between 1990 and 2000, a 40-percent increase during this 10-year period. It then grew by more than1.4 million people (28 percent) by 2012 to more than 6.55 million people.[2] It is projected to nearly double to 12.8 million by 2050.[3] Arizona also is in the dry Southwest with limited precipitation in much of the state, so water has been a concern as the state's population has grown. In fact, the pursuit of water has shaped Arizona since humans came to the region. Canals built more than 1,000 years ago show evidence of the Hohokam Indians' efforts to irrigate and develop agriculture in the arid

---

[1] U.S. Census Bureau, "Annual Estimates of the Population for the United States, Regions, States, and Puerto Rico: April 1, 2010 to July 1, 2011," NST-EST2011-01, December 22, 2009.

[2] U.S. Census Bureau, "Population Change and Distribution: 1990 to 2001," April 2001, p.2; U.S. Census Bureau, "State & County QuickFacts," undated.

[3] Arizona Department of Administration, "Arizona Population Projections 2006–2055," 2006.

lands of central Arizona's Verde Valley. In the 1800s, controlling water—storing, channeling, and using it for irrigation—was key to farming, ranching, mining, and other growth and development within different parts of Arizona.[4] More recently, water projects helped transform the state from a collection of mining camps to thriving villages and cities.

A state forecast of water demand and supply to the year 2110 found statewide water demand is predicted to grow to 8.1 million acre-feet by 2035 and to 10.6 million acre-feet by 2110. In contrast, currently developed water supplies are in estimated to be in the range of 6.5 million acre-feet to 6.8 million acre-feet statewide (not all of developed supplies are being used). While portions of the state likely will have sufficient water to meet future needs, others will have shortfalls (even with additional conservation and greater reuse of effluent) unless other sources of water are successfully developed. For example, the Upper San Pedro Region, where Fort Huachuca is located, is forecast to have water demands of between 39,500 to -50,500 acre-feet in 2035 and between 56,800 and 68,600 acre-feet in 2110. Current developed sources in the region are 33,700 acre-feet.[5]

Water sources that could potentially be developed statewide to meet future needs include additional groundwater; surface water (both in-state rivers and the Colorado River); reclaimed water and other water such as brackish or poor-quality groundwater; mine and agricultural drainage; desalinated water; and water made available through weather modification. However, the state has not developed a comprehensive plan to obtain additional water since there is so much variability across its various regions. Moreover, many hydrologic, technical, legal, and economic aspects to developing these sources must be addressed before they can be developed successfully. Therefore, regions within the state face many challenges and uncertainties regarding future water supplies to meet growing demand.[6]

Background about the regional climates and geography is useful for understanding Arizona's water situation. Arizona can be divided into three physiographic regions, as shown in Figure 6.1, that vary in climate, population, and other features. In the north, the Colorado Plateau covers approximately 140,000 square miles over Arizona, Utah, Colorado, and New Mexico. This area is largely desert with some areas of forest, and elevations range from 3,000 to 4,000 feet. The Grand Canyon is part of the Colorado Plateau; the plateau also includes the city of Flagstaff. Annual precipitation in this area is low—10 inches on average.[7]

---

[4] For an example of how the management of water helped shaped such development, see Duane A. Smith, "The River of Sorrows: The History of the Lower Dolores River Valley," National Park Service, undated.

[5] Arizona Water Resources Development Commission, "Arizona Water Resources Development Commission Final Report," October 2011.

[6] Arizona Water Resources Development Commission Final Report, 2011; and Tony Davis, "Arizona's drinking-water needs will force trade-offs," *Arizona Daily Star*, February 23, 2014.

[7] Annabelle Foos, "Geology of the Colorado Plateau," Akron, Ohio: University of Akron, 1999.

Figure 6.1. Physiographic Regions in Arizona

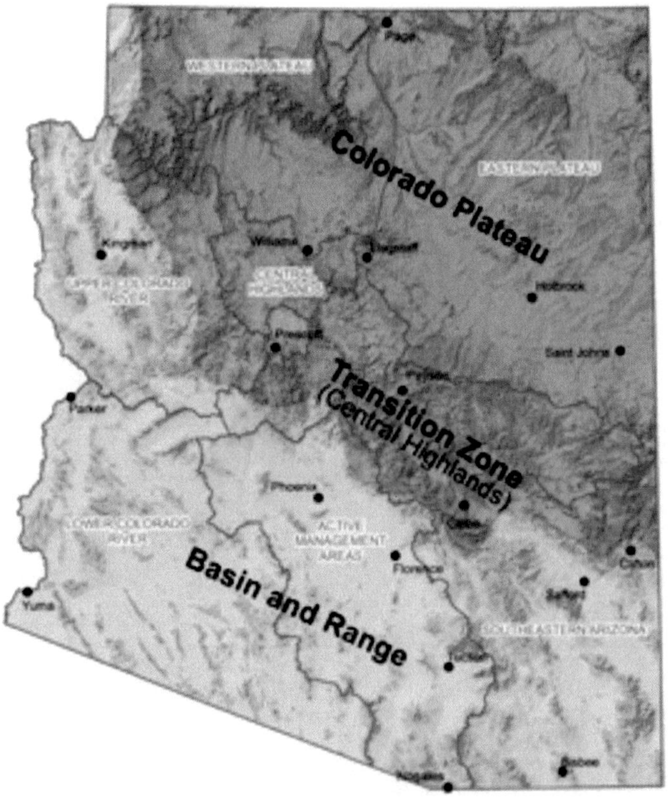

SOURCE: Arizona Department of Water Resources, "Geography of
the Southeastern Arizona Planning Area," undated.

The second key region is termed the Basin and Range Province. This area is vast: It lies between the Sierra Nevada mountains in the west and the Colorado Plateau in the east and stretches from the northern end of Nevada into Mexico. It occupies 170,000 square miles. In Arizona, it cuts across the southwestern portion of the state. The Basin and Range province contains much of Arizona's population, as shown in Figure 6.2, including the cities of Phoenix and Tucson. Fort Huachuca also lies in this region, in the southeast corner of the state. The Basin and Range's average precipitation is four inches to eight inches in the lower basins and from about 16 inches to 30 inches in the upper mountains. However, because of the arid environment, almost all the precipitation in the basins and most of the precipitation in the mountains is lost to evapotranspiration. Only about 5 percent of the precipitation that falls recharges the basin-fill aquifers.[8]

The Transition Zone serves as a boundary between these two regions and shares characteristics with both.

---

[8] U.S. Geological Survey, "Ground Water Atlas of the United States: Arizona, Colorado, New Mexico, Utah," 1995.

Figure 6.2. County Population in Arizona

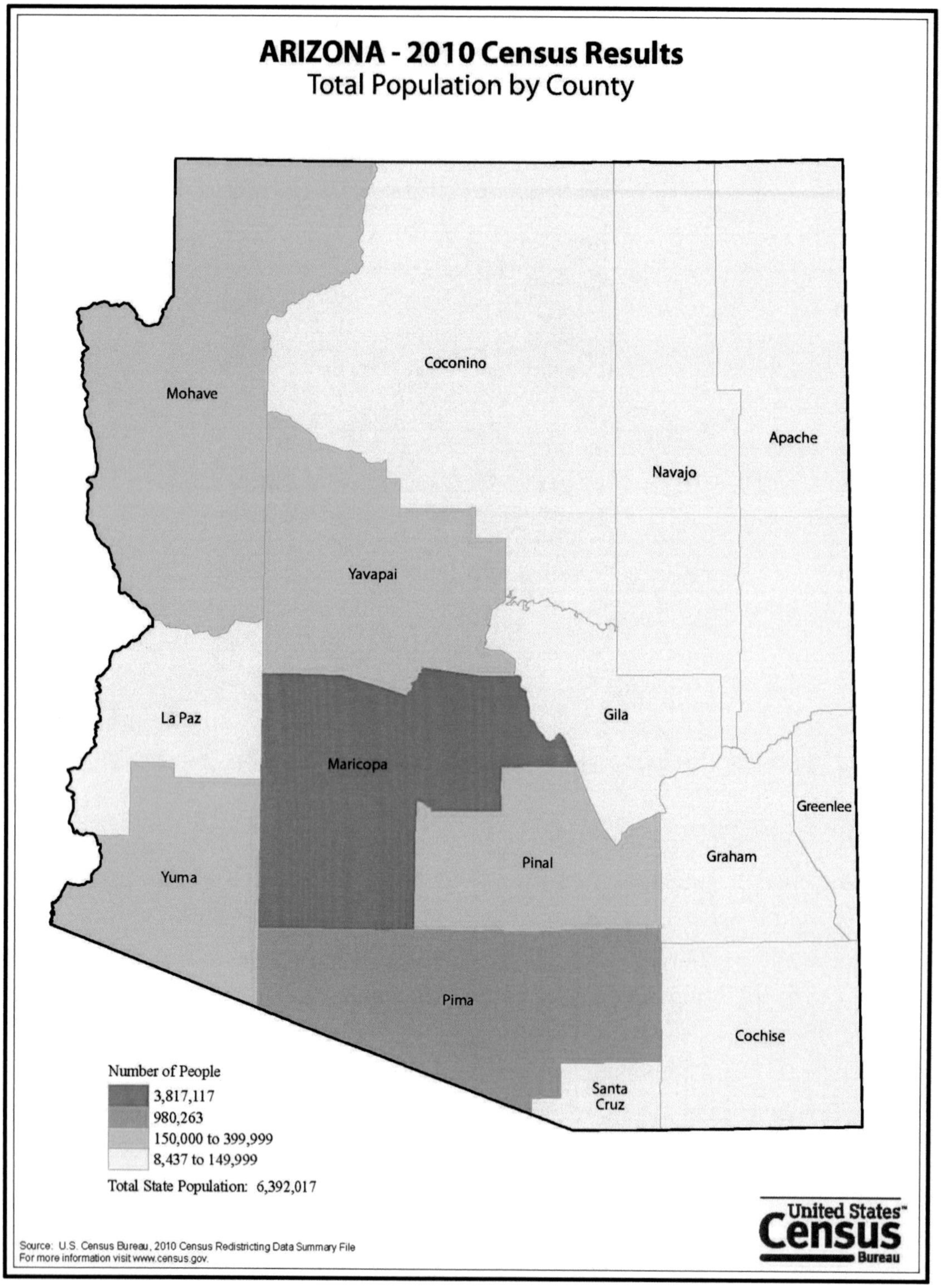

SOURCE: U.S. Census Bureau, "Arizona 2010 Census Results, Total Population by County," undated.

Arizona Water Supply and Demand

Arizona uses about eight MAF of water each year—roughly 75 percent for agriculture, 20 percent for municipal uses, and 5 percent for industrial uses. Agriculture use has been declining in recent decades, from a peak of more than 4 MAF in 1976. Although per-capita use of water also has been declining, the total municipal water use has grown over time with the population. Most municipal water—60 percent—is used for outdoor irrigation and landscaping, a key target area for conservation efforts.[9]

Arizona relies on four sources of water to meet its demand. Groundwater provides 40 percent; the Colorado River alone provides 39 percent; other surface water from lakes, rivers, and streams account for 19 percent; and effluent accounts for the remaining 2 percent.[10] Although it is relatively dry with average annual precipitation of 13 inches,[11] Arizona contains 28 major rivers. However, most of these rivers are tributaries of the Colorado River, and many are dry channels, which only have flows during some times of the year.[12] Over-pumping of water from wells has eliminated a large percentage of the natural perennial flow of rivers.[13]

Despite recent growth, municipal water use has stabilized in some areas. As Figure 6.3 shows, the population of Phoenix has grown from approximately 1 million people in 1990 to 1.5 million in 2007. While total water use grew alongside population in the early 1990s, it declined after a peak in the early 2000s and has stabilized in recent years. As Figure 6.4 shows, this period saw a decline in per-capita use, from approximately 250 gallons per person per day in 1990 to approximately 195 gallons per person per day in 2007.[14]

---

[9] Pitzer, Eden, and Gelt, 2007, p. 7.

[10] Arizona Department of Water Resources, "Securing Arizona's Water Future," undated, p. 1; Bonnie G. Colby, David A. de Kok, Gary Woodard, et al., "Arizona's Water Future: Challenges and Opportunities," University of Arizona Office of Economic Development and Water Resources Research Center, 2004, pp. 59–62.

[11] U.S. Department of the Interior and U.S. Geological Survey, "Arizona," National Atlas of the United States of America, undated.

[12] Pitzer, Eden, and Gelt, 2007, p. 4.

[13] Joe Gelt, "Land Subsidence, Earth Fissures Change Arizona's Landscape," *Arroyo*, Vol. 6, No. 2, Summer 1992.

[14] City of Phoenix, "Historical Population and Water Use," undated.

Figure 6.3. Population Growth and Water Use in Phoenix, Arizona

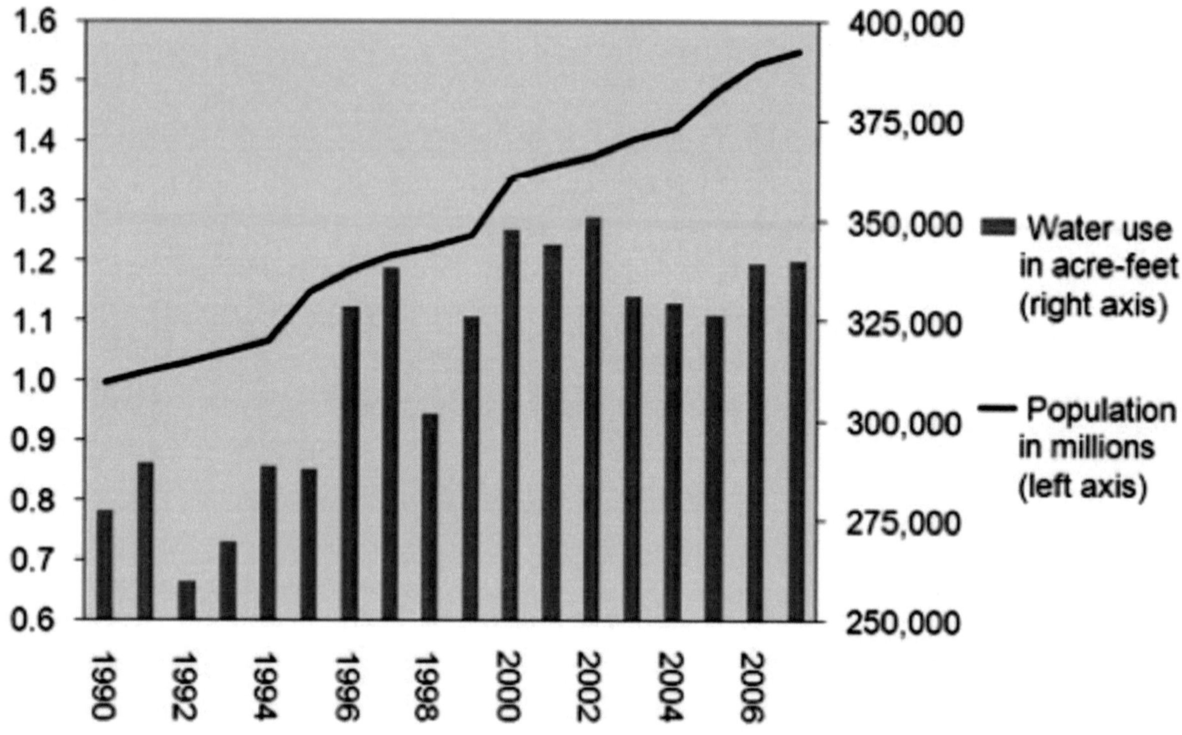

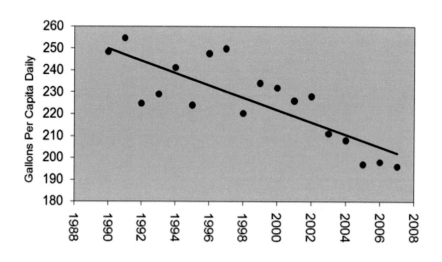

SOURCE: City of Phoenix, "Historical Population and Water Use," undated.

Figure 6.4. Per Capita Water Use in Phoenix, Arizona

SOURCE: City of Phoenix, "Historical Population and Water Use," undated.

The water use across Arizona varies greatly. As shown in Figure 6.5, several cities are using fewer than 100 gallons per capita per day, while others are using nearly 250 gallons. These differences are attributable to development patterns such as lot sizes as well as differences in conservation measures and water pricing structures.[15]

Figure 6.5. Per Capita Water Use in Several Cities in Arizona

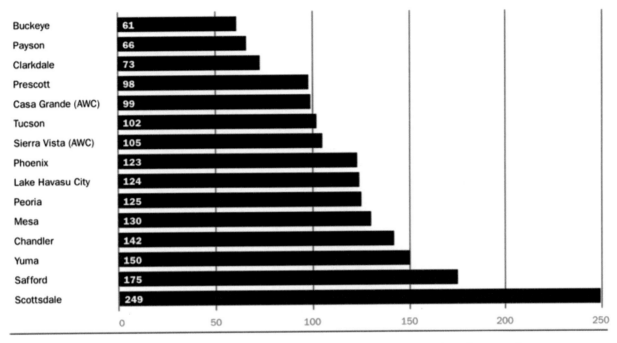

SOURCE: Western Resource Advocates, "Arizona Water Meter: A Comparison of Water Conservation Programs in 15 Arizona Communities," October 2010, p. 5.

As is discussed more below, because of problems with over-drafting the groundwater, Arizona created Active Management Areas (AMAs) to better manage groundwater use and eliminate severe groundwater overdraft. Although approximately 80 percent of Arizona's population lives in AMAs, and AMAs account for approximately 75 percent of the state's water consumption, they cover only 13 percent of the state's area.[16] Rural growth outside the AMAs, and the water resources needed to support this growth, are of increasing concern in Arizona, as rural communities have expanded dramatically. Fort Huachuca is located is such a rural area. Many residential developments are being created in deserts, and growing populations, both seasonal and permanent, are straining infrastructure and developed water supplies. A 2003 study by the Arizona Department of Water Resources notes:

> As a result of high growth rates, physically and legally limited water supplies, drought, economic constraints, and relatively little water resource management and planning, water supplies are stressed in some parts of rural Arizona. In some

---

[15] Western Resource Advocates, "Arizona Water Meter: A Comparison of Water Conservation Programs in 15 Arizona Communities," October 2010.

[16] Arizona Department of Water Resources, "Rural Programs," undated.

areas of the state, stakeholders and leaders have expressed the need for additional resources to support local water resource planning.[17]

Private water companies will be serving many new developments. The Arizona Corporation Commission (ACC) regulates those companies. The ACC is analogous to the public utilities commissions in many states but with broader authority. In 2012, the ACC regulated more than 400 individual water systems operated by nearly 350 companies.[18]

Additionally, many areas are without adequate surface water and will rely on groundwater to meet their needs. Groundwater outside of AMAs is not well regulated, raising concerns about over-extraction and groundwater depletion.[19]

The Rural Arizona Watershed Initiative helps watershed groups outside AMAs in their water-resource planning efforts. They are encouraged to develop management strategies suitable for their watershed with input from local stakeholders. The Department of Water Resources provides non-regulatory technical assistance.

The Upper San Pedro Partnership (USPP) is an active collaborative watershed group. It is a coalition of nearly two dozen organizations, including Fort Huachuca, that works to identify, prioritize, and implement comprehensive policies and projects to meet the area's water needs while protecting the San Pedro Riparian National Conservation Area (SPRNCA). A key goal of the USPP is to ensure the long-term viability of Fort Huachuca.[20]

Inter-State Water Agreements

As is the case in Colorado, Arizona is involved in a number of compacts and treaties that help allocate and manage water resources. We briefly discuss two types here. First, we discuss the Colorado River Basin Compact again because it is a key water source for Arizona. Then we discuss some of Arizona's agreements with Mexico. This discussion illustrates how for many states these river basin agreements also can involve the United States' neighbors, Mexico or Canada, because rivers cross into these other countries.

Colorado River Basin Compact

The Colorado River is a key water source for Arizona. As was discussed in Chapter Five, the 1,450-mile Colorado River flows from the Rocky Mountains, across Colorado and Utah and along the borders of Arizona, California and Nevada, into Mexico to the Gulf of California. The Colorado River's tributaries make up the Colorado River Basin, which also covers the states of

---

[17] Arizona Department of Water Resources, "Rural Water Resources Study/Rural Water Resources 2003 Questionnaire Report," October 2004, p. 1.

[18] Arizona Corporation Commission, "Utilities Division," undated.

[19] L. William Staudenmaier, "Between a Rock and a Dry Place: The Rural Water Supply Challenge for Arizona, *Arizona Law Review,* Vol. 49, 2007, pp. 321–338.

[20] Upper San Pedro Partnership, "About Us," undated.

Wyoming and New Mexico. (For a map of the Colorado River Basin and the seven states that share it, see Figure 5.6.)

The shared use of the Colorado River Basin among these seven states—Arizona, California, Colorado, New Mexico, Nevada, Utah, and Wyoming—required negotiation and has been controversial over the last century. Arizona has argued and fought with California and other states in and out of court to ensure it has access to water from this key source.

The Colorado River Basin Compact is an agreement among the seven states to divide water from the Colorado River equitably, establish the relative importance of different uses of water, ensure development of the basin, and reduce existing and future controversies. The compact requires the Upper Basin states (Colorado, New Mexico, Utah, and Wyoming) to deliver 7.5 MAF of water per year on a 10-year rolling average to the Lower Basin states (Arizona, California, and Nevada). If the 10-year rolling average falls below this level, the Lower Basin States may make a "call" on the compact, forcing the Upper Basin states to reduce water consumption.[21] In 1928, Congress divided the Lower Basin allocation, giving California 4.4 MAF, Arizona 2.8 MAF, and Nevada 0.3 MAF each year.[22]

Arizona has fought legally to ensure that it maintains its 2.8 MAF. In fact, when California tried to claim senior water rights over some of Arizona's claims to this Colorado River water, the case went to the U.S. Supreme Court, whose ruling confirmed Arizona's 2.8 MAF right against California's claim of senior water rights and Arizona's rights to tributaries to the Colorado River within Arizona. The court also confirmed that during Lower Basin surplus years, the surplus is split between Arizona and California.

Water Agreements with Mexico

The border between Arizona and Mexico requires special considerations related to water resources. The United States and Mexico created the International Boundary Water Commission (IBWC) in 1889 to address water rights treaties and agreements between the two countries. Many of the IBWC's efforts are highly relevant to Arizona's water resources. For instance, in 1944, the IBWC facilitated the Treaty for the Utilization of Waters of the Colorado and Tijuana Rivers and of the Rio Grande. The treaty grants Mexico 1.5 MAF of water from the Colorado River.[23] Salinity in the lower Colorado River was a concern at the time, as it affected agriculture and drinking water in Mexico, but water quality considerations did not become part of the 1944 treaty. Salinity grew in subsequent decades, in part because Arizona began discharging wastewater into the Colorado River.[24] After several rounds of unsatisfactory treaties, the two

---

[21] Watson, and Scarborough, 2009, p. 2.

[22] Pitzer, Eden, and Gelt, 2007, p. 12.

[23] U.S. Department of State, "Utilization of the Waters of the Colorado and Tijuana Rivers and of the River Grande," U.S. Government Printing Office, 1946.

[24] Allie Alexis Umof, "An Analysis of the 1944 U.S.-Mexico Water Treaty: Its Past, Present, and Future," *Environs: Environmental Law and Policy Journal,* Vol. 32, No. 1, Fall 2008, p. 78.

countries signed Minute 242 in 1973, which guarantees a specific level of water quality to Mexico.[25]

Other concerns have arisen in recent decades. The Nogales International Wastewater Treatment Plant in Nogales, Arizona, treats water on both sides of the border. A needed upgrade was blocked in the early 2000s by a lawsuit claiming that the facility did not comply with U.S. EPA's water quality standards. To resolve the problem, the two countries negotiated a $62 million upgrade to the plant that began in May 2007. Additionally, the U.S.-Mexico Transboundary Aquifer Assessment Act calls for collaborative studies of aquifers, including Arizona's Santa Cruz River Valley and San Pedro aquifers, which are affected by the Nogales wastewater treatment plant.[26]

Intra-State Water Projects

As in Colorado, Arizona has numerous dams, reservoirs, conveyance canals and other water transfer and storage infrastructure projects to manage and transfer water. Such infrastructure, as was seen in the Colorado case study with the C-BT Project, can be key to enabling certain types of water markets. For instance, if water is to be sold or leased, conveyance infrastructure is needed to move it from one location to another. Here we briefly discuss the largest water infrastructure project in Arizona, the Central Arizona Project.

Central Arizona Project (CAP)

Arizona was a reluctant party to the Colorado River Compact out of great concern that California would take more than its fair share of water. However, it also sought to secure its own water sources and needed federal approval to build the Central Arizona Project (CAP) to deliver Colorado River water to users across the state. It ratified the Colorado River Compact in 1944 in part to win federal approval of the CAP. However, approval did not come until 1968, in part because of legal uncertainties related to water rights. Construction began in 1973 the first water deliveries were made nearly 25 years later in 1985 (see Figure 6.6). Today, the system consists of a 336-mile canal from the Colorado River to Tucson, with numerous aqueducts, tunnels, pumps, and pipelines. The project is part of the state's critical infrastructure, because by 2007 "more than 4 million people and 300,000 acres of irrigated farmland in Central Arizona depend on the CAP for delivery of over half their water supply."[27]

---

[25] International Boundary and Water Commission United States and Mexico, "Minute No. 242, Permanent and Definitive Solution to the International Problem of Salinity of the Colorado River," August 30, 1973.

[26] Pitzer, Eden, and Gelt, 2007, pp. 20–21.

[27] Pitzer, Eden, and Gelt, 2007, p. 11.

Figure 6.6. Central Arizona Project

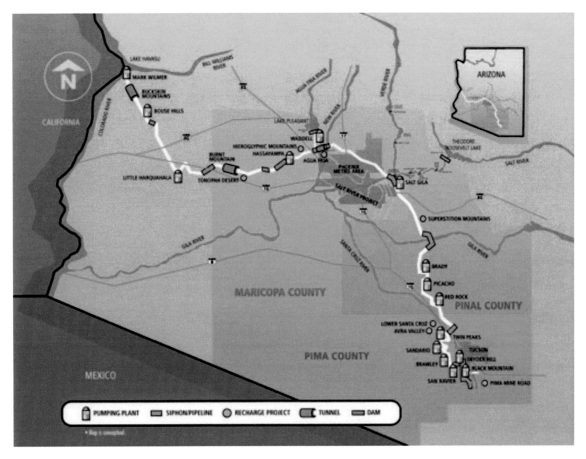

SOURCE: Arizona Department of Water Resources, undated.

Arizona Water Rights and Regulations

Next we discuss some of the features of Arizona's water laws, regulations, and policies that have implications for water markets and partnerships and Fort Huachuca, and that differ from those in Colorado and some other states. For instance, unlike most states, Arizona has an aggressive program to manage the use of groundwater in AMAs to address groundwater overdraft and has legally defined reuse for effluent, both of which we discuss below.

Surface Water Rights

Arizona has a dual system of water governance that largely separates surface water and groundwater rights. As with other states in the West, Arizona uses a doctrine of prior appropriation to govern surface water rights. Prior appropriation was written into Arizona law in 1864, and today the Arizona Department of Water Resources records surface water rights through a water-permit program.

A key event in Arizona's water rights history occurred in 1990, when the state recognized official water rights for environmental purposes and granted permits to support fish and other

251

wildlife habitat. Since then, the state has issued dozens of water permits for environmental purposes.[28]

Because water rights have been so controversial in Arizona, the state has been using a judicial process called general stream adjudication to resolve all the water rights along two major river systems, the Gila River Basin and the Little Colorado River, for which water rights and claims are many and complex. We briefly discuss these processes next.

The first is the adjudication process of the Gila River Basin, which stretches from east to west across southern Arizona. The process began as petitions in 1974 and 1979 to resolve water rights claims in the Salt, Verde, Gila, and San Pedro rivers. This basin drains most of central and southern Arizona, and today includes 85 percent of Arizona's population (including the Tucson and Phoenix metropolitan areas) and most of its agriculture and industry. Fort Huachuca also is within this water basin.

In 1979, the Arizona Legislature amended its adjudication statutes to transfer these claims to the Maricopa County Superior Court, and in 1981, the Arizona Supreme Court consolidated the adjudications into a single proceeding. By late 2012, the adjudication concerned 83,500 claims by more than 24,000 claimants. The parties include the United States, the State of Arizona, Native American tribes, municipalities, public and private utilities, agricultural irrigation districts, industrial corporations, and individual farms, ranches, and other private water users.

Fort Huachuca also is a water rights claimant in the Gila River Basin adjudication process. Fort Huachuca lands were acquired through two Executive orders, three public land orders, voluntary conveyances, condemnation, land exchanges, and leases. The determination of Fort Huachuca's implied federal reserved water rights requires a determination of the primary purposes of the military reservation. The Fort Huachuca case is contested. The litigants include the Arizona Water Company, ASARCO, Babacomari Ranch Company, Bella Vista Water Company, Inc., BHP Copper Inc., City of Sierra Vista, Freeport-McMoRan, Gila River Indian Community, Pueblo Del Sol Water Company, St. David Irrigation District, San Carlos Apache Tribe, State of Arizona, Tonto Apache Tribe, and Yavapai-Apache Nation. The Cities of Mesa and Phoenix withdrew as litigants.[29] If Fort Huachuca loses its claim, it could limit its ability to do its mission because its water rights and use would be curtailed.

---

[28] Pitzer, Eden, and Gelt, 2007, p. 5.

[29] Michael Jeanes, "The Water Case Turns 30," *Maricopa Lawyer*, Maricopa County Bar Association, Vol. 28, No. 5, May 2009; Edward Ballinger, Jr., "In Re the General Adjudication of all Rights to Use Water in the Gila River System and Source, Case Regarding Fort Huachuca, Order Granting the Special Master's Motion for Adoption of the April 4, 2008 Report Regarding Fort Huachuca," Civil No. w1-11-605, Superior Court of Arizona, Maricopa County, September 7, 2011.

The Gila River Basin adjudication process has not yet resulted in a final decree, but it has resulted in significant revisions to Arizona's water code and multiple Arizona Supreme Court decisions.[30]

The adjudication process in the Little Colorado River, a tributary of the Colorado River in northern Arizona, began similarly. The Phelps Dodge Corporation filed a petition in 1978 with the State Land Department to determine the water rights of the Little Colorado River. Under the same 1979 legislative amendments, the Little Colorado River Adjudication was transferred to Apache County Superior Court.[31] By fall 2012 this adjudication process concerned the claims of 3,100 water users in northeastern Arizona.

Groundwater Rights

Before 1980, the state applied a "reasonable use" doctrine to groundwater extraction, allowing owners of land to extract groundwater within their property rights. This approach led to overdraft for many decades, and the state legislature responded in 1980 with the Groundwater Management Act. This legislation has three main purposes: control severe overdraft, allocate limited groundwater resources more efficiently and augment groundwater through supply development.[32]

As a way of accomplishing these purposes, the act created Active Management Areas (AMAs) around major basins and sub-basins to address areas with the most severe groundwater overdraft. Today there are five AMAs, as shown in Figure 6.7: Phoenix, Pina, Prescott, Tucson, and Santa Cruz. These AMAs cover 80 percent of Arizona's population and 70 percent of the state's groundwater overdraft.[33]

The act governs groundwater management in these areas through six provisions:

1. "Establishment of a program of groundwater rights and permits.
2. A provision prohibiting irrigation of new agricultural lands within AMAs.
3. Preparation of a series of five water management plans for each AMA designed to create a comprehensive system of conservation targets and other water management criteria.
4. Development of a program requiring developers to demonstrate a 100-year assured water supply for new growth.
5. A requirement to meter/measure water pumped from all large wells.

---

[30] Joseph M. Feller, "The Adjudication that Ate Arizona Water Law," *Arizona Law Review*, Vol. 49, 2007, pp. 405–440.

[31] Maricopa County Superior Court, "Arizona's General Stream Adjudications," undated.

[32] Arizona Department of Water Resources, "Overview of the Arizona Groundwater Management Code," undated, p. 1.

[33] Arizona Department of Water Resources, undated, p. 2.

6. A program for annual water withdrawal and use reporting. These reports may be audited to ensure water-user compliance with the provisions of the Groundwater Code and management plans. Penalties may be assessed for non-compliance."[34]

The Department of Water Resources develops long-term plans, conservation strategies, regulations, and enforcement strategies to reduce overdraft in each AMA. The act also established an Assured Water Supply program that requires developers within AMAs to demonstrate that an assured water supply will be "physically, legally, and continuously available for the next 100 years" before the developer can sell parcels.[35]

The Groundwater Management Act also established two Irrigation Non-expansion Areas (INAs) of Joseph City and Douglas; the Harquahala INA was later added. The number of acres of land that can be irrigated is restricted in these areas.[36]

---

[34] Arizona Department of Water Resources, undated, p. 2.

[35] Pima County Geographic Information Systems, "Arizona's Assured Water Supply Rules," undated.

[36] Arizona Department of Water Resources, "Active Management Areas (AMAs) & Irrigation Non-expansion Areas (INAs)," undated.

Figure 6.7. Map of Active Management Areas in Arizona

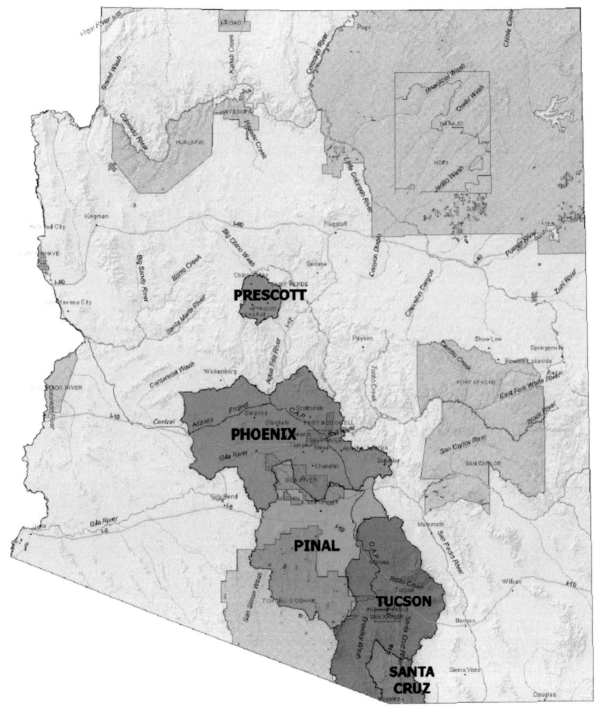

Effluent Rights

Effluent, or treated wastewater, is legally distinct from groundwater and surface water in Arizona. Effluent is owned by the entity that generates it, but then is treated as surface water

255

once it is discharged into surface water bodies. As such, it is subject to surface water rights determined by Prior Appropriation, with one exception: the entity that generated it is not required to discharge it and instead can reuse the water, even if it uses natural channels to convey the water and even at the expense of downstream water rights holders. This legal distinction was tested when in the 1980s, Phoenix began using approximately 70,000 AF of effluent at the Palo Verde Nuclear Generating Station instead of discharging it into the Salt River. Those with downstream water rights lost a suit to compel discharge. This has had significant effect on the water market for effluent, as described below.[37]

Tribal Water Rights

Another important group of water users who have significant and long-standing water rights are Native American tribes. Tribes' claims to water can complicate water allocation in many parts of the United States. In fact, tribal water rights play a key role in Arizona's water governance:

> More than one-fourth of the state's land is held in trust as reservations for the benefit of American Indians. The determination of water rights and use by Indian communities could have a significant impact on water supplies in Arizona.[38]

Arizona's Native Americans use less water per capita than non-Native Americans. Their use may be only a small fraction of the water to which they hold rights. Native Americans hold senior water rights, as established in 1908 with the "Winters Doctrine." This doctrine assigns water rights according to when a reservation was established. An agriculture purpose was assigned in 1963 following a court decision and calculated based on the amount of land that could be irrigated at reasonable costs. This could be significant because Native American reservation lands cover 28 percent of the state, as shown in Figure 6.8. Native American water rights may be significantly larger than the rights currently identified—perhaps larger than the state's total surface water supply—because many water rights claims remain unresolved. This uncertainty creates significant difficulties for long term water planning.[39]

---

[37] Susanna Eden, Robert Glennon, Alan Ker, et al., "Agricultural Water to Municipal Use: The Legal and Institutional Context for Voluntary Transactions in Arizona," *The Water Report*, December 15, 2008. p.12.

[38] Pitzer, Eden, and Gelt, 2007, p. 18.

[39] Pitzer, Eden, and Gelt, 2007, p. 18.

Figure 6.8. Tribal Lands in Arizona

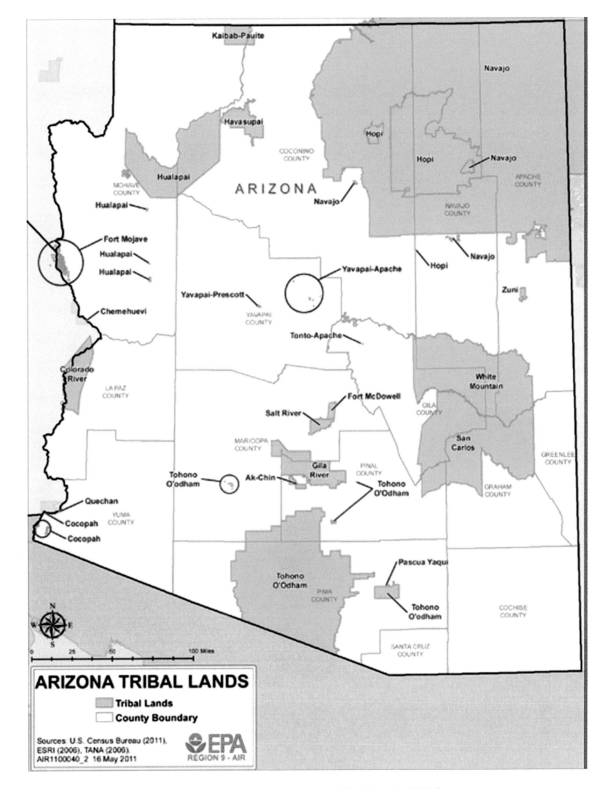

SOURCE: U.S. Environmental Protection Agency, "Arizona Tribal Lands, 2011."

Water Market Activities in Arizona

Within Arizona, many market mechanisms have been employed. The following discussion provides a sample of how that has been done in Arizona. Arizona depends on both its native surface and groundwater sources as well as Colorado River water conveyed by the CAP and the Salt River Project (SRP). These projects also provide the means to transfer both surface and groundwater through leases and sales. The Groundwater Management Act provides a major incentive for water transfers with its requirement that new land development cannot occur unless there is a guaranteed water supply to meet anticipated needs for at least 100 years. An auction was used in at least one instance to purchase water in support of development. And the Arizona Water Bank is a well-known approach to recharging depleted aquifers with excess CAP water, in effect banking it for future use (for Arizona as well as for California and Nevada if requested). Finally, several cities employ block pricing, which is a means to encourage water conservation through a price signal. These market mechanisms could be an opportunity for Fort Huachuca or another Army facility in Arizona in the future. However, today it would be challenging for Fort Huachuca to participate given its location and current water situation, which is described below.

Leases and Sales

Overall, Arizona led western states in the amount of water transferred using one-year leases, multi-year leases, and sales during 1987 to 2005. Colorado dominates in terms of *numbers of transactions*. However, in terms of the *amount of water committed,* Arizona, California, and Texas are the top three. Within-sector trades (agriculture-to-agriculture and urban-to-urban) are the most common trades that occurred in the 1987–2005 time frame.[40]

Arizona has five categories of water transfers: Colorado River water (which represents the majority of surface water transfers), non-Colorado River surface water, groundwater, effluent, and Native American reserved water rights. Outside of the Colorado and Salt rivers, surface use and thus surface transfers make up a minor portion of the state's total water consumption.[41] Table 6.1 provides a summary of the water transfers by type of water transferred.

---

[40] This analysis looked at annual leases, multi-year leases and sales. Two different measures for the amount of water transferred are used because different mechanisms, annual leases, multi-year leases and sales are employed to transfer water. While annual leases commit water volumes for one year, sales and multi-year leases commit water volumes several years. So in this analysis, the volume of water transferred is measured in two ways: the annual volume transferred and the total committed volume. See Brewer, Glennon, Ker, and Libecap, 2008, pp. 91–112.

[41] National Research Council, "Central Arizona: The Endless Search for New Supplies to Water the Desert," in *Water Transfers in the West: Efficiency, Equity, and the Environment*, Washington, D.C.: National Academies Press, 1992.

Table 6.1. Summary of Water Transfers in Arizona (1987–2004)

Water Type	Number of Transactions	Acre-feet (thousands)
CAP	65	5,984
Groundwater (Type II)	47	20
Reclaimed	15	83
Groundwater (non-AMA)	11	20
Surface water	9	1,314
Groundwater (Type I)	1	<2

SOURCE: Bonnie Colby and Katharine Jacobs (eds.), "Arizona Water Policy: Management Innovations in an Urbanizing, Arid Region," Resources for the Future, Washington, D.C., 2006.

Water transfers have occurred in Arizona as early as 1948, when the city of Prescott purchased farmland to use the groundwater for municipal use. In the early 1970s, Tucson also began purchasing farmland for access to water. These purchases, later referred to as water farming, were controversial among local farmers who claimed that the pumping led to water declines. These purchases typically were for land relatively near the city and within the same county. The water was in the same hydrologic basin. So, while there was controversy, it was localized. With the passage of the Groundwater Act of 1980, however, water farming was done on a larger scale and later was enabled by the CAP canal for conveyance. Before completion of the CAP, groundwater transfers within AMAs accounted for the largest portion of all types of water transfers in the state and were primarily transfers from the agricultural to the urban sector. In the late 1980s these purchases fueled controversy because of the more extensive economic, social, and environmental costs. Additional legislation and the availability of CAP water to meet urban needs reduced these kinds of transfers. Groundwater transfers today are more complicated than surface-water transfers because of the many different kinds of groundwater rights that exist; some rights are transferable out of the basin of origin (e.g., those grandfathered rights in AMAs that had not been used for irrigation, called Type II), while others are not (e.g., those outside of AMAs, and those within AMAs that had been previously used for irrigation but not since 1965, called Type I).[42]

In addition to surface water and groundwater, effluent has been sold in Arizona. Over the objections of downstream water rights holders, the Arizona Supreme Court, in *Arizona Public Service Company v. John F. Long*, ruled in 1989 that effluent is not considered surface water or groundwater and thereby belongs to the entity that generated it. This decision was subject only to the beneficial use requirement and effectively made effluent a source of transferable water. Cities in Arizona have been able to transfer effluent in various ways to generate revenue. Prescott Valley, Arizona, auctioned off its effluent in the form of groundwater credits to a developer (described in more detail later and in Chapter Four). Effluent has been transferred by the city of

---

[42] Colby and Jacobs, 2006; National Research Council, 1992.

Tucson, which sold effluent to farmers in the nearby Cortaro-Marana Irrigation district. Tucson also requires new golf courses to use it for irrigation. Phoenix, too, has sold effluent, notably to the Palo Verde nuclear power plant for use in its cooling towers. And effluent has been traded in Native American water-rights settlements in various ways (as trade for potable water or groundwater credits that may be used for CAP water), for example, with the Gila River Indian community and the Tohono O'odham tribe. In fact, of the 15 cities' utilities studied in Arizona, all were using effluent beneficially, most commonly for irrigation, high-water landscapes, and aquifer recharge.[43]

Water Auctions in Arizona

Auctions are one means of selling water. As was discussed in Chapter Four, the Prescott Valley, Arizona, auction held in 2007 is considered one of the most successful water-rights auctions. In the decade or so leading up to the auction (and since), Prescott Valley Town was in a high-growth area of Arizona, constrained by limited water-rights availability. To raise money for water infrastructure, the town held a traditional auction to sell groundwater credits that resulted from discharging newly available treated wastewater effluent into the aquifer.[44]

The town of Prescott Valley is in the Prescott AMA of Arizona, as discussed earlier, an area in which groundwater resources have special regulatory and management controls. In the Prescott AMA, the primary management goal is to equalize groundwater withdrawal with replacement by the year 2025.[45] As of 1999, Arizona water law required new developments to secure access to reliable water sources for the next 100 years, without which developers could not build. The town of Prescott Valley was especially constrained because of the area's high growth and limited options for new supplies of water (the town does not have access to water from the CAP, while its larger high-growth neighbor Phoenix does have access).

The town discharges its treated effluent into the dry bed of the Agua Fria River, where it helps to recharge the underground aquifer; qualifying the town for groundwater credits.[46] Since the new water rights made available by the town are in effect paper rights in the form of groundwater credits, they are more flexible than wet water (i.e., more easily transported). The

---

[43] Colby and Jacobs, 2006; National Research Council, 1992; Western Resource Advocates, "Arizona Water Meter," October 2010.

[44] The treated wastewater resulted from an upgrade to the town's wastewater treatment facility. There is irony in that the needed infrastructure was a pipeline to carry groundwater from another aquifer (within the same management area) to support already-permitted development in Prescott Valley.

[45] The smallest in land area of the five, the Prescott AMA is about 485 square miles in Yavapai County, as it is within the Verde River Watershed. It contains two sub-basins: the Little Chino sub-basin and the Upper Agua Fria sub-basin, each of which contains "economically significant amounts of groundwater found at depths ranging from just below the soil surface to a depth of over 500 feet." From Arizona Department of Water Resources, "Active Management Areas and Irrigation Nonexpansion Areas," undated; U.S. Soil Conservation Service, "Managing Arid and Semi Arid Watersheds: Prescott Active Management Areas."

[46] Friederici, 2007.

amount of water offered was 2,724 acre-feet, enough to support an additional 12,000 homes or other industrial, commercial, and recreational uses. So the offering of these water rights was attractive to developers and other potential water users in the water-poor area. However, establishing a price for these water rights was not easy, since the value is in part a function of anticipated growth and development in the area. By employing an auction to sell off the newly acquired water rights at "market" prices, the town hoped to generate enough income to fund its infrastructure requirements. After some trial and error, the auction was restructured and held late 2007. The successful bidder was a New York-based financial firm (Water Asset Management's subsidiary, Water Property Investors, LLC, offered the highest bid and successfully bid for the water at the cost of $24,500 per acre-foot). Water Property Investors, LLC was required to put the water to beneficial use in the town. About one-third of the water rights have since been sold to developers by the investment firm for undisclosed prices despite the slowing in the residential market.[47]

The price paid by the investment firm reflects the higher-scarcity value of water in the Prescott Valley Town and region; at these prices, water conservation becomes more economically viable. For comparison, the cities in the Phoenix area negotiated deals with nearby Native American communities to buy the use of tribal water for the next 100 years for $1,500 to $1,800 per acre-foot (recall they have access to the CAP infrastructure).[48]

At the time, this was the largest sale for the highest price paid for water rights, yielding more than $67 million for Prescott Valley Town. The auction allowed the town to raise money from its water resource, a resource whose true market price is not well known. An open process was used to derive the market price, which reflected the scarcity of water in the area and which also motivated conservation. To hold this auction, the town had to gain regulatory approvals, including clarification of water rights, and also benefitted from partnering with the private sector to sell these water rights successfully.[49]

Arizona Water Bank

The Arizona Water Bank was established in 1996 to enable the storage of excess Colorado River Water within the CAP system for future use during periods of low supply.[50] Until the Arizona

---

[47] In 2009, a developer from Denver purchased 200 acre-feet for an undisclosed price and a real estate investment firm from Scottsdale, Arizona, purchased 700 acre-feet for an undisclosed price. See Scott, 2012.

[48] Friederici, 2007.

[49] "Arizona Water Rights Auction Tops $20m," November 2007; "Arizona Town Holds Successful Effluent Water Auction," 2008; Friederici, 2007; Christia Gibbons, "Prescott Valley Water Auction Could Impact Phoenix," *Phoenix Business Journal*, October 29, 2006.; Scott, 2012.

[50] The AWBA procures excess CAP water supplies from the CAWCD. Water is allocated according to a priority system established by the U.S. Secretary of Interior as follows: municipal and industrial, Indian, non-Indian agriculture and miscellaneous. Therefore, excess CAP water is the most junior priority. The three Arizona counties within this system are Maricopa, Pinal and Pima. See Arizona Senate Research Staff, "Arizona State Senate Background Brief: Arizona Water Banking Authority," Phoenix, Ariz., November 15, 2010.

Water Bank was created, Arizona did not use its full 2.8 million acre feet allotment of Colorado River water (in part because farmers were able to pump groundwater more cheaply than the cost of CAP water). Most of Arizona's unused water would have gone to Southern California.[51]

The administering authority of the Arizona Water Bank, the AWBA, purchases excess water from the CAP at a price that the Central Arizona Water Conservation District sets each year.[52] The purchase is considered consumptive use and is not subject to non-use forfeiture. The AWBA is permitted to store this water within state owned underground facilities and aquifers (direct recharge), or irrigation districts are allowed to use the water in lieu of pumping groundwater (indirect recharge). In exchange for storing the water, the AWBA receives credits that it can then transfer at some future time either to the Central Arizona Groundwater Replenishment District (CAGWRD) or the Arizona Department of Water Resources (ADWR) for use for their municipal and industrial customers along the Central Arizona Project canal should they require it during dry periods (The amount transferred is reduced by a 5 percent conservation requirement.).[53] Therefore the Arizona Water Bank enables the use of Arizona's allotment of Colorado River water to replenish groundwater aquifers, implementing Arizona's groundwater management plan, and to firm water supplies for municipal and industrial users during periods of low water supply.

In addition to these benefits, the AWBA can contract with authorized entities in California and Nevada[54] to store Colorado River water for them for a fee, thereby replenishing Arizona's groundwater aquifers. Under these agreements, whenever either California or Nevada would need their stored water, they would draw the stored amount from the Colorado River. In 2005 the AWBA began storing water for the Southern Nevada Water Authority.[55] The AWBA also became responsible that year for working with Native American tribes to fulfill the state's obligation in firming their water supplies.

In 2010, 211,712 acre-feet of CAP water was recharged to the aquifers (19,000 acre-feet was on behalf of Nevada). Of the 2.78 MAF of Colorado River water used by Arizona in 2010, 1.13 MAF was used directly on the river, 1.44 MAF was used by CAP customers and .21 MAF by the AWBA. As of 2010, the cumulative total amount of Colorado River water stored by the bank was approximately 3.78 MAF.[56]

---

[51] Arizona Water Banking Authority, "Executive Summary," undated.

[52] Subsequent legislation in 1999 allows the AWBA to store effluent in addition to Colorado River water if there is excess capacity after the surplus Colorado River water is stored; to firm water supplies for non-CAP municipal and industrial users within the service area along the canal such as the Salt River Project, Maricopa Water District, and Roosevelt Water Conservation District; and to establish a lending process for water storage with other entities.

[53] The ADWR could also decide not to sell, but to retire these credits.

[54] Up to 1.25 million acre-feet for Nevada.

[55] Arizona Water Banking Authority, homepage, undated.

[56] Arizona Water Banking Authority, "Activities," undated.

Block Pricing

Block pricing is one means of encouraging conservation by increasing price. The key elements of block pricing, which were discussed in greater detail in a previous section, are the fixed fee (lower fees encourage conservation), block size (ideally based on historical water usage by type of customer), and price change from block to block (must be noticeable to water users to encourage conservation). Water rate structures were studied and compared in 15 Arizona communities. These communities, many of which had increasing block pricing, were Buckeye, Casa Grande, Chandler, Clarkdale, Lake Havasu City, Mesa, Payson, Peoria, Phoenix, Prescott, Safford, Scottsdale, Sierra Vista, Tucson, and Yuma. Their price structures, which include a fixed fee and cost per marginal gallon of water, are shown in Figure 6.9 (as of 2010). Notably, Prescott, Buckeye, and Tucson send strong conservation signals to customers, although Tucson allows for a fairly generous allotment of water in its first block, but increases rates significantly for exceeding this limit. Prescott prefers to use moderate marginal prices increases in smaller increments. At the other end of the spectrum lie Yuma, Lake Havasu City, and Sierra Vista (adjacent to Fort Huachuca), which do not send much of a conservation signal at all.[57]

Figure 6.9. Marginal Water Rates for Nine Arizona Communities (2010)

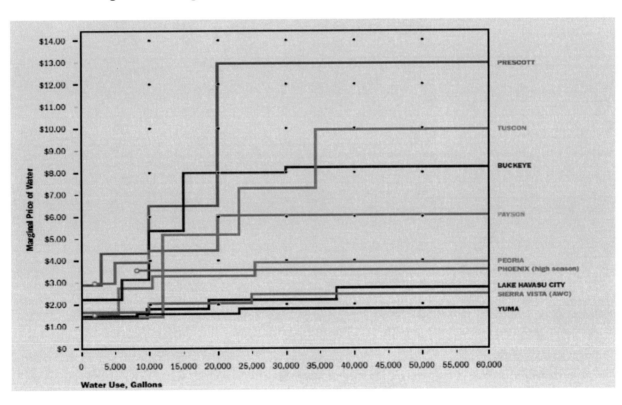

SOURCE: Western Resource Advocates, "Arizona Water Meter," October 2010.

---

[57] Western Resource Advocates, "Arizona Water Meter," October 2010.

263

Fort Huachuca Water Case Study

Next we present an overview of water activities at Fort Huachuca, which has one of the most progressive and successful installation water conservation programs within the Army. The program has been driven by federal lawsuits regarding concerns that the installation's pumping of groundwater was hurting endangered species, as is explained below. Between 1995 and 2010, Fort Huachuca reduced groundwater pumping by 60 percent through a range of water-conservation projects, including installing artificial turf on physical training (PT) fields, installing waterless urinals, implementing more efficient irrigation systems, and exploiting rainwater for irrigation of grassy areas.[58]

We begin by providing some background information about this fort and its mission. Then we discuss some of the water resource and rights challenges that the fort faces, given the water pressures in southeastern Arizona. Next, this section briefly provides an overview of federal water conservation requirements. Then we describe the water supply and infrastructure at Fort Huachuca. In this section, there is a detailed description of the wastewater treatment facility situation because it illustrates the key challenges that installations face in managing and maintaining water infrastructure. This is followed by a brief overview of Fort Huachuca water demand. Lastly, we give an overview of the fort's water program, highlighting some water efficiency, reclaimed water, and water reuse activities, as well as its funding challenges.

Background on Fort Huachuca

Fort Huachuca was founded in 1877 and is located in Cochise County, in southeast Arizona, about 15 miles north of the Mexican border. The city of Sierra Vista is located south and east of the post, and Huachuca City is to the north and east. Fort Huachuca also is located at the edge of Huachuca Mountains with the cantonment area located on a significant slope, which causes major stormwater runoff issues during the rainy season in July and August. The cantonment area contains more than 1,890 buildings.

In 2010, Fort Huachuca had a population of about 6,500 active duty Soldiers, 7,400 family members and 5,000 civilians. Fort Huachuca may have over 18,000 people on post during the day.

Fort Huachuca has a rich history. It is known as the fort from which Geronimo was captured and was the home of the Buffalo Soldiers, the 10th Cavalry Regiment. Construction of buildings on the main cantonment began in the 1880s. The adobe houses still onsite are historic buildings dating back to the late 1800s.

Historically, Fort Huachuca was established as the home of the "Signal Corps" because of its location at the edge of the mountains. The metal-bearing mountains serve as a "bowl" that attenuates all interference signals. This means that stray signals are kept out, while those

---

[58] Caprioli, 2010.

generated internally stay within the post. The Fort's central importance to intelligence persists today. For example, network security is a key Army mission based at Fort Huachuca, and the Network Enterprise Technology Command (NETCOM)/9th Signal Command is located here. Extensive C5ISR[59] activities are conducted at Huachuca. The post is located on approximately 73,100 acres of land, but it operates and uses an "electronic mitten" region over 2,600 square miles in size that includes the area on post. The restricted air space extends to the ground surface, and this unique airspace enables unmanned aerial training for both the Army and the Air Force.

Ongoing population growth near Fort Huachuca has raised concerns about potential interference with the post's testing and training missions. For instance, "Maintaining low levels of electronic interference and broad lines of sight in the immediate vicinity of the installation is critical to preserving the installation's training and testing mission."[60] To help prevent incompatible development near the post, Fort Huachuca has been participating in the Army's Compatible Use Buffer (ACUB) Program. Through this program, Fort Huachuca has partnered with The Nature Conservancy and other non-Army organizations to protect land from incompatible development, mostly through conservation easements.[61] These easement deals have restricted development by maintaining open space on nearby ranchlands and have retired agricultural land use. By September 30, 2012, 6,142 acres had been preserved. Mission and community benefits of these easements have included reducing current and future water use (which will be discussed more later); preserving on-installation maneuver, helicopter, and night flying training capability that generates noise pollution or requires minimal light pollution; reducing electromagnetic interference; preserving working lands and local character; providing habitat for endangered species; and supporting regional planning objectives.[62]

Fort Huachuca has more than 50 tenant organizations, including multiple commands that are not part of IMCOM and not directly under garrison command. Such tenants include the Intelligence Electronic Warfare Test Directorate (IEWTD), MEDCOM, the Joint Interoperability Test Command, Information Systems Engineering Command (ISEC), the Electronic Proving Ground and the Fort Huachuca Military Intelligence School (MIS). These tenants also frequently change. Increasingly, the post is likely to have fewer permanent tenants and more temporary, training-related tenants. For example, many members of the intelligence and unmanned aerial

---

[59] C5ISR stands for Command, Control, Communications, Computers, Combat Systems, Intelligence, Surveillance, and Reconnaissance.

[60] U.S. Army Environmental Command, "Fort Huachuca ACUB," Army Compatible Use Buffer Program fact sheet, undated.

[61] A conservation easement is a deed restriction landowners voluntarily place on their property to protect ordinarily in perpetuity the conservation values of the land.

[62] For more details on these benefits, see U.S. Army Readiness and Environmental Protection Integration (REPI) Program, "U.S. Army: Fort Huachuca: Arizona," fact sheet, undated. This site also is the source for the status of these transactions, including the acres preserved.

vehicle (UAV) communities will need to do some training onsite. The changing nature of the post's population has implications for water, which are explained below.

Water Rights Challenges at Fort Huachuca

To understand Fort Huachuca's water issues requires some background information about the water resources in the region and competition for water, legal challenges, and how the Endangered Species Act (ESA) has become a main driving force for water conservation there.

Water Competition in the Upper San Pedro River Basin

Fort Huachuca is located in the Sierra Vista sub-watershed within the Upper San Pedro River basin in the southwest of Arizona. Fort Huachuca borders the San Pedro Riparian National Conservation Area (SPRNCA) to the east. Also east of Fort Huachuca, the San Pedro River flows north across the Mexican border and then through SPRNCA; the Babocomari River borders the northern part of the post (see Figure 6.10).

The Upper San Pedro basin contains approximately 1,875 square miles and consists of the San Pedro River Valley and the surrounding mountains. The San Pedro River drains most of the basin's surface water. It enters Arizona at the boundary with Mexico near Palominas, Arizona, and flows 62 miles northwest before leaving the basin. The San Pedro River is frequently dry and flows only as a result of local precipitation, except for an 18-mile stretch that is created by groundwater forced to the surface by bedrock.[63] Since the 1990s, the region largely has been in a period of drought, with a few years of adequate rainfall in that period. This means that the contributions of precipitation to the flow of the San Pedro River have not been good. This interrelationship between the regional groundwater aquifer and the San Pedro River has been a key concern for the region, which is explained below.

---

[63] Arizona Department of Water Resources, "Upper San Pedro," undated.

Figure 6.10. Map of Fort Huachuca in Relationship to its Neighbors

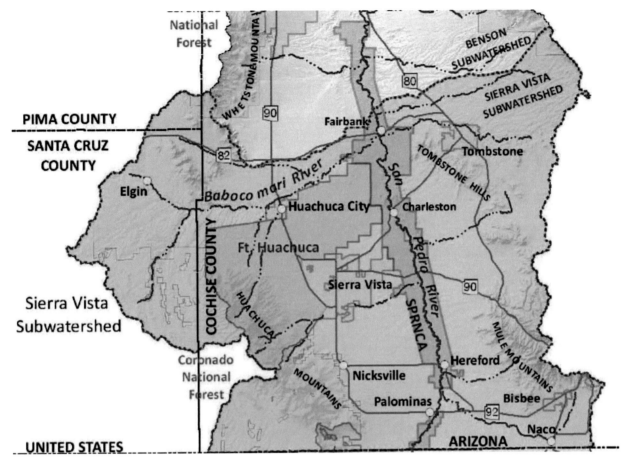

SOURCE: John Ruble, "Successfully Reducing Your Groundwater Footprint,"
PowerPoint presentation, DPW, Fort Huachuca, March 21, 2011.

Fort Huachuca originally was established in 1877 in part because of the local water resources. Huachuca Canyon, for example, has surface water most of the year, apart from the driest months of May and June immediately before the rainy season. Garden Canyon also has ephemeral surface water sources. Because of its water resources and southern location, many tropical birds can be found in the region; around 15 to 18 species of hummingbirds and the lesser long-nose bat, also a nectar-feeder, can be found locally. The area has 2,500 acres of forest, and the balance is grassland. Partly because of the numerous bird species and warm climate, the region has experienced a large growth in retirees and also has a large tourism industry. This population growth has created pressures on the water resources in the region.

Endangered Species Act Water Issues

The Upper San Pedro River (USPR) basin is an environmentally sensitive ecosystem and home to at least two endangered species, the Huachuca water umbel and the southwestern willow flycatcher, (see Figure 6.11). The water umbel is a semi-aquatic plant found in isolated sites in northern New Mexico and southeastern Arizona. The flycatcher is found across several states in

267

the region, but is threatened in large part by the loss of riparian vegetation and water diversions and groundwater pumping that threaten riparian vegetation.[64] Key habitat for both of these species occurs within the San Pedro Riparian National Conservation Area.

Figure 6.11. Huachuca Water Umbel and Southwestern Willow Flycatcher Are Endangered Species in the Upper San Pedro River Basin

SOURCES: Center for Biological Diversity, "Huachuca Water Umbel," undated; EarthJustice.org, "Court Rejects Fort Huachuca's Groundwater Pumping Plan," May 31, 2011.

Fort Huachuca's water comes from pumping groundwater within the USPR Basin, which has led to issues with the ESA for more than a decade. Declining groundwater could potentially decrease the amount of water in the San Pedro River, which could potentially affect the species' habitat. The ESA directs all federal agencies to work to conserve endangered and threatened species and to consult with the U.S. Fish and Wildlife Service (USFWS) on whether its activities could affect a species that is listed as threatened or endangered. If it appears that the agency's action may affect a listed species, that agency prepares a biological assessment (BA) to assist in its determination of the project's effect on a species. In places where it is needed, the USFWS prepares a biological opinion on whether the proposed activity will *jeopardize* the continued existence of a listed species. When USFWS makes a jeopardy determination, it also provides the consulting federal agency with reasonable and prudent alternative actions.[65] The USFWS has issued Biological Opinions in 1999, 2002, and 2007 regarding Fort Huachuca's groundwater pumping. Through Fort Huachuca's consultation with USFWS in 2006 and 2007, the fort identified its effect on the San Pedro River and developed a BA that outlined specific conservation requirements for the fort to achieve to reduce groundwater use to acceptable levels.

The USFWS 2007 Biological Opinion found that Fort Huachuca was not jeopardizing imperiled animal and plant species that depend on the San Pedro River. This opinion was based in part on the fort taking a variety of measures to mitigate its potential effect, including continuing to reduce groundwater use, participating in the Upper San Pedro Partnership, and

---

[64] U.S. Fish and Wildlife Service, "Southwestern Willow Flycatcher," Reno, Nev.: Nevada Fish and Wildlife Office, last updated September 28, 2012.

[65] For more details about the process that a federal agency goes through to respond to the ESA, see U.S. Fish and Wildlife Service "Endangered Species, Section 7 Consultation: A Brief Explanation," undated.

recognizing that the fort is responsible for some off-post groundwater use by Soldiers, family members, and civilians who work on the post but live off-post (explained further below).

The Center for Biological Diversity (CBD) and Maricopa Audubon Society have filed several lawsuits against Fort Huachuca, the DoD, and the USFWS claiming that Fort Huachuca's groundwater pumping has negatively affected the San Pedro River ecosystem. A ruling by the United States District Court for the District of Arizona in June 2011 struck down the 2007 biological opinion issued by the USFWS. As of August 2012, DPW personnel were working on a new programmatic BA. The rough draft has been submitted to USFWS for feedback. These reviews take up to 135 days, after which a biological opinion is issued.

Partly as a result of its ongoing legal challenges, Fort Huachuca has implemented an aggressive water-conservation program and efforts to help recharge the groundwater basin. In addition, Fort Huachuca has to calculate the fort's effect on water use in the region from both on-post and off-post fort-attributed water demand. Fort Huachuca is accountable for water usage "attributable" to its activities off the installation, including water used by Huachuca personnel and families as well as associated contractors and service providers.

Gila River Water Rights Challenge

Similarly, Fort Huachuca's water rights also are being challenged in the Gila River adjudication process, which could also affect the fort's mission. In 1974, the Salt River Project filed a claim in state court to water rights to the Salt and Gila Rivers. The San Pedro River is a tributary to the Gila River. This triggered a string of claimants filing for overlapping water rights, with about 83,000 individual claims eventually coming forward. Despite this large number, no water rights cases have been adjudicated in 38 years. In 1989, the fort initially filed a water rights claim to about 10,000 acre-feet per year of the Gila River flow.

Thus, the Gila River Adjudication has been pending since 1974. While this controversy continues at the time this report was written, the crux of the matter is the legal process of identifying and quantifying of water rights among thousands of parties to the case. As part of this process, Army and Department of Justice lawyers are actively pursing the Fort Huachuca's water rights. To this end, the Army has been successful in establishing a federal reserved water rights claim: in September 2011, the Special Master in charge of the adjudication issued a report that the fort was reserved federal land established for "military purposes," and that the parcels of land in question had establishment date of either 1881 or 1883. Given the nature of the installation and the early dates of establishment, the Special Master determined that the fort is thereby entitled to a quantity of water under the doctrine of implied federal reserved water rights. The determination of the exact quantity of water to which the fort is entitled, however, is still ongoing.

Given the scarcity of water in Arizona, everyone, including cities such as Huachuca, feels the need to protect its water rights. However, despite the water rights legal challenges, some in the region see Fort Huachuca as a net positive for water conservation; this is because the fort uses "only" 1,000 acre-feet, while other uses of the same land, such as suburban development, might withdraw much more. In fact, an alternative futures assessment in a Harvard University study concluded that the long-term sustainability of San Pedro River is actually *supported* by Fort Huachuca; private development would be expected to place a much bigger stress on the water supply and draw much more water.[66]

Federal, DoD, and Army Water Conservation Requirements

Fort Huachuca, like other Army installations, must meet federal, DoD, and Army water-conservation requirements. Signed in 2007, Executive Order 13423 directs each agency beginning in FY 2008 to reduce water consumption intensity, relative to the baseline of the agency's water consumption in FY 2007, through life-cycle cost-effective measures by 2 percent annually through the end of the fiscal year 2015 or 16 percent by the end of FY 2015.[67] The 2009 Army Energy Security Implementation Strategy also specifically identified water conservation as a related priority.[68] The 2010 *Department of Defense Strategic Sustainability Performance Plan, FY2010*, outlined annual targets for meeting the following goals:[69]

- Potable water consumption intensity by facilities reduced by 26 percent of FY 2007 levels by FY 2020
- Industrial and irrigation water consumption reduced by 20 percent of FY 2010 levels by FY 2020
- All development and redevelopment projects of 5,000 square feet or greater maintain pre-development hydrology to the maximum extent technically feasible.

Executive Order (EO) 13514 requires all federal agencies to set a baseline for potable water use at FY 2007 and reduce potable water use intensity (WUI) by 2 percent per year based on FY 2007 baseline through FY 2020, which is a total reduction of 26 percent.[70] Because of the pressure to reduce its groundwater use, Fort Huachuca already has an aggressive water conservation program and has made progress on these goals, as is discussed below.

Fort Huachuca Water Supply and Infrastructure

Next we discuss where and how Fort Huachuca gets the water it uses on post. Then we discuss the fort's wastewater treatment system and some of the challenges that it faces.

---

[66] Carl Steinitz et al., "Alternative Futures for Landscapes in the Upper San Pedro River Basin of Arizona and Sonora," U.S. Forest Service Gen. Tech. Rep. PSW-GTR-191, 2005.

[67] See Executive Office of the President, 2007.

[68] U.S. Department of the Army, 2009, p. 12.

[69] U.S. Department of Defense, 2010, pp. II-14-15.

[70] Even though such water conservation goals are Army-wide requirements, each installation is expected to strive toward meeting them at the individual installation-level. Namely, the Army has spread these goals equally to all installations rather than require more water conservation from some installations than others.

Fort Huachuca Water Sources

Fort Huachuca was one of the first Western settlements in the Upper San Pedro Basin. The post has some of the oldest reserved surface water claims in Arizona. The installation's surface water comes from ephemeral springs that are fed by snowmelt and runoff from the Huachuca Mountains. These surface waters were originally the only source of potable water at the post. However, by 1983, the fort had discontinued the use of springs as a source of potable water and switched to well water.

By 2009, Fort Huachuca had

> eight municipal water supply wells, ranging in depths from 710 to 1,230 feet. Two of the wells (800 gallons per minute (gpm) pump capacity) are located on the East Range and six wells (500–700 gpm pump capacity) are located between the main gate and the east gate. The east range wells are not chlorinated and are used just for non-drinking applications (flushing, showering).[71]

Despite a constant state of drought conditions for five or more years beginning in 2004, the six potable-water wells have been sufficient for Fort Huachuca's needs. The post has never needed to implement its "drought plan" for alternative water sources. Another five non-municipal wells produce less than 5 af per year, which supports military testing and research activities across the post. The fort owns and operates its own water and wastewater treatment systems.[72]

The potable water production wells at Fort Huachuca pull high-quality water from the aquifer. This water only requires the addition of chlorine and fluorine, prior to delivery in the distribution system. Testing of the supply occurs monthly. There are no plans to privatize any of the water infrastructure or its operations and maintenance (O&M). As it stands, the post does not have to pay for water to reuse and recharge, and post staff have not been instructed to privatize this commodity. Additionally, Fort Huachuca personnel's responsibility for water means it is easier for the personnel to justify related expenses and to control ESA concerns.

All of the housing at Fort Huachuca is privatized. Regardless, these facilities need to meet the same requirements for water use and recharge as the other Fort Huachuca facilities. Namely, all impervious surface water must be recharged. To meet these requirements, many of these dwellings must be rebuilt; in the meantime, the private owners and operators are maintaining the existing facilities until they can be rebuilt to specifications.

Fort Huachuca Wastewater Treatment Facility

In this subsection, we discuss the operations of Fort Huachuca's wastewater treatment facility. We discuss this facility in detail because it offers insights into a number of different topics. First, it illustrates the installation's challenges in trying to operate and maintain large-scale water

---

[71] Fort Huachuca Directorate of Public Works, "Water Resource Management Plan: Fort Huachuca, AZ," Fort Huachuca, Ariz., April 3, 2009.

[72] Fort Huachuca Directorate of Public Works, 2009.

infrastructure that is outdated and needs to be rebuilt to current needs. However, given the current fiscal climate, the installation cannot find the capital to invest in a new wastewater treatment facility. Second, it shows the creative uses of effluent, the significance of effluent uses, and how it is a valuable commodity for Fort Huachuca. Third, this example illustrates a creative public-to-public partnership in the water area. Lastly, this example shows how aggressive water conservation can create some unexpected consequences for wastewater infrastructure.

Fort Huachuca owns and operates its own wastewater treatment facility, which handles all wastewater from the post. Fort Huachuca's very successful water conservation efforts have, however, had unintended adverse effects on its wastewater treatment facilities. Two related problems have resulted: A decrease in the total volumes of water that are processed through the facility and an increase in concentrations of chemicals dissolved in the water. Fort Huachuca's wastewater facility was designed to treat 2 million gallons of water per day (gpd), or about 2,240 acre-feet per year,[73] a design level set in 2002–2003. As of August 2012, the facility was generally receiving less than 500,000 gpd (approximately 560 acre-feet per year) of wastewater influent and, as a result, was routinely having to supplement the flow with other sources to keep the system functioning.[74] This included the direct feed of potable water into the wastewater system, as much as feeding the system with 100,000 gpd of potable water for a week at a time. Additionally, the facility was only running eight to 10 hours per day, five days a week, further complicating successful operations and decreasing the overall efficiency of the system.

The concentration of effluent is both a function of the total population, which has stayed roughly constant or even increased, and the total volume of water, which has drastically decreased. Low-flow toilets and waterless urinals, for example, illustrate how waste and solute levels might stay the same, but water flow could decrease at the same time. The result at Fort Huachuca is a waste stream that frequently is as high as three times the "normal strength" of municipal effluent. "Total nitrogen" is problematic,[75] and elevated concentrations of nitrates are a frequent concern in the treated water leaving the facility. Other nitrogen species generally are present, but it is the nitrate levels that are monitored for compliance with both the Safe Water Drinking Act (SWDA) and the Clean Water Act (CWA). Currently, Fort Huachuca's wastewater effluent meets all SDWA requirements. However, nitrate levels are a special problem. Biological, chemical, and physical methods can be deployed to combat these issues and attempt

---

[73] One acre-foot is equal to 325,851.429 gallons.

[74] For example, as much as 450,000 gallons had been pulled from the bottom of the treatment facility basin to supplement flow. Also, significant quantities of treated wastewater, up to 58 percent per day, were being re-circulated.

[75] "Total nitrogen" includes nitrites (NO_2^-), nitrates (NO_3^-), and TKN, the "total Kjeldahl nitrogen," which measures the total organic nitrogen, ammonia (NH_3), and ammonium (NH_4^+) in effluent. NH_4^+ in human waste is converted first to NO_3^- and then to NO_2^-. NO_2^- acts as a fertilizer and, when present in high concentrations in discharged effluent, can lead to overgrowth of microorganisms and eutrophication.

to remove some of the problematic solutes. However, biological species can become overloaded or even experience toxic levels of solutes under these conditions.

Another potential consequence of water conservation and increased concentrations is that the longevity of the collection system, including sewage lines, can be compromised. The lines may corrode in the presence of elevated levels of sulfuric acid, which is formed when hydrogen sulfide (H_2S) and oxygen react in pipes. H_2S can form when anaerobic bacteria break down organic material in sewer systems.

Mitigation Strategies at the Wastewater Treatment Facility

Several mitigation strategies were being implemented or planned as of August 2012. Siemens designed the original plant, and the company has been working with Huachuca staff to implement a number of fixes, all of which are admittedly non-optimal. In fact, they could be considered band-aid solutions because they do not deal with the fundamental mismatch in influent supply with facility size, leading to water volume and concentration issues.

One set of concepts for improving performance involved augmenting the breakdown or removal of solutes. For example, live enzymes were being added to the wastewater to enhance the natural biological breakdown of chemicals of concern. This bacterial additive, mixed into the waste stream prior to entering the plant, had reduced the amount of ammonia (NH_4^+) in the influent by about 20 to 30 percent. Increasing aeration, so that oxygen can break down solutes and so that microorganism populations are better maintained, also was being employed to try to reduce concentrations. Other concepts—such as commercial ammonia strippers—are available, but require ongoing maintenance and are expensive. Simple, inexpensive fixes, such as sink traps for food service operations that decrease the organic load so that it cannot contribute to the already high concentrations, were being identified and implemented across the post.

A second family of modifications would be to augment the flow of the "underhydrated" facility, to dilute the waste stream by a number of approaches. For example, modifications to recirculate the effluent had been inexpensive (on the order of $2,000) and effective, and had enabled the installation to push more costly modifications down the road by two years or more.[76] Modifying the facility's equalization basins to achieve a more "steady state" flow and loading level was proving to be more expensive—all told, a multimillion-dollar project.[77] In the meantime, the operators had begun adding 100,000 gpd of potable water in an attempt to stabilize flow volumes and decrease concentrations of nitrates. However, operators thought that realistically they would need to have about 700,000 gallons per day total at a minimum to achieve proper function and attain the discharge limit of 10 milligrams nitrate per gallon. This would mean that perhaps as much as 200,000 to 300,000 gallons per day of supplemental flow would be required. These parameters were being tested as of August 2012.

---

[76] For example, a more complex aeration and circulation scheme had been estimated to cost $230,000.

[77] The estimate for the equalization basin project was $9.5 million.

A typical ratio of potable water supplied to a system, relative to what returns as effluent, is about 3:2. That means that about 60 percent of the water volume supplied to homes and businesses returns as wastewater to be treated. Areas of loss include irrigation, washing cars, supply pipes, and other evaporation and leakage pathways. Fort Huachuca had decreased losses in its system from those in previous years, so that by late 2012 it was experiencing a 60 percent or even higher return rate—up to as high as 68 percent. Consequently, these type of losses are not a likely place to reduce losses and increase volumes returned to the effluent treatment plant and not a focus of DPW personnel strategies to improve operations at the facility.

In Arizona, a facility can detain stormwater for some usages, but it cannot directly add this captured water to its wastewater treatment influent (i.e., mix sewage and stormwater) to augment the flow volumes and decrease concentrations. As a result, this option was not open to Fort Huachuca for dilution or for increasing the volume of flow.

Another concept in dilution being considered was to partner with Huachuca City to make use of nearby Huachuca City's water effluent. The lagoons of the city's own wastewater treatment facility were located in the flood plains of the Babocomari River, and a consent decree was issued for these to be decommissioned.[78] Fort Huachuca had begun to negotiate a deal to bring in the effluent from the city. Several years earlier, the city had applied for a U.S. EPA grant for the project, but this had fallen through. As part of this process, Huachuca staff had assisted Huachuca City staff in working through the National Environmental Policy Act (NEPA) process and documentation. In the summer of 2012, the project was looking like it might come to fruition as a MCA (Military Construction, Army) project.

As of August 2012, the environmental assessment (EA) for the Huachuca City effluent diversion had been completed, and the public comment period had elapsed. Because the town administrative staff is quite small, Huachuca staff had prepared and issued a request for proposals (RFP) for a contractor to execute the project. The project would entail delivering the wastewater by pipe and installing and integrating proper electronics and controls to manage this new waste stream. The fort's end of the pipe has already been constructed, and would require some minor retesting at the time of being connected to the city's pipe. This multipurpose public-to-public partnership meant that Fort Huachuca was (1) improving the operations of its own wastewater treatment facility (by making use of excess capacity), (2) increasing reuse or recharge and working toward meeting Endangered Species Act (ESA) goals, and (3) helping Huachuca City obviate its need for a new wastewater treatment facility. This example shows how creative partnerships with local governments for sharing water infrastructure are doable and can have benefits for both the installation and local government. However, Huachuca City's sewage volume is relatively small, at about 90 acre-feet per year.[79] This means that use of this

---

[78] The state of Arizona issued a "Notice of Opportunity to Correct" to the city's wastewater facility, but this case was resolved and closed due to the plan to reroute the effluent to Fort Huachuca.

[79] Recall that the typical volume at the Fort Huachuca facility was about 560 acre-feet per year.

water will have a small effect on the total volumes moving through the treatment facility at Fort Huachuca.

Ultimately, the best solution for Fort Huachuca's Wastewater Treatment Plant problems would be the construction of a new appropriately-sized facility. Such a facility could make use of some of the upgrades (e.g., equalization basins), but would address the fundamental supply-sizing mismatch. Increases in energy efficiency also would be expected with a more appropriately sized facility that could run consistently. However, a new, state-of-the-art wastewater treatment plant was expected to cost on the order of $65 million to $85 million, and would require MILCON funding. As of August 2012, Fort Huachuca had nothing new in MCA dollars through FY 2017 or 2018, so this was not expected to be an option in the near term. Lack of funding for a new facility that would meet the wastewater challenges and save energy illustrates a fundamental challenge that installations face, not being able to find funds for needed capital investments in large-scale infrastructure, which is one of the main reasons why the DoD has been privatizing water and wastewater treatment plants.

Uses of Treated Water from the Wastewater Facility

The wastewater treatment facility is not permitted to discharge any water. Instead, all water must be routed to reuse or recharge purposes. All wastewater is treated to U.S. EPA Class B or B+ quality. This treated effluent is used to irrigate the golf course and the remaining excess is used to recharge the aquifer.[80] Recharge amounts are expected to increase when Huachuca City's effluent is redirected to the facility.

During summer 2012, DPW personnel were examining a major recharge basin project, which was expected to cost about $6 million. This gravity-fed basin would receive treated water, hold the water for testing,[81] and then allow the water to filter back through the ground into the aquifer.

Another output of the wastewater facility is the biosolids that are collected. Polymer is added to this waste and, after dewatering, the resulting material is buried within 24 hours. Tipping fees associated with this waste are about $100,000 per year, and it is taken to the county landfill about 15 miles away. The biosolids are difficult to dry with the existing system, and about 80 percent of the weight is water even after dewatering (i.e., only 20 percent is dry solid material).

Fort Huachuca Water Usage

On-post water usage has decreased over the years because of Fort Huachuca's aggressive water conservation activities while the post population has gone up. Between 1992 and 2010, the total water savings have been 2,065 acre-feet/year (673M gallons) and the associated energy cost

---

[80] Timothy Faulkner, "Sustainable Water: Reaching Net Zero," *U.S. Army Journal of Installation Management,* Spring 2011, p. 22.

[81] As noted elsewhere, treated effluent nitrate levels are a concern for recharge at treatment levels that are not problematic for reuses such as irrigation.

avoidance $263,000 (see Figure 6.12).[82] The peak year for water withdrawals was in 1989, in which the post withdrew a net 32,070 acre-feet of water. Significant water reduction efforts then were instituted, so that by about 10 years ago, the post was drawing much less: about 10,000 to 12,000 acre-feet per year (afy) of attributable water.[83] As of August 2012, this deficit has been reduced to about 4,000 afy. This success has largely been a result of on-post projects and activities, which the water personnel note has resulted in water use on post that is "under control."

Figure 6.12. Groundwater Pumping and Population Trends at Fort Huachuca

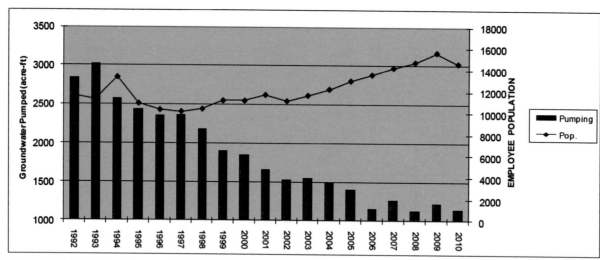

SOURCE: John Ruble, "Successfully Reducing Your Groundwater Footprint," PowerPoint presentation, DPW, Fort Huachuca, 2011.

Estimating Off-post Water Usage

To understand the water usage on- and off-post, the Huachuca personnel gather population data and employment information annually on September 30. This enables them to understand and analyze the demographics and water use implications. There are multiple categories of attribution for water use: direct, indirect, and induced populations associated with Fort Huachuca that all contribute to net water usage. Next, we give an example of this calculation from a few years ago to illustrate the methodology and the complexity involved in trying to accurately estimate water use. These are not exact numbers, only rough-order-of-magnitude numbers used to illustrate the process. Given the dynamic nature of the different tenant populations at the post, the real numbers change from year to year.

For this example, Fort Huachuca had about 1,000 on-site military houses and barracks for year-round and visiting populations. About 20 percent of the directly attributable population

---

[82] Ruble, 2011.

[83] This is a net withdrawal rate, which meant that their pumping rate from the aquifer exceeded their recharge rate to the aquifer by this amount.

lived on post. Directly attributable water usage includes water use by all individuals employed at Fort Huachuca. This means the "direct population" includes military personnel, civilian DoD employees, and contractors. It also includes the more than 20,000 students who spend part of the year at Huachuca; these students alone total up to about 3,000 to 4,000 full-time equivalents annually. The directly attributable population, or "direct population," is therefore about 15,000.[84] The "indirect population" includes the family and household members of military, civilian, and contractor personnel who live on and off the fort. The "induced population" includes individuals not directly working at or for Fort Huachuca but those, for example, who work in service industries that support the larger regional infrastructure who would not be employed in these positions were it not for the existence of Fort Huachuca. This total of all direct, indirect, and induced population was estimated to be about 33,000 to 34,000 people.[85] In practice, a large part of the local population was attributable, with the exclusion of non-military retirees in the region and some influx of tourism due to local birding interest, among others.

On-Post Water Demand

Fort Huachuca has assessed sources for the installation's on-post water demand, and it can be broken down into four water use categories:

- Domestic water consumption
- Commercial water consumption
- Landscape irrigation
- Unaccounted for uses and losses.

Domestic water consumption refers to water used for household purposes, such as bathing, drinking, food preparation, washing dishes and clothes, cleaning the house, flushing toilets, car washing, and watering lawns. This category includes dormitories, temporary lodging facilities and multi-family and single-family housing. Most of the water is used for bathing, cleaning, and cooking. In 2005, this category was about one-third of the post's total water demand (see Table 6.2).[86]

Commercial water consumption is associated with installation facility operations at Fort Huachuca, including water use in restrooms, HVAC systems, dining facilities, and ice machines/commissary. It also includes water consumption at recreational facilities, in construction activities, for fire training, and in vehicle maintenance and washing. Construction

---

[84] In accounting for off-post water use, the post gets a "debit" for withdrawal and a "credit" for recharge in the form of returns to the municipal wastewater facilities for the approximately 80 percent of directly attributable water use that is offsite.

[85] Note that the Huachuca calculations are conducted in a manner that eliminates double counting (e.g., spouses who both work and live at Fort Huachuca).

[86] For more details on the breakdown of these different usages, see Fort Huachuca Directorate of Public Works, 2009, pp. 5–8.

activities and HVAC systems are by far the two biggest users in this category. In 2005, this category was about 24 percent of the post's water demand (see Table 6.2).

Table 6.2. Summary of Fort Huachuca Water Consumption in 2005

Major Use	Annual Consumption (af/year)	Percent of Total
Domestic Water Consumption	467.60	33.33%
Commercial Water Consumption	332.20	23.68%
Landscape Irrigation	106.24	7.57%
Unaccounted for Uses and Losses	496.96	35.42%
Total	1,403	100%

SOURCE: Fort Huachuca Directorate of Public Works, 2009, page 8.

Landscape irrigation is a large consumptive use of water at Fort Huachuca. Water is used to irrigate the golf course, cemetery, and recreational fields. However, Fort Huachuca has been much more aggressive than most other Army posts at Xeriscaping, replacing some athletic fields with artificial turf, and other activities to reduce landscaping water requirements (which is discussed more below). In 2005, such irrigation accounted for only about 8 percent of the post's water consumption (see Table 6.2). Note that Military Family Housing irrigation uses were included under the Domestic Water Use category and are not included in this 8 percent.

The last category of water demand is Unaccounted for Uses and Losses, which is mostly water losses from leaks. Fort Huachuca has an older water distribution system that is prone to leakage, a common problem with military installations. In 2005, this category was about 35 percent of total water demand at Fort Huachuca (see Table 6.2 again). As part of its water-conservation activities, Fort Huachuca has been aggressively trying to identify and fix such water leaks, which is explained in more depth below.

Fort Huachuca Water Program

In this section, we provide an overview of the Fort Huachuca water program. We begin this section by briefly mentioning staff that has water responsibilities. Then we provide an overview of the installation's Water Program goals, approach, and philosophy. Next, a brief overview is presented of some of Fort Huachuca's water efficiency projects. Because it is a major issue with water market trends, we discuss how Fort Huachuca is taking advantage of alternative water sources. The post is doing some innovative things in this area. Lastly, we briefly discuss some of the issues with funding for the Water Program, as this is a fundamental problem for installations.

Fort Huachuca Water Staff

The main responsibility for water-related issues is shared between the Environmental and Natural Resources Division (ENRD) and the Engineering Division within the Directorate of

Public Works (DPW). The Engineering Division has both an Energy Branch and a Water Conservation Branch, which sometimes reside in Environment Divisions at some other garrisons. There is significant cross-division collaboration. For example, ENRD has a hydrologist on staff; 75 percent of this person's time is spent on water conservation, drainage, and erosion issues. Much of this is compliance-relevant work for the ESA issues. A civil engineer in the Engineering Division, in turn, works with this hydrologist on these issues. Additionally, legal personnel—including an environmental attorney—help with water issues. O&M personnel also provide support for some water issues.

The wastewater treatment facility has several staff members, including a lab analyst for wastewater treatment process control and the water compliance manager for the facility.

The DPW personnel also sometimes call upon outside experts to help with water issues. For example, the development of a BA required involvement by two attorneys from the Army Environmental Law Division, one outside biologist, and others, including a staff member at ACSIM to coordinate the activities. Additionally, DPW personnel noted that senior commanders have been very helpful and that they understand the importance of water issues. In fact, new commanders take a course to learn about relevant issues, including ESA requirements, compliance, and reporting issues. However, one of the "biggest challenges in water" for Fort Huachuca was a lack of dedicated staff for water issues; most personnel who were supposed to be full time for water had responsibilities for some other functions, such as emergency operations. Other water personnel saw their time spread across many duties and had many competing demands for their time. For example, DPW personnel do not usually have the time to conduct in-depth, life- cycle cost analyses related to their larger project ideas, so they generally are unable to justify expensive projects that are not part of other infrastructure improvements and that also are directly relevant to ESA requirements. Because they lack the personnel to conduct a systematic, site-wide water design, their projects are "one-offs" that do not necessarily identify the most cost-effective projects or the most important ones. Additionally, budget constraints had eliminated a number of positions related to the water program and produced a 15 percent cut across IMCOM for FY13 relative to the FY 12 TDA. On the other hand, the DPW personnel interviewed for this study felt that they had "good control" over the projects they were able to undertake, and that they were able to do a substantial amount of improvement given the level of resources.

Overview of Fort Huachuca's Water Program

In April 2009, Fort Huachuca released its Water Resource Management Plan to promote the efficient use of water and to meet the water conservation requirements established in the 2006 BA and the USFWS 2007 Biological Opinion (BO). This plan also is based on guidance in EO 13423, OSD and Army regulations as discussed earlier. This plan lays out steps for Fort Huachuca to meet its attributable groundwater demand of 1,418 acre-feet per year (af/year), based on its population as part of the Sierra Vista sub-watershed, by the year 2016. The fort's proposed mitigation strategy included:

> 1) conservation measures totaling 116 acre-feet of reduced pumping and 836 acre-feet of increased artificial recharge on post by 2016, and

> 2) conservation easements purchased by the Fort that eliminate 1073 acre-feet of pumping near the San Pedro River.[87]

Fort Huachuca has developed a Groundwater Resource Management System, along with the water resource management plan, to lay out strategies to improve water use efficiency both on- and off-post. The post's Groundwater Resource Management System program includes the management plan, goals, metrics, community education, and feedback as well as legal, policy, education, technology, community outreach and partnership elements.[88]

The post also has developed and implemented the Fort Huachuca Irrigation and Water Management Policy. Key parts of this policy include restrictions on exterior water use (such as establishing a strict watering schedule and placing limits on new turf); prohibitions on wasteful water use practices (such as decorative water features and unattended water hoses); and rapid responses to water leaks and the reporting of wasteful water use practices.[89]

The four pillars of Fort Huachuca's Water Program implementation are education, conservation, reuse, and recharge.[90] In fact, this integrated approach is the key to Huachuca's water successes to date. For instance, in addition to successful water-conservation efforts, which have greatly reduced the demand for water, and specifically for potable water, Fort Huachuca has been very aggressive and innovative in the area of water reuse and recharge (discussed more below). As one Fort Huachuca staff member noted, "Everything we do with water is multipurpose." Water recharge, water management, and flood control often are addressed in a single project. For example, all water that is treated in the fort's wastewater treatment facility is reused or recharged to the aquifer. As one staff member noted, "every opportunity" is taken to improve water utilization on post.

Fort Huachuca has made real progress, not just on reducing the use of groundwater, but also in groundwater basin recharge and water reuse. The post also has been aggressively obtaining conservation easements to retire agricultural pumping within the region and prevent residential development on such lands. Figure 6.13 shows Fort Huachuca water trendlines for these different activities from 1994 through 2010. Groundwater pumping has decreased by more than 64 percent between 1989 and 2010, from 3,207 to 1,142 afa. In that same time period, conservation

---

[87] Fort Huachuca Directorate of Public Works, 2009.

[88] Faulkner, 2011, pp. 19–20.

[89] Ruble, 2011.

[90] Faulkner, 2011, p. 21.

easements have retired 1,073 afa of agricultural pumping and approximately 200 afa of avoiding future residential pumping on these properties.[91]

Figure 6.13. Fort Huachuca Water Trendlines

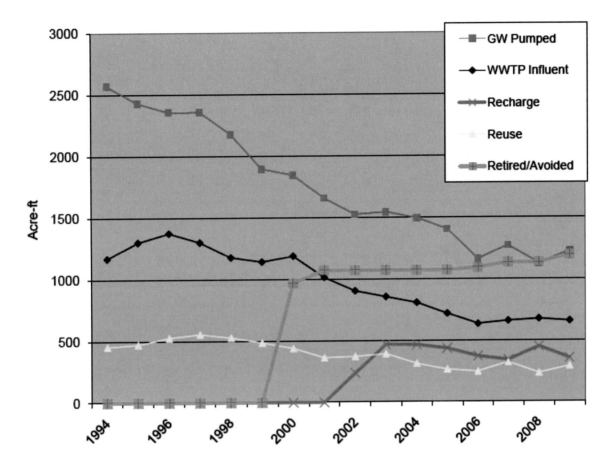

SOURCE: John Ruble, "Successfully Reducing Your Groundwater Footprint," PowerPoint presentation, DPW, Fort Huachuca, 2011, p.18.

Water Efficiency Projects

Over the past 20 years, Fort Huachuca has implemented a range of projects and best management practices to reduce water consumption. Here is a list of some of them:

- Installing artificial turf at different locations across the post, including child-care facilities and physical training fields
- Leak detection and repair or replacement of pipes
- Installing water meters for recharge basin flow measurements, calibrated water well meters for flow usage, and repaired water system controls
- Replacing hot-water heating and cold water cooling lines south of Central Energy Plant and other areas

---

[91] Ruble, 2011.

- Treated effluent being used to irrigate Fort Huachuca golf course: installed an effluent reuse water holding pond and pumping facilities for irrigating the golf course
- Installing or retro-fitting nearly 500 waterless urinals
- Replacing evaporative coolers in 10 test areas and other buildings with air conditioning systems
- Broadly implementing low-flow technologies, such as horizontal axis washing machines
- Repairing HVAC controls within North, South, and West Central Energy Plants to reduce energy and water consumption
- Installing digital mammography and replacing Mobile X-Ray
- Installing a detention basin to detain and recharge stormwater
- Implementing rainwater harvesting, such as rooftop capture at Barnes Field House
- Connecting overflow pipe from 1.5 MG reservoir to sewer to capture spillage during reservoir cleaning. After processing at the wastewater treatment plant, this water will be recycled for irrigation and other non-potable uses
- Repairing and restoring existing rainfall recharge basins.[92]

Some selected water projects are worth highlighting because they did something especially progressive or challenging for the Army, such as using artificial turf, and they provide useful lessons for other installations in water-conservation project implementation.

One innovative water conservation strategy was the implementation of strict design guides for landscaping, allowing only native and low water use plants. General use of artificial turf and Xeriscaping was implemented post-wide when possible. For example, the child-care facility used artificial turf. Training facilities have also been designed to use less water and less grass. For example, the physical training (PT) field at Huachuca uses artificial turf rather than a natural grass surface. Not only does this decrease water use and maintenance costs (e.g., snow melts more readily in winter on these surfaces, and mowing is not required[93]), but water from the field can be captured and recharged. The turf also has proven to be much safer and reduces injury in training. Despite some lack of Army-wide buy-in on these training surfaces, the advantages were clear to Huachuca personnel, including little or no O&M (only raking and re-pelleting every few years), no watering, and fewer injuries. As the Fort Huachuca garrison commander, COL Timothy Faulkner, wrote in 2011, "Artificial turf has the ancillary benefit of saving grass maintenance, fuel cost and provides the Soldiers the great hand-to-hand combat training areas with proper cushioning."[94]

Similarly, the obstacle course on site was Xeriscaped to reduce maintenance and eliminate the need to water in the dry seasons. With the addition of proper drainage and a water-retainment system, the course could be used year-round. Previously, the rainy season would cause standing-

---

[92] This list mainly came from: U.S. Department of the Army, *Installation Management Water Portfolio: 2011–2017*, April 2011, and Faulkner, 2011.

[93] Construction at a grade of 1 percent or less, and with sufficient under drainage, allows these surfaces not to washout during monsoons, further reducing maintenance.

[94] Faulkner, 2011, p. 22.

water problems, limiting use of the course and leading to injuries (from mud and holes on the course). While the initial response to such changes can be negative, the DPW personnel at Huachuca have found that the units have not only accepted these changes, but also prefer them.

Another water conservation effort has been phasing out the use of evaporative coolers ("swamp coolers") on the post. In fact, these coolers largely had been replaced. This meant small increases in electricity bills, but significantly reduced water use. Given the arid climate, and typical dew points of 50 percent to 60 percent—and as low at 3 percent immediately before the rainy season—evaporative cooling was not appropriate.

Fort Huachuca personnel learned some important lessons about the role that spare parts and maintenance can play in water-conservation project implementation. One recommendation that arose was to ensure that spare parts for older, less water-efficient equipment be removed from stock. DPW personnel had found that, although they had moved to all 1.2 to 1.4 gallon-per-flush toilets as well as some dual-flush toilets, O&M personnel had, in some cases, performed repairs with old 4 gallon-per-flush spare parts and inadvertently had changed the post back to less water-efficient operations.

Residential Communities Initiative (RCI) privatized housing contractors also have been helping to control water use. While water meters were installed in every RCI home, no charge was associated with using a baseline amount of water. In a revenue-neutral mechanism, water use above the baseline level incurred a fee, and water use below the level resulted in an incentive payment, creating incentives to use water efficiently. Fort Huachuca did not collect any money, but incentives were appropriate. Additionally, the regional and local water education programs, such as "Water Wise and Energy Smart," helped motivate water conservation in housing areas.

Fort Huachuca personnel noted that their relatively aggressive water efficiency efforts to date means only so much more can be done to conserve water without eventually influencing the quality of life. For example, on Fort Huachuca, the per-capita water use on post is 52 gallons per day, increasing to about 70 gallons per day when industrial and commercial water use is included. For comparison, private water utilities often report upwards of 80 gallons per day for private users, and more than 140 gallons per day when including industrial uses, including golf courses. Some Arizona cities, in fact, were using 200 gallons per day or more (see Figure 6.5). This high level of water efficiency on Fort Huachuca means that, while their water efficiency efforts will continue, the focus increasingly will be on making use of reclaimed water, as described below. And as noted, the irony of more aggressive water conservation is that it "hurts" operations at the water-treatment facilities.

In addition to nearing the limits of water efficiency that potentially will begin affecting quality of life, the reductions in water use have been so effective that they had uncovered unexpected problems associated with drastically reducing water volumes through an existing water treatment system, as discussed earlier. DPW personnel noted that these problems are "almost hidden" until the regime of very high water conservation becomes established. They referred to themselves as "the poster children" for the problems that other installations might

283

encounter down the road, and should be aware of, when they embark on aggressive water-efficiency policies.

Taking Advantage of Alternative Water Sources

Next we discuss how Fort Huachuca has tried to take advantage of reclaimed water and how it reuses wastewater and conducts basin water recharging. Then we discuss stormwater runoff, an important water issue that Fort Huachuca faces that is difficult to address given its funding constraints.

Reclaimed Water

Wastewater effluent treated onsite and stormwater collected onsite are both valuable sources of reclaimed water for subsequent use on post. Water reuse (for such purposes as irrigation and greywater sanitary applications) is an option for treated effluent, but not all reuse is an option for collected stormwater. On the other hand, water can be recharged from either of these sources. So the first priority with treated effluent is generally reuse; excess treated wastewater has the potential for being used as recharge. Water recharge, however, was mainly sourced from collected stormwater, because of the demand for the treated water for reuse purposes.

Fort Huachuca personnel also noted that greywater use for water reuse applications actually was not beneficial to their system overall. This was because it actually increased problems of effluent treatment by reducing overall volumes at the treatment facility. Therefore, they had no incentive to reuse greywater, for example, in cases such as reusing laundry water for flushing toilets. The take-home message was that if a facility already is conserving aggressively but lacks funding to retrofit the system to handle these lower volumes of water, reusing greywater may actually exacerbate problems.

Fort Huachuca staff noted the attractiveness of the concept of capturing stormwater, reusing the stormwater, and then recharging the stormwater after treatment. Some 12,000 to 14,000 acres of impervious surfaces are located on post. If the water runoff from all of these surfaces were used, for example, as an input for the wastewater treatment facility, it eventually would be returned to recharge the aquifer but would, in the interim, enhance the performance of the wastewater treatment facility. This was not an option for this water, which was required to be directly recharged and was not allowed to be captured for reuse in Arizona. However, an example of an apparently legal option to increase flows through the sanitary system without directly routing stormwater to the treatment plant involves the rooftop capture of rainwater being reused for flushing toilets and then making its way to the wastewater treatment facilities. Unfortunately, the state requirement that storm and sanitary systems be completely separate complicated the execution of these concepts.

Water Reuse

In addition to conservation efforts, water reuse figures prominently in Huachuca's water planning strategy. Water reuse supplies all of the water needs for the golf course on post, which is open to both Huachuca personnel and the public. As a result of the volume decreases at the treatment facility, the golf course has been the main end user of wastewater from the facility. The Huachuca course is a "desert course," which means that it has many non-irrigated areas (such as the first 75 yards), artificial turf is used, and the greens are the only irrigated areas requiring water. Irrigation must occur at night.[95] Nitrate had been an issue of concern to Huachuca personnel with treated effluent being used for irrigation, but in this case, it fertilizes the course. This reuse is relatively straightforward and non-problematic; the only complaint is the management consideration of nighttime watering.

Although this facility at times has been criticized for it's water use, the ENRD staff note that because Huachuca-affiliated individuals and the public both use the course, its existence likely preempts the need to build additional private golf courses that might not be as aggressively reusing water. For example, the fort's course has a 250 acre-foot per-year limit, while an existing private course in town—part of which potentially could be considered part of Huachuca's "attributable pumping"—uses 400 to 500 acre-feet per year of potable water.

As of August 2012, the reduction in treated wastewater volumes meant that most if not all of the treated water onsite was destined for reuse. However, the possibility of increasing the volumes of this water by collection of Huachuca City effluent means that this might not always be the case. Consequently, in the future, Huachuca personnel might need to decide at some points of the year whether to use the water for recharge.

Water Recharge

The concept of water recharge involves capturing stormwater, or using treated wastewater, and allowing it to slowly drain back into the aquifer from the surface, using the natural filtering system of the layers of rocks and earth between the surface and the aquifer to purify the water. The slow drainage can occur from natural wetlands or from human-made collection basins. These projects are not actually "water conservation" projects, but a mitigation approach in which the water from the aquifer is replenished at a greater rate than would have occurred without the project. These projects also can be important for stormwater management, which is another significant concern for Fort Huachuca (discussed below). Namely, such projects have a dual benefit of recharging the aquifer and helping to manage stormwater runoff. While conservation reduces withdrawals, this approach keeps water onsite and returns it to the water table by recharging wells. Water recharge projects have been implemented at various locations at Fort Huachuca. On the east range, a number of recharge basins are in remote locations. Such

---

[95] This is a requirement for all treated effluent at this level and is a condition of the permit to use treated effluent on the golf course.

detention basins, referred to as "pit parks," are located across the post. These reduce flooding and allow the recharge of approximately 1 acre-foot to 2 acre-feet per year for every basin.

Conservation easements are another way that Fort Huachuca is increasing its recharge credits. For example, 3,900 acres of conservation easements were acquired on the east side of the post just outside of the fence. Another conservation easement is along the Babocomari River border. In a conservation easement purchased at the southeast edge of the post, a deal was made with the county in which the fort gets the recharge credit but the county assumes ownership of the land. As discussed earlier, such a land deal has mission benefits by preserving open space and preventing incompatible development. It also has the water recharge benefits because it takes land out of agricultural use and can prevent new development water use.

A third example of recharge projects involves those that are more integrated with the post's infrastructure, such as smaller collection basins alongside buildings, roads, parking lots, and sidewalks. As noted, all water from all such impervious surfaces must be recharged to the water table. The idea in this case is to recharge the aquifer as well as to reduce and "peak shave" the flooding that occurs during the rainy season. The first such project was an "experiment" at the Burger King renovation site, which presented an opportunity for incremental additions to water recharge in the parking lot. The success of this project has led to others, such as roadside collection basins that have been integrated with Xeriscaping along some of the roads that slope toward one of the problematic drainage pathways that creates a river of drainage during rainy season. This not only allows for recharge but also reduces the erosion that is occurring along this drainage pathway and is threatening to undercut a road further downhill.

Another recharge example is the parking lot at the post commissary that had previously experienced serious flooding problems during the monsoon season. It was being rebuilt in August 2012 to better manage stormwater runoff and includes intermittent, smaller collection basins between parking rows and a larger basin at the location of the worst flooding in the lowest-lying part of the parking lot (see Figure 6.14).[96] The commissary parking lot was being designed with "curb cuts" to hold water. Instead of a raised curb that would keep most rain water on the impervious surfaces and out of landscaped areas, the lot is curbless, and 18-inch cuts, open to the ground at the bottom, were intermittently placed between parking rows. The cuts were filled with large rocks; perforated half-circle pipes (made of high-density polyethylene, or HDPE) were placed at the bottom of these curb cuts to retain water and allow for it to drain into the ground more slowly and eventually recharge the aquifer. Again, Xeriscaping was part of the design planning, including small trees to mark the sidewalks that allowed passage of pedestrians and shopping carts around the collection basins; these trees would require no watering once

---

[96] Once a year or so, an extreme weather event would cause an overflow of the grates in the parking lot and a large volume of standing water would result.

established.[97] This project was expected to collect between six to eight acre-feet of water per year. Importantly, the fort personnel were taking advantage of leveraging funding for already-planned construction and adding water recharge projects as an incremental cost.

Figure 6.14. Fort Huachuca Commissary Parking Lot Area
Designed to Better Manage Stormwater Runoff

SOURCE: Photo by Beth Lachman.

As noted, the possibility remains that treated effluent might become another source of recharge, especially if the Huachuca City effluent is piped to the post. In this case, the Fort Huachuca staff would need to consider a few factors, including the total volume of treated effluent relative to reuse demand. It might be the case, for example, that the golf course reservoirs would be filled first, especially in the dry months of May and June. After that, it might

---

[97] The contractor developing the project agreed to keep the landscaped plants alive for one year until established as part of their contract.

make sense to fill trenches and water the parade fields. Beyond that, recharge might be the next priority.[98]

On the other hand, stormwater capture for reuse was not an option. So the DPW personnel have no incentive to capture rooftop water for reuse, and such capture actually would exacerbate problems at the effluent treatment facility. Rooftop capture for recharge, however, would be attractive if the staff could increase volumes being treated at their wastewater treatment facility with other waste streams.

Because Fort Huachuca has significant offsite attributable net water withdrawal—80 percent of total attributable use—recharge from the city of Sierra Vista is important for this accounting. The downtown Sierra Vista water treatment facility recharges about 2,700 acre-feet per year. The post gets credit for about half of this amount, or 1,350 acre-feet per year.

Stormwater Runoff Issues

Because of the climate, topography, and precipitation patterns of the region where Fort Huachuca is located, the cumulative effects of stormwater runoff can be significant. Namely, the rainy season in July and August bring significant quantities of rainfall in a very short period. Water recharge projects (described above) have the co-benefit of mitigating these problems, but they certainly have not eliminated them. For example, the commissary parking lot project is expected to remove less than 1 percent of the volume of water that drains from impervious surfaces into a seasonal ditch known as the "Hatfield Channel." This channel is a lingering issue that DPW personnel have been unable to find sufficient funding to mitigate, starting with the resources necessary to develop a broad, integrated stormwater management strategy. The Hatfield Channel is downhill from a significant amount of nearly continuous impervious surfaces on the post. This includes rooftops, roads, and parking lots. In its lower section, the ditch is about 30 feet wide and 8 feet to 15 feet deep due to years of erosion from the stormwater runoff from all of these surfaces. It also is located about 6 feet short of undercutting the major installation roadway that it runs along. The ditch is a perfect example of the need for a "grand design" for stormwater management (and recharge) on the post. To understand how to manage the problem, a complete study of the water flows over the entire installation would need to be performed to prevent one solution from creating another problem. Such a plan would probably incorporate a wetlands basin located near the wastewater treatment facility. The estimated cost of the design alone would be $300,000; the entire project likely would cost $30 million. Because Sustainment, Restoration and Modernization (SRM) funds can only be allocated to "design-build" projects and not design alone, Fort Huachuca personnel have been unable to find sufficient resources to

---

[98] The approximately 300 feet between the surface and the aquifer provides a sufficient filter for most of the trace material that remains behind in the treated water. However, nitrate can be a problem and therefore would need to be removed at a higher level than is currently required for use on the golf course if it were to be specifically targeted to recharge basins.

initiate the project. If such funds could be obtained for the design upfront, the implementation could be phased as a multi-year project.

Funding for the Water Program

DPW personnel at Fort Huachuca cited funding for projects as a major challenge in general and for water-related projects in particular. The funding has been lacking to cover high up-front capital costs of infrastructure improvements, from fixing leaky roofs to repairing pavement. However, as noted, the ESA provides motivation and justification for water-efficiency and recharge projects. The Huachuca staff noted that, in many instances, this has included an ability to justify increased costs associated with these projects with little or no resistance. Additionally, the personnel's approach is to leverage capital investments for existing projects whenever possible and to add water-relevant modifications at incremental costs. The PT field drainage systems, the water-capture project in the Burger King parking lot, and other projects were funded in this way. This approach has been the key to winning acceptance of, and even support for, program goals.

SRM funding has been the primary source of water program funding. Although this funding is intended primarily as a revenue stream for maintaining existing facilities, the Huachuca personnel have leveraged it and actively sought out incremental cost increases to O&M activities that have large water efficiency or recharge benefits. The commissary parking lot, for example, was being funded in this way. The ESA has allowed them to specifically prioritize SRM projects that are water-relevant. This has meant that projects are not subject to a strict cost-benefit analysis before they are undertaken; any project that helps meet water saving mandates and leverages existing funding streams is viewed as an opportunity. Because sustaining the mission (i.e., staying open and operational) trumps all other mandates, the Fort Huachuca water personnel have a somewhat unusual level of support for their projects. This has also meant that personnel can act relatively quickly to take advantage of a water saving opportunity (e.g., year-end funding) if they have planned ahead.

Another funding challenge is that water project costs can be substantial. For example, the cost of a typical artificial-turf PT field is on the order of $170,000, but it varies from field to field. Fort Huachuca personnel knew this project reduced costs for water and O&M and also has health and safety benefits resulting from training on softer surfaces; however, no specific cost-benefit analysis was performed. In this case, SRM funding was not the source, but funding was obtained because a senior Army commander saw the numerous benefits in the project and identified an alternative source.

The commissary parking-lot project similarly added about $200,000 to a $600,000 project. The motivation for the baseline expense, without the water recharge, was to fix a flooding issue. Other drainage approaches had not worked, so this was a justifiable SRM expense. The incremental costs of a project that will recharge 6 to 8 acre-feet of water and therefore would avoid the need for a drainage system entirely was justified on ESA grounds. The purchase of conservation easements, with similar justifications, costs around $20,000 per acre-feet per year. Fort Huachuca has been able to obtain some OSD funds for conservation easements through OSD's Readiness and Environmental Protection Integration (REPI) Program.[99]

In some instances, Fort Huachuca water personnel have been unable to find funding for projects. For example, the "Hatfield Detention Basin" was estimated to cost $550,000. IMCOM rejected the proposal for funding this project because it was not viewed as a true "water conservation" project. Further attempts to secure year-end funds were unsuccessful. While the project would result in O&M benefits, such as erosion reduction, Fort Huachuca personnel would rather not use SRM funds for this project because of its size and because it does not leverage another infrastructure improvement.

In the case of the effluent project in partnership with Huachuca City, the funding came from the Army in a show of support that "went all the way up to Headquarters" when U.S. EPA funding fell through. The project was viewed as one that was in direct support of ESA mandates, so it was easily authorized. Working with the Upper San Pedro Partnership[100] provided a mechanism for execution and a vehicle for the contract. The funding was a federal MCA earmark, making use of Operations and Maintenance Army (OMA) dollars, but ultimately the project would reduce the net cost to taxpayers. This was because Huachuca City could avoid the local taxpayer expense of construction of a new facility, and Fort Huachuca could improve its own operations, reducing O&M costs and avoiding a very costly construction of its own, which ultimately also would be a taxpayer expense.

In the past, Fort Huachuca has used both ESPCs and UESCs for energy projects, and water projects have been included in such projects. For example, installing more water-efficient restroom water fixtures has been included in ESPCs. Water savings in ESPCs have totaled as much as 93 acre-feet/year of water use.[101] In addition, Fort Huachuca's 2009 Water Resource Management Plan suggests the use of such mechanisms, as well as ECIPs, to help fund water best management practices.

Summary

Fort Huachuca lies adjacent to the city of Sierra Vista in southeast Arizona about 15 miles north of the Mexican border. Given the water challenges within this state and Fort Huachuca's aggressive water conservation and management activities, this post, like Fort Carson, can serve as a useful military installation model for water management.

---

[99] More details on the OSD Readiness and Environmental Protection Integration (REPI) Program can be found at U.S Department of Defense, "REPI (Readiness and Environmental Protection Program)," undated.

[100] For more information, see Upper San Pedro Partnership, homepage, undated.

[101] Ruble, 2011.

The Arizona Water Context

Arizona is a dry state that has seen significant population growth during the last few decades. Most of the water use in Arizona is for agriculture (75 percent) followed by municipal (20 percent), and industrial (5 percent). Arizona is heavily reliant on both Colorado River water and groundwater to supply its needs. The state is party to a number of compacts and treaties for waters that it shares with other states and Mexico. The region in which Fort Huachuca lies has low precipitation, much of which is lost to evapotranspiration, so only a small portion recharges the aquifers. Arizona, like Colorado, has numerous reservoirs, conveyance canals and other water transfer and storage infrastructure. Notable projects are the Central Arizona Project, which delivers Colorado River water across the state, and the Salt River Project.

Surface water rights in Arizona are based on the prior-appropriation doctrine. Long-standing adjudications also are in process to clarify water rights in two river basins. The Gila River Basin adjudication includes Fort Huachuca. Since 1980, groundwater has been more actively managed in select regions of the state designated Active Management Areas. Outside of these regions, where Fort Huachuca lies, reasonable use is applied. Much development has occurred in some areas outside of the AMAs, causing concern regarding over-extraction and groundwater depletion. In an effort to improve water use in some of these areas, collaborative watershed-based groups such as the Upper San Pedro Partnership have developed. This partnership includes Fort Huachuca and many other organizations that seek to identify, prioritize, and implement comprehensive polices to meet the water needs of the San Pedro River watershed. Finally, effluent rights in Arizona are legally distinct from surface water and groundwater rights and are owned by the entity that generates the effluent and is treated as surface water once discharged.

Water markets in Arizona are facilitated by the conveyance infrastructure that facilitates water transfer. Water transfers of CAP water, groundwater, reclaimed water, and other surface water routinely occur in Arizona. The Groundwater Management Act requirement that new development in AMAs must have a 100-year guaranteed water supply, as well as the groundwater credits system for aquifer recharge created by the act, also have helped to create incentives for water transfer, and use of alternative water sources. A notable case occurred in Prescott Valley, which auctioned groundwater credits derived from its effluent to developers to raise money for new infrastructure. This auction yielded more than $67 million for Prescott Valley. Arizona also has an operational water bank that improves the state's use of Colorado River water and the conjunctive management of surface water and groundwater. Many cities also use increasing block pricing to encourage water conservation.

Fort Huachuca Water Management

Fort Huachuca is located in the Sierra Vista sub-watershed within the Upper San Pedro River basin. The Upper San Pedro River (USPR) basin is an environmentally sensitive ecosystem and home to at least two endangered species, the Huachuca water umbel and the southwestern willow flycatcher. These species depend on riparian areas along the river. Fort Huachuca's water comes from pumping groundwater within the USPR Basin, which has led to issues with the ESA for over a decade since declining groundwater could potentially decrease the amount of water in the San Pedro River, which could potentially affect the species' habitat. The USFWS has issued Biological Opinions in 1999, 2002, and 2007 regarding Fort Huachuca's groundwater pumping. Through Fort Huachuca's consultation with USFWS in 2006 and 2007, the fort identified its effect on the San Pedro River and developed a plan that outlined specific conservation requirements for the fort to achieve to reduce groundwater use to acceptable levels.

The USFWS 2007 Biological Opinion found that Fort Huachuca was not jeopardizing imperiled animal and plant species that depend on the San Pedro River. This opinion was based in part on the fort taking a variety of measures to mitigate its potential effect, including continuing to reduce groundwater use, participating in the Upper San Pedro Partnership, and recognizing that the fort is responsible for some off-post groundwater use by Soldiers, family members, and civilians who work on the post but live off-post. The Center for Biological Diversity (CBD) and Maricopa Audubon Society have filed several lawsuits against Fort Huachuca, the DoD, and the USFWS claiming that Fort Huachuca's groundwater pumping has negatively affected the San Pedro River ecosystem. These legal challenges are ongoing.

Fort Huachuca uses groundwater sources from both potable and non-potable wells. The installation owns and operates its water infrastructure, which includes the wells, conveyance infrastructure, and a wastewater treatment facility (a water treatment facility is not needed because of good groundwater water quality). The wastewater treatment facility sends treated wastewater either to reuse or aquifer recharge.

The four pillars of Fort Huachuca's water program are education, conservation, reuse, and recharge, which are implemented through several water plans and policies. In part because of the ongoing legal challenges and federal and Army policies, Fort Huachuca has implemented an aggressive water conservation program and efforts to help recharge the groundwater basin. Fort Huachuca has reduced groundwater pumping by 60 percent in the 15 years prior to 2010. The installation has effectively implemented a range of water conservation projects that reduced water use for irrigation, domestic use, and system leaks. In addition to conservation efforts, water reuse figures prominently in Huachuca's water management. Water reuse through treated effluent supplies all of the water needs for the golf course on post. Water recharge projects also have been implemented at different locations around Fort Huachuca. Water recharge involves capturing stormwater, or using treated wastewater, and allowing it to slowly drain back into the aquifer from the surface, using the natural filtering system of the layers of rocks and earth between the surface and the aquifer to purify the water. For example, a number of recharge basins, such as in the parking lot area at the commissary, have been installed throughout the post. Fort Huachuca has faced and addressed some technical, legal, and financial challenges to using such alternative water sources. For example, water management personnel were able to justify and acquire SRM funds to build the commissary parking lot stormwater management and recharge basin project because of former flooding problems. The fort also is pursuing a partnership with the Huachuca City to have the city send its wastewater to the Fort Huachuca plant, which would increase the amount of treated wastewater for reuse and recharge. Partnerships resulting from its ACUB program also contributed to the reduction in groundwater pumping in the region due to several easements that retired agricultural pumping and protected lands from development (and additional water use).

Fort Huachuca's very successful water conservation efforts have, however, had unintended adverse effects on its wastewater treatment facilities. Conservation has reduced the flows to the wastewater facility to 20 percent to 25 percent of its design capacity. The effects of this reduction are less efficient operations and high chemical concentrations in effluent. Additionally, the longevity of the collection system is compromised because of the elevated levels of chemicals. Installation personnel have worked with the wastewater facility designer to install workarounds for these conditions. Ultimately, the best solution for Fort Huachuca's Wastewater Treatment Plant problems would be the construction of a new appropriately sized facility that would also save energy. However, a new, state-of-the-art wastewater treatment plant was expected to cost on the order of $65 million to $85 million, and would require MILCON funding, which is not likely to happen any time soon given current federal budget constraints. Lack of funding for a new facility that would meet the wastewater challenges and save energy illustrates a fundamental challenge that installations face—the inability to find funds for needed capital investments in large-scale infrastructure.

Even with all the water conservation pressures and management support, as the wastewater treatment facility discussion illustrated, funding for water projects has been a major challenge at Fort Huachuca. The funding to cover high up-front capital costs of infrastructure improvements was lacking, from fixing leaky roofs to repairing pavement. However, the ESA provides motivation and justification for water efficiency and recharge projects, including the ability to justify increased costs associated with some construction projects. Fort Huachuca has been able to use both traditional and some non-traditional funding sources for water projects, including SRM, OMA, REPI, ESPCS, UESCs, and ECIPs. Additionally, the post personnel have been able to leverage capital investments for existing projects whenever possible and to add water-relevant modifications at incremental costs.

7. Findings and Recommendations

In this chapter, we present the findings and recommendations of this study. We have grouped them into four categories: water markets, installation water rights, water partnerships, and other issues.

The focus of our study was on water-market mechanisms and partnership opportunities for leveraging industry investments in installation water systems, but what we found has broader implications for basic water concerns at installations, including water rights, having access to water supplies, and water security.

Water Market Findings

We begin this chapter by discussing the key findings regarding our analysis of the evolving water markets.[1]

Wastewater Is Becoming a Valuable Asset

A significant fundamental finding is that effluent from a wastewater treatment plant has become an asset. Many local governments, companies, and other organizations throughout the United States have found that effluent is a new and useful water source. As discussed in Chapter Two, water recycling, in which treated wastewater is being used for many different purposes, is becoming increasingly common, especially in parts of the country where traditional water sources are over-appropriated. Reclaimed (recycled) water is used for landscape and crop irrigation, stream and wetlands enhancement, industrial processes, fountains, decorative ponds, recreational lakes, and toilet flushing. Military installations such as Fort Carson and Fort Huachuca have taken advantage of treated effluent to irrigate golf courses. Processed effluent also is used to help protect groundwater supplies from seawater intrusion along such places as the Florida and California coasts and to help recharge aquifers, as Fort Huachuca has done in Arizona. The most successful water auction in the United States, in Prescott Valley, Arizona, was for groundwater credits derived from the city's effluent. Effluent likely will become an important commodity in more of these water market deals in the future.

Recycled water has a number of advantages, including reducing the reliance on increasingly scarce and expensive potable water sources, reducing discharges of wastewater into oceans and rivers, and providing a consistent local drought-resistance water supply. It also can be used to help minimize groundwater overdraft problems when injected into groundwater basins. However, as discussed in Chapter Two, the use of treated effluent also has some disadvantages

---

[1] The details on water markets were discussed in Chapter Four.

that can limit its use, most notably the cost for treatment and delivery. Saline content, nutrients, and other chemicals in the water are other potential problems in using treated wastewater. In addition, water recycling has second-order effects since the use of recycled water in effect alters patterns of water use. These changes can affect the ecosystem and downstream users who may have rights to water that comes from treated effluent that historically has been discharged into rivers.

Rainwater and stormwater runoff also is becoming a valuable alternative source of water, with many of the same types of uses, including irrigation and flushing toilets. Cisterns, rain gardens, and rain barrels are used to take advantage of this water source. However, such sources do not have the same consistency as a drought-resistance water supply. Consequently, capturing rainwater and stormwater runoff is not as valuable from a water market perspective. Within many states regulatory challenges also exist in the ability to co-mingle stormwater runoff with wastewater.

Water Markets Provide Future Opportunities for Installations

As discussed in Chapter Four, we found that water market mechanisms still are relatively new and not well developed. Army installations have limited opportunities to use water markets today. However, many innovative experiments are taking place to transfer water, or to more effectively use price to guide water use, and the markets are evolving. As water-market mechanisms continue to evolve and grow, Army installations may have opportunities in the future. These opportunities are likely to occur first in growing communities experiencing water stress, areas with good conveyance infrastructure, or areas with poor water quality. Future participation in such water markets could help provide long-term water security and financial benefits to Army installations. Basic water leasing and selling, water auctions, and water banks are potential opportunities for installations to lease (or sell) excess water, including effluent, to fund installation water efficiency and infrastructure investments. There also could be opportunities to procure water if existing installation water supplies are not sufficient. Obviously, these opportunities will vary by location, and are still limited for much of the country, but they likely will grow. Water banks, especially those that are used to balance available water supply during seasonal fluctuations or periods of drought, can help installations in another important way. They potentially could be used as a hedge against drought or other water emergency. Obviously, participating in a regional water bank can present challenges and risks, such as concerns about ceding control over water. But the benefits, including working with key local agencies regarding drought issues, could be worth considering such issues. Moreover, in addition to the recognition that water is an asset, the attractiveness of Army water investments on financial terms may improve as more of these markets establish an economically efficient price for water.

However, installations' ability to take advantage of water-market mechanisms is dependent on their understanding those mechanisms, having the expertise to participate in them, and having the water rights to the traditional and alternative water sources at their installations.

Block pricing can affect the price that an installation pays for water. Installations that buy their water from a public utility may be subjected to increasing or decreasing block pricing, changing the economics of water use on post. For installations that charge tenants for their water use, increasing block rates potentially could be implemented to motivate tenants to conserve water. Finally, water-quality trading also could provide some potential opportunities to either buy or sell water quality credits. In summary, water market mechanisms have the potential to help installations have sufficient water for their missions, help with other aspects of water security, save money, or spur investments in water efficiency and infrastructure projects.

Recommendations Regarding Water Markets

We have a number of recommendations for the Army to help ensure that its installations can take advantage of current and future water markets. The Army should continue to monitor water market trends and how water market mechanisms are evolving over time. Similarly, the Army should monitor state and local government policies and issues that pertain to water allocation, especially in locations near major Army installations. Water managers at Army installations should understand and track what is happening with local and regional water policy, especially in times of drought or other water shortage problems. Active markets, or those with high water prices, may suggest to the Army that water resources are becoming scarce, affecting supply reliability and the value of its own water resources. Installation managers also need to understand the regional and local water use patterns, sources, and infrastructure. They also should consider the trends in the use of water regionally to potentially take advantage of future water market mechanisms to acquire or market installation water. Tracking such issues is important for installations to be able to main their water rights and to have access to crucial water supplies and for water security. Installation managers need to have such knowledge even when their potable water and/or wastewater systems are privatized. Army installation managers cannot rely on the privatized utility company to track policy, supply, use, infrastructure, market, and other water trends that could jeopardize or help ensure an installation's long-term water rights, access to water, or other water security concerns.

In addition to having the installation managers tracking such trends, Army headquarters should invest more staff expertise into water issues because of the increasing scarcity of this resource, future threats to water access at installations, and water's critical role in the Army performing its installation missions. Specifically, Army headquarters should consider developing more detailed expertise in water rights, the value of water, water supply issues, and evolving water market mechanisms. In addition, in its installation strategic planning, the Army should include the long-term value of water and water market mechanisms.

Installations themselves should ensure the full cost of water (commodity price, all infrastructure, long-term supply) is used when assessing water investments and water use, and ensure that water users are charged the full cost of water use. With some additional expertise, installations could consider experimenting with an increasing block-pricing approach for non-garrison organizations.

Lastly, the Army should try to ensure that installations maintain all of their existing water rights, including traditional surface and groundwater rights and new alternative water sources, such as wastewater and stormwater runoff. Army installations cannot take advantage of future water-market mechanisms or have long-term water security without having clear rights to their water sources. Understanding and maintaining such rights is especially important with privatized water systems in which the Army no longer controls the water assets. We discuss such issues further below.

Findings About Installation Water Rights

As one would expect, we found that water rights are becoming more important because of the increasing competition for this scarce water resource.[2] As a region, state, or local area faces shortages of water, legal water rights are generating greater tension and challenges. States and local governments are looking to ensure that all water sources, even small ones, are appropriately allocated. Municipal and private interests, whether cities, towns, ranching, agriculture, oil and gas companies, developers, and other industries are more likely to legally fight to maintain their own water rights and challenge other water-rights claims.

As was just discussed, an installation or other organization has to have clear water rights to its water sources to participate in any water-market activity. As we conducted our analyses and talked with numerous water experts and installation personnel, we found that many installations probably do not fully understand their water rights. There are a number of reasons.

First, many installation managers do not understand the value of water, because it still costs so little and it is difficult to quantify the potential long-term value of water. Hence, these managers do not pay much attention to water rights. Also, with the privatization of installation water systems, installations tend to have less expertise and focus on water issues. Second, the issue of water rights is a complex one that is dependent on when and how the different parcels of land that form an installation were acquired. Historical context and understanding is important. Often water rights, especially in Western states with prior appropriate systems, need to be traced and understood back over many decades, even back over a century. For example, Fort Bliss has some water rights on the McGregor Range in New Mexico that can be traced back to the 1880s. Installation personnel change over time, and often the knowledge and historical context for understanding installation water rights is lost. Third, installation personnel especially new

---

[2] Water rights are explained at the end of Chapter Two.

personnel, often lack the training, expertise, and experience with the detailed technical and legal aspects of understanding, documenting, and tracking water rights.

Some installations, such as Fort Carson, also are at risk of losing some water rights. Large installations often have many different types of water rights, including small sources, such as a remote lake or well on a training range used for limited training purposes. As water becomes more scarce, states and local governments are looking to appropriate any water source they can, even small ones. If an installation has not maintained or submitted proper documentation on its water rights, such as not reporting beneficial use for a water source, it is at risk of losing some of those rights. Installations need to ensure that they document and maintain access to all water rights, even the small and remote water sources, because they may become a valuable asset for a future water market or important for future installation missions and water security. Such water rights include surface water, groundwater, effluent from installation sewer systems, and stormwater runoff.

State and local governments play a key role in planning and managing water sources and rights. State and local government water managers are actively managing water resources to achieve economic, environmental, and social goals and priorities, largely based on input from water users. However, states have authority to enforce and grant water-rights claims and to take away those rights. It also is important to distinguish between paper and real water rights. An installation may have legally documented paper water rights, but if it does not have access to the actual water, it has a problem. As water becomes scarcer, situations increasingly occur in which water-rights holders have used more than their fair share of a water source such as a stream, and state and local authorities have not stepped in to address this problem. If the state or local agency that manages water allocation does not address a violation, then access to water may be lost.

Installation Water Rights Recommendations

Army installation personnel need help to make sure they have done everything they can to establish, document, maintain, and enhance their water rights. Army headquarters should develop water rights guidance to help installation managers understand the long-term value of water rights and what they need to do to establish, maintain, and enhance those water rights. In fact, since this study was conducted, the Secretary of the Army issued Directive 2014-08, "Water Rights Policy for Army Installations in the United States"[3] in May 2014, which set new Water Rights Policy. Also that month, the Office of the Under Secretary of Defense provided guidance on water rights policy by issuing a Memorandum, "Water Rights and Water Resources Management on Department of Defense Installations and Ranges in the United States and

---

[3] U.S. Department of the Army, "Water Rights Policy for Army Installations in the United States," Washington, D.C.: Secretary of the Army Directive 2014-8, May 12, 2014.

Territories."[4] Army headquarters also should provide detailed practical how-to water rights guidance for garrison commanders, installation managers, lawyers, DPW personnel, and other personnel who have water responsibilities. Pursuant to Directive 2014-08, ACSIM has been working on such water rights implementation guidance, which was to be issued in late 2014. Installations should be following such guidance.

ASA(IE&E), in collaboration with ACSIM, should consider starting an Army working group on installation water rights and ensuring access to long-term water sources to identify and address installation water rights concerns and needs. Such a group could have periodic teleconferences with interested and knowledgeable personnel. In early 2013, ASA(IE&E) conducted a survey about installation water rights with installation staff. These survey results could be used to help identify key personnel and topics for this working group. Since this study recommendation was made, ASA(IE&E) and ACSIM have formed a working group on installation water rights and ensuring access to long-term water sources. The first meeting was held June 25, 2014.

Army installations should be collaborating with state and local governments on water rights. Installation personnel should engage in relevant federal, state, and local water planning processes, including drought planning, to understand water development, allocation, and emergency plans to learn how such actions might impact their installation water rights and access to water sources. For example, the water planning activities regarding the water flow on the Chattahoochee River in Georgia could impact Fort Benning. More importantly, if installation personnel are engaged in such planning processes, they can make sure the installation's interests are represented in the process and help minimize any potential impact on their water rights. Similarly, they should be collaborating with any state or other government regulators that have authority over water rights to help clarify their installations' water quantity and ensure their priority rights. Participating in such processes is also a proactive approach to help learn about and limit potential encroachment on installation water rights and sources.

Findings About Water Partnerships

Traditional installation water partnerships, such as ESPCs and UESCs, can be useful tools to acquire capital investments for installation water conservation and efficiency projects.[5] However, such partnership mechanisms are used more frequently for energy than for water projects, often because the required return-on-investment rates are more difficult to achieve with water projects. However, water and energy projects can be bundled together in an ESPC/UESC task, so that a short-term payback energy investment, such as replacing light bulbs, can be combined with a

---

[4] U.S. Department of Defense, "Water Rights and Water Resources Management on Department of Defense Installations and Ranges in the United States and Territories," Washington, D.C.: Office of the Under Secretary of Defense, May 23, 2014.

[5] Chapter Three provides an overview of the different types of water partnerships.

longer-term water conservation investment payback, such as low-flush toilets. In addition, UESCs have historically been implemented by energy utilities, which are less likely to know about and consider water-conservation investments. Further encouragement of water utilities to conduct UESCs is needed. Finally, more water projects could be conducted using such tools, but these tools cannot be used for certain types of water investment projects, such as large-scale water infrastructure investments and stormwater management projects.

Collaborations and partnerships play an important role in regional, state, and local water planning and management, and they are likely to increase. Such water collaborations and partnerships focus on managing and sharing water resources and supply, long-range water and drought planning, watershed management and planning, and developing and sharing water infrastructure. Such water collaborations and partnerships can be between two cities to share management of drinking water and wastewater treatment systems. They also, as in the case of the Great Lakes Partnerships, can be among states to help manage key regional water resources. More regional collaborations and partnerships with multiple entities to manage water resources, such as state, regional, and local water planning processes, are developing because of a number of reasons, including the shared nature of using water supplies, the fact that water is a public good, multiple stakeholder interests, increasing water demands, and increasingly scarce supplies of water.

State, regional, and local government and water utility partnerships are important to installations to help them save money by lowering costs, maintain water rights, invest in shared infrastructure, and have access to water supply, especially during droughts. Similarly, regional partnerships, involving multiple organizations, to help plan, manage, and/or share water resources are likely to grow and are an important opportunity for installations' long-term water security.

Because of the passage of NDAA 2013 Section 331 in January 2013, installations have expanded authority to partner with state and local governments. This authority gave new opportunity for installations to partner with nearby cities and towns to share water infrastructure and services. We already found some examples of such collaboration, including Fort Carson sharing costs of replacing a shared water pipeline with a state prison in Colorado and Fort Huachuca partnering with Huachuca City to process the city's wastewater. Local government collaborations are an emerging opportunity for water infrastructure investments.

Recommendations About Installation Water Partnerships

Army headquarters organizations, such as ACSIM, should encourage installation energy and water personnel to incorporate more water projects into their ESPCs and UESCs. Installation personnel also should be encouraging ESCO and energy utility personnel to assess and incorporate more water projects when they are developing energy ESPC and UESC proposals.

Army headquarters and installations also should try to reach out directly to water utilities to encourage them to develop and submit water UESC projects.

Installation personnel should try to take an active role in state, regional, and local governments regional and local water planning and collaborative management by collaborating and partnering with these organizations whenever they can. For example, installation water personnel could volunteer to be on a regional drought planning or water supply planning working group. In fact, Fort Carson personnel could volunteer to help in the development of the next Colorado Drought Management Plan. Similarly, Fort Benning personnel could become involved in the state water management process for the river that supplies their drinking water by volunteering to help in the Middle Chattahoochee River planning process, at least providing public input to the Middle Chattahoochee Regional Plan and maybe even trying for representation on the Middle Chattahoochee Water Planning Council. Installation personnel also should be trying to engage more often with state and local government water managers to better understand water supply, demand, and management issues so that mutually beneficial financing and partnership opportunities can be identified and pursued. Installations could potentially save money by pursuing more public-to-public partnerships to share water infrastructure and/or supply, as in the example of Lake Oswego and Tigard, Oregon.

Army headquarters should ensure that installation strategic planning and guidance emphasize the importance of collaborating and partnering with state and local governments regarding water issues, from water and drought planning to shared infrastructure projects. Through such collaboration and partnering, installations can help maintain water rights and access to water sources, enhancing installations' long-term water security. Through such partnerships in shared water infrastructure, installations also have the potential to achieve some economies of scale and save costs.

Lastly, ACSIM or IMCOM should help installations conduct pilot experiments in partnering with a local government to increase investment in water infrastructure and save long-term costs. The passage of NDAA 2013 Section 331 has energized communities and military installations about partnership opportunities, so it is a good time to pursue installation water-infrastructure partnerships with nearby cities and towns.

Other Findings and Recommendations

Lastly, in this study we had some findings and recommendations in two other areas: water utility privatization and water scarcity at installations with successful water conservation programs.

Water Utility Privatization

When an Army installation's water systems have been privatized, it limits the installation's control and flexibility over water resources and has the potential to impact an installation's future water system investments, partnerships, and market opportunities.[6] Since wastewater is a valuable commodity for future water markets, Army installations should have the right to sell or use such effluent in water investment deals or on post in the future. Recent and future installation privatization deals appear to protect an installation's water rights to use the wastewater it generates because of a clause in the contract language. However, if an installation's wastewater treatment facility was privatized a while ago, it may have lost the water rights to the installation's wastewater if this right was not specified in the contract. For example, one installation with privatized water utilities wanted to consider using its wastewater for irrigation, but learned that it would cost significantly more because of the charges that the privatized utility would charge for the reclaimed water. In addition, at some installations, the privatized water utility has limited incentives to invest in water conservation, especially when the utility sells water to the installation. Water privatization also can make it more difficult to partner with a local government in shared water infrastructure. Lastly, installations are likely to lose water management personnel and water expertise as installations rely more and more on the water-utility contractors. Such personnel and expertise are potentially needed for installations to deal with future water scarcity, security and market opportunities, and ensuring that installation water needs and concerns are addressed.

To address these issues, we have several recommendations for installation water-utility privatization. First, for installations that have privatized their wastewater system, the installation legal staff should check to make sure effluent rights were protected in the privatization contract and if not, try to amend the existing privatization contracts to incorporate such water rights. The installation should be able to maintain the flexibility and right to use such effluent for irrigation, groundwater recharge, other non-potable water uses, and as a market commodity, without additional charges being imposed by the privatized utility. The Army already pays for wastewater processing. It should not have to pay again to use this wastewater resource.

Second, before privatizing an Army installation's water systems, the Army should assess a number of different factors, including the importance of having long-term water expertise, water rights, and water market opportunities at an installation. Such issues are more important for installations in arid areas. In such areas, the Army may want to reconsider privatization or ensure that these issues are sufficiently addressed when privatizing the water utilities. The Army also should assess all of the different water rights implications, the potential long-term water security pros and cons, and the implications for water conservation. Alternative approaches for funding the needed water infrastructure investments, including different partnership opportunities (such as partnerships with local governments), also should be considered and assessed to ensure that the Army is getting the best deal for all water issues.

Third, when privatizing water utilities, any of the downsides just assessed should be minimized ahead of time as much as possible. For example, the Army may take steps to ensure

---

[6] Utility privatization is discussed in Chapter Three.

that water rights, security, and market staff expertise are maintained at an installation even with the privatization of all the water utilities, especially if the installation is in a part of the country that experiences or is likely to experience water scarcity problems in the future.

Lastly, the Army should conduct a study of the implementation of previous Army water utility privatization deals to develop lessons learned and incorporate them into future deals so that they are conducted more effectively and efficiently to ensure installation long-term water rights, security, conservation, and market opportunities. Many previous problems with water utility privatization, such as the lack of incentives to invest in water-conservation efforts, could potentially be addressed in contract language.

Future Water Shortages and Successful Installation Water Conservation

Some proactive installations, such as Fort Huachuca and Fort Carson, which have made significant investments in water conservation, run the risk of being penalized by state and local government during times of drought or other water emergency. The Army should try to ensure that installations' successful efforts in water conservation are considered during periods of regional and local water scarcity. The Army could proactively address this possibility at different organizational levels before a water crisis arrives. Army headquarters or IMCOM regions should consider working with state agencies as they prepare statewide water management and drought emergency plans, especially in states such as Texas with a large Army presence. At a minimum, the Army should be reviewing those state plans and calculating the impact on Army installations for enterprise and regional water security considerations, so that installations can proactively prepare for the effects of regional water issues within their own strategic planning. At the local level, installation personnel should proactively be reviewing and working with state and local governments in their regional and local water scarcity planning since, as a stakeholder, the Army needs to provide input to ensure that its needs are considered.

Appendix A. Select Water Banks in the United States

This appendix provides detailed background on the following water banks mentioned in Chapter Four:

- The California Drought Water Banks of 1991, 1992, and 1994
- Arizona Water Bank
- Central Kansas Water Bank Association
- Kittitas Water Exchange
- Arkansas River Water Bank Pilot

The California Drought Water Banks of 1991, 1992, and 1994

The California Drought Water Banks of 1991, 1992, and 1994 provide good examples of water banks that are used to balance water supply.[1]

A drought water bank was established in California in 1991 after five years of continuous drought. In early 1991, water deliveries to agriculture were cut by 75 percent to 100 percent in some areas, while urban areas saw reductions on the order of 15 percent to 45 percent.[2]

The 1991 Drought Bank was significant in that it was the first time voluntary, market-based water transfers occurred on a large scale in California. While the bank was not created explicitly by legislation, the emergency legislation passed streamlined the water rights transfer process (by allowing for transfers outside the suppliers' area and protecting these rights during drought). Executive orders and actions also aided the water bank.

The primary purpose of the bank was to enable voluntary transfers of water from suppliers (agriculture) in Northern California to demanders (urban, municipal, and agriculture) in the central and southern parts of the state (within the State Water Project (SWP) customers and the Central Valley Project (CVP) service areas) at a time when traditional water supply sources were stressed. The bank activities served to protect existing water rights holders while providing water to critical needs without damaging fish and wildlife.

California's Department of Water Resources (DWR) administered the drought banks, which were activated during the 1991, 1992, and 1994 drought years and which served as a clearinghouse for short-term trades. Water supplies could come from irrigation entitlements and

---

[1] Peggy Clifford, Clay Laundry, and Andrea Larsen-Hayden "Analysis of Water Banks in the Western States," Washington Department of Ecology Water Resources Program, Report 04-11-011, July 2004; Lloyd Dixon, Nancy Moore, and Susan Schechter, *California's 1991 Drought Water Bank,* Santa Monica, Calif.: RAND Corporation, MR-301-CDWR, 1993; Harvard University, "Innovations in American Government Awards: California Drought Water Bank," 1995; Michael O'Donnell, and Bonnie Colby, "Water Banks: A Tool for Enhancing Water Supply Reliability," Tucson, Ariz.: University of Arizona, Department of Agricultural and Resource Economics, January 2010.

[2] Dixon, Moore, and Schechter, 1993, p. 1.

appropriative water-rights holders. California's DWR purchased water for the bank from: 1) surface water made available by groundwater substitution, 2) surface water stored in local reservoirs that was surplus to local needs; and 3) water made available by farmers fallowing or forgoing irrigation of designated farmland. Demanders were required to submit bids quantifying critical needs for water. Priority was given to water needed for drinking, health, sanitation, fire protection, and critical agricultural needs. The DWR negotiated contracts with suppliers at a range of prices established by a committee of potential demanders and then sold the water to demanders for a fixed price according to critical-needs allocation rules that took into account economic, social or environmental losses.[3]

Because San Francisco Bay and Sacramento's San Joaquin River Delta are environmentally sensitive areas, the bank also sought to ensure minimum stream flows.

Activity levels of the 1991, 1992, and 1994 banks varied but were high at times. In 1991, the bank purchased nearly 825,000 acre-feet of water and sold 435,000 acre-feet.[4] The purchase price was $125 per acre-foot, and the selling price was $175 per acre-foot. Fifty-one percent of the water came from no-irrigation contracts, 32 percent came from sources using groundwater substitution and 17 percent came from stored water sources.[5] To supply this water, 166,000 acres of agricultural land were fallowed.[6] Analyses determined that statewide, the banks contributed jobs and economic opportunity, although some economic activity shifted away from the agricultural areas that supplied water to the southern part of the state. Third-party effects from the water transfers were determined to be minimal. In 2001, 2002 and 2003, the California DWR began administering a dry-leasing program. These are similar in purpose to the drought water banks, but do not require drought conditions to be active.

Arizona Water Bank

The Arizona Water Bank (AWB) provides an example of a water bank that balances use of water supply sources and improves conjunctive use.[7] It was established in 1996 to enable the storage of

---

[3] For the 1991 bank, DWR negotiated a range of prices with sellers. Subsequent banks developed water pools that were price uniformly.

[4] Subsequent banks only made purchases after demanders were identified to avoid the need to carry-over water supplies and lowered the purchase price. Other lessons incorporated into subsequent banks were to avoid no-irrigation contracts if possible, by using groundwater substitution and stored water contract sources more frequently.

[5] Dixon, Moore, and Schechter, 1993, p. 5.

[6] David Zilberman, Ariel Dinar, Neal MacDougall, Madhu Khanna, Cheril Brown, and Frederico Castillo, "Individual and Institutional Responses to the Drought: The Case of California Agriculture," *Journal of Contemporary Water Research and Education*, Vol. 121, No. 1, 2002.

[7] The Arizona Water Bank Authority was created and modified by two Arizona bills. House Bill 2492 in 1996 created the bank, while the enactment of House Bill 2463 in 1999 amended the earlier legislation. Creation of the bank was not controversial, since it was created within the limits of existing laws and revenue sources (which were redirected to the bank). See Clifford, Laundry, and Larsen-Hayden, 2004, pp. 32–36; and O'Donnell and Colby, 2010.

excess Colorado River Water within the Central Arizona Project system for future use during periods of low supply.[8] Until the AWB was created, Arizona did not use its full 2.8 million acre-feet allotment of Colorado River water. The state likely would not have used this amount until 2030 (excess Central Arizona Project (CAP) water was available in part because farmers were able to pump groundwater more cheaply and did not choose to purchase CAP supplied water). Most of Arizona's unused water would have gone to Southern California.[9]

The administering authority of the Arizona Water Bank, the Arizona Water Banking Authority (AWBA), purchases excess water from the Central Arizona Project (CAP) at a price set each year by the Central Arizona Water Conservation District (CAWCD).[10] The purchase is considered consumptive use and is therefore not subject to non-use forfeiture. The AWBA is permitted to store this water within state-owned underground facilities and aquifers (direct recharge), or irrigation districts are allowed to use the water in lieu of pumping groundwater (indirect recharge). In exchange for storing the water, the AWBA receives credits that it then can transfer at some future time to either the Central Arizona Groundwater Replenishment District (CAGWRD) or the Arizona Department of Water Resources (ADWR) for use for their municipal and industrial customers along the CAP canal should they require it during dry periods (the amount transferred is reduced by a 5 percent conservation requirement.).[11] Therefore, the Arizona Water Bank enables the use of Arizona's allotment of Colorado River water to replenish groundwater aquifers, to implement Arizona's groundwater management plan, and to provide water supplies for municipal and industrial users during periods of low water supply. The AWBA became active in 1997. Figure A.1 shows the rapid increase in water stored in Arizona's groundwater storage facilities due to the Arizona Water Bank Authority.

---

[8] The AWBA procures excess CAP water supplies from the CAWCD. Water is allocated according to a priority system established by the U.S. Secretary of Interior as follows: municipal and industrial, Indian, non-Indian agriculture and miscellaneous. Therefore, excess CAP water is the most junior priority. The three Arizona counties within this system are Maricopa, Pinal, and Pima. See Arizona Senate Research Staff, "Arizona State Senate Background Brief: Arizona Water Banking Authority," Phoenix, Arizona, November 15, 2010.

[9] Arizona Water Banking Authority, "Executive Summary," undated.

[10] Subsequent legislation in 1999 allows the AWBA to store effluent in addition to Colorado River water if there is excess capacity after the surplus Colorado River water is stored. It also firms water supplies for non-CAP municipal and industrial users within the service area along the canal such as the Salt River Project, Maricopa Water District, and Roosevelt Water Conservation District and establishes a lending process for water storage with other entities.

[11] The ADWR also could decide to retire and not sell these credits.

Figure A.1. Cumulative Storage in Arizona's Groundwater Storage Facilities, by Storer

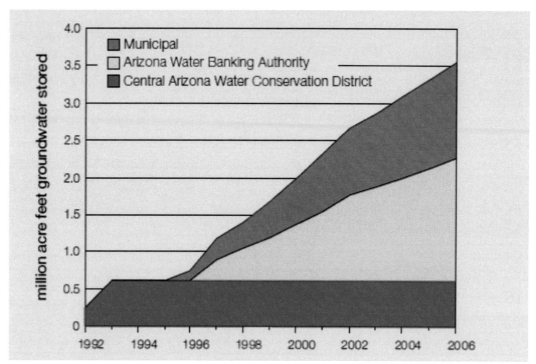

SOURCE: Taylor Shipman and Sharon B. Megdal, "Arizona's Groundwater Savings Program," *Southwest Hydrology,* May/June 2008, p. 11.

In 2005, the AWBA also became responsible for working with Native American tribes to fulfill the state's obligation in firming[12] their water supplies, which also provides drought protection for Native American tribes. Additionally, the AWBA can contract with authorized entities in California and Nevada[13] to store Colorado River water for them for a fee, thereby replenishing Arizona's groundwater aquifers. Under these agreements, when either California or Nevada would need their stored water, they would draw the stored amount from Lake Mead and the Colorado River. In 2005, the AWBA began storing water for the Southern Nevada Water Authority.[14] The agreement was renegotiated in 2013 after the Southern Nevada Authority lacked funds (due to the economic downturn and the investment required for a third intake from

---

[12] *Firming* a water supply refers to making the supply more secure, or reliable. This often is done in periods when supplies are limited by supplementing water supplies with water stored in a reservoir or a water bank. For example, the Arizona Water Bank's unused Colorado River water is used to ensure that water is available for Arizona municipal uses, industrial uses and for Indian water rights settlements when other supplies are limited. See Arizona State Senate, "Fact Sheet For H.B. 2835: Arizona Water Settlements Act Implementation," Phoenix, Arizona, March 6, 2006.

The Windy Gap Firming Project in Colorado is increasing reservoir capacity in order to make water supplies to municipalities on the Front Range more secure.

[13] Up to 1.25 million acre-feet for Nevada.

[14] Arizona Water Banking Authority, homepage, undated.

Lake Mead), and Arizona did not have surplus Colorado River Water (due in part to increased demand from municipalities and Indian tribes).[15]

The AWBA is a separate state agency that receives administrative, technical, and legal support from the Arizona Department of Water Resources. The AWBA pays the delivery and storage cost for transporting excess Colorado River water through the CAP canal. It has three primary sources of funding: pumping fees charged in each of the three groundwater active management areas that it serves (monies are to be applied within the AMA it was collected in); an ad valorem tax charged in the three CAP counties (again, monies are to be applied within the county the tax was collected in); and general-appropriation funds determined by the state legislature (these have not been available since 2001). In addition, California and Nevada provide funds should they request that the AWBA store water for them.

In 2010, 211,712 acre-feet of CAP water was recharged to the aquifers (19,000 acre-feet was on behalf of Nevada). Of the 2.78 million acre-feet (MAF) of Colorado River water that Arizona used in 2010, 1.13 MAF was used directly on the river, CAP customers used 1.44 MAF and the AWBA used .21 MAF. As of 2010, the cumulative total amount of Colorado River water stored by the bank was approximately 3.78 MAF.[16]

Table A.1 presents the cumulative total acre-feet stored by the AWBA as of December 2012 by type of funds used to procure the water and AMA. Three AMAs are participating in the bank: Phoenix, Pinal, and Tucson.[17]

---

[15] Arizona Water Banking Authority, homepage, undated; Henry Brean, "Water Authority Puts Water Banking Agreement with Arizona on Hold," *Las Vegas Review-Journal*, April 18, 2013.

[16] Arizona Water Banking Authority, "Activities," undated.

[17] Arizona Water Banking Authority, "Arizona Water Bank Authority Annual Report 2012," July 1, 2013.

Table A.1. Cumulative Total and Location of Long-Term Storage Credits Accrued Through December 2012, by Location and Funding Source

AMA Name	Acre Feet
Phoenix AMA	
4¢ Ad valorem Tax]	1,329,925
Groundwater Withdrawal Fee	293,632
General Fund	42,316
Indian Firming	0
Shortage Reparations	20,642
Interstate Water Banking - Nevada	51,009
AMA Total	1,737,523
Pinal AMA	
4¢ Ad valorem Tax	187,465
Groundwater Withdrawal Fee	394,896
General Fund	306,968
Indian Firming	0
Shortage Reparations	60,507
Interstate Water Banking - Nevada	439,851
Other- (GSF Operator Full Cost Share)	14,125
AMA Total	1,384,620
Tucson AMA	
4¢ Ad valorem Tax	390,334
Groundwater Withdrawal Fee (a)	98,788
General Fund	54,546
Indian Firming	28,481
Shortage Reparations	1,227
Interstate Water Banking - Nevada	109,791
AMA Total	650,768
Totals by Source of Funds	
4¢ Ad valorem Tax	1,907,724
Groundwater Withdrawal Fee	787,317
General Fund (b)	403,830
Indian Firming	28,481
Shortage Reparations	82,375
Interstate Water Banking—Nevada (c)	600,651
Other- (GSF Operator Full Cost Share)	14,125
TOTAL	3,824,505

SOURCE: Arizona Water Banking Authority, "Arizona Water Bank Authority Annual Report 2012," July 1, 2013.
NOTES: a) Includes 234 acre-feet of credits purchased from the Tohono O'odham Nation pursuant to § 45-841.01.
b) By resolution passed in 2002, the AWBA established on-river firming as the highest priority of use for credits accrued through expenditure of general fund appropriations. Pursuant to the AWBA Agreement to Firm with MCWA dated February 4, 2005, a total of 230,280 acre-feet of credits were transferred to the AWBA long-term storage subaccount for the MCWA in 2005. An additional 25,894 acre-feet of credits have been reserved under Exhibit C the Amended Agreement to Firm, dated December 8, 2010, for a total of 256,174 acre-feet. By resolution passed in 2008, the AWBA established a replacement account for 4th priority Colorado River M&I users.
c) Includes 50,000 acre-feet of credits transferred to SNWA.

In sum, the Arizona Water Bank enables the use of Arizona's allotment of Colorado River water to replenish groundwater aquifers; implements Arizona's groundwater management plan;

provides water supplies for municipal and industrial users during periods of low water supply; fulfills the state requirement to improve the reliability of Indian tribes' water supplies; and provides an alternative management option for Nevada and California in the sharing of Colorado River Water.

Central Kansas Water Bank Association

The Central Kansas Water Bank Association is an example of a water bank that enables conservation and redistributes water use away from sensitive, water-poor areas. The first water bank in Kansas was established in 2005 as a means of promoting water conservation while providing an alternative management approach for reallocating limited water resources either to areas with growing need or away from supply-poor areas to improve stream flow. Located in central Kansas within the Big Bend Groundwater Management District, the Central Kansas Water Bank Association is a groundwater bank created to alleviate stress on the Rattlesnake Creek Basin using voluntary means. Because of the hydrology in the area, the groundwater and surface water flows are inter-linked, and groundwater use near the creek affects the creek's flow. The water bank makes it possible to reduce groundwater usage near the creek by "depositing" the water rights in the bank and providing opportunities to lease water rights from other areas. The Central Kansas Water Bank Association also serves other purposes, such as:

- Facilitating term water leases (up to a 10-year term) so that water use can be redistributed away from sensitive, water-stressed, areas (such as the Rattlesnake Creek Basin) within the same geographical and hydrological unit
- Providing a means for areas with new water demands to attain water when water rights are not available
- Offering a safe deposit function for depositors who can save a portion of unused water right for future use
- Promoting conservation by requiring deposits and leases to conserve a minimum of 10 percent.

The area covered by the Central Kansas Water Bank Association is shown in Figure A.2. While the Rattlesnake Creek basin is of primary concern, the geographic coverage needed to be large enough to allow for a reasonable number of transactions.

Figure A.2. Geographic Coverage of the Central Kansas Water Bank Association

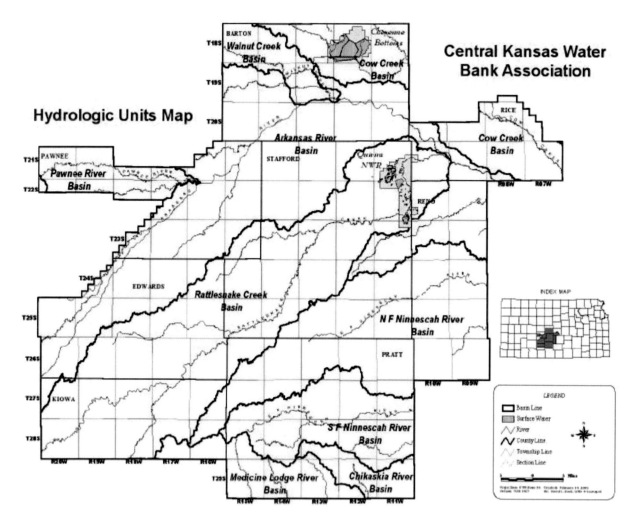

SOURCE: Susan Stover et al., "Central Kansas Water Bank Association Five Year Review and Recommendations," January 20, 2011, p. 5.

The Central Kansas Water Bank Association is a private, not-for-profit groundwater bank. The bank provides an electronic bulletin board that helps to bring depositors and lessees together. The bank manages water deposits, leases, and safe deposits but does not own, buy, or sell water rights. Deposits can be made annually for up to five years in duration while leases can be for a maximum of ten years (this is based on the length of time the bank has been chartered for). Water rights deposited in the bank constitute "due and sufficient cause for non-use" and so therefore are not forfeited. A fee schedule is used to generate funds to operate the bank.

Water deposited in the bank can be used anywhere within the same hydrological unit covered by the geographic area of the bank for beneficial use, expanding both the location and uses of the water right. The actual amount of water leased is restricted to the quantity that had been used historically (rather than the authorized quantity) for the same level of consumptive use (calculated using standard factors). A conservation requirement that varies by area is then applied. The water can be leased for a period of up to ten years at a price agreed upon by the depositor and the lessee. Because it is a term-permitted water right, it would be junior to an established water right in the area.[18]

Chartered in 2005 for a seven-year trial period, the groundwater bank was evaluated (as required in the originating statute) in 2010 and 2011. The evaluation committee noted the low participation in the bank. In the five years of bank operations reviewed, only three deposits were made totaling more than 600 acre-feet of water. Prices set by the depositor ranged from $250 to $1,000 per acre-foot. One lease was made in 2009, but no water was withdrawn. In addition 10 safe deposits were made and no withdrawals were made.[19] The review committee determined that low participation was due to the cumbersome procedures; expensive fees; complex formulas for determining the water quantity available for lease; large reductions taken on deposits and leases for conservation and consumptive use; and the uncertainty due to the temporary nature of the bank (which originally was chartered for seven years). Despite this, evaluators felt the bank could be a useful mechanism for conserving and transferring water and recommended making water banks a permanent water-management option in Kansas. Specific modifications to the bank structure and procedures to streamline the water rights transfer process and encourage more transactions were recommended.[20]

Kittitas Water Exchange

The first pilot of water banking in Washington state was the Kittitas Water Exchange in the Yakima Valley. It was developed to ensure stream flows as well as to provide a source of water for future development (note that 15 of 23 water banks surveyed in 2004 had some environmental objective). Lying within the Yakima River Basin, the area covered by the Kittitas Exchange has extensive irrigation using USBR reservoir water (from the Yakima), threatened and endangered species issues (trout and salmon) and congressionally-required minimum stream flow requirements to sustain Yakima Nation fishing rights. Compounding these demands, the river basin has experienced low total water supply in the years 2001, 2004, and 2005 as well as pressures from population growth.

Water rights in Kittitas County were fully appropriated by 1945, to get seniority water rights had to be purchased before 1905. In dry years, even senior water-rights holders did not get their entire share of water, and since 1990, junior rights holders had not received their full allotment.[21]

---

[18] Kansas Department of Agriculture, "Voluntary Incentive Programs," undated; and Tracy Streeter, "Testimony before the House Agriculture and Natural Resources Committee on HB 2516," January 31, 2012.

[19] Susan Stover et al., "Central Kansas Water Bank Association Five Year Review and Recommendations," January 20, 2011, p. 10.

[20] For more details on the recommendations, please see Stover et al., 2011.

[21] All rights to Yakima Basin surface water have been adjudicated under the priority system: Those predating 1905 are considered senior water rights. Junior water rates, those dating after 1905, are curtailed in low water years to ensure senior water right holders get the water they are entitled to. Surface water not appropriated by mid-1905 was

Because of these pressures and the associated uncertainties with the water supply, Upper Kittitas County and the state issued a moratorium in 2009 on new groundwater wells in the area. Later, the moratorium was lifted, and new wells were allowed only if they could be water-neutral. This meant that a well could be developed only if an equivalent amount of water rights (those with the same amount of consumptive use) were placed in a basin-specific mitigation bank. In other words, any newly developed groundwater well in Kittitas County had to purchase mitigation water to offset effects on stream flows and senior water rights holders to receive a permit. About the same time, a USGS study determined that wells, even deep aquifer wells, did indeed affect surface water in the Yakima Basin. The final report was released in 2011.[22]

Several key legislative actions made water banking a possibility in Washington state. The state had statutory authority to develop water banks since 2003 under House Bill 1640. House Bill 1640 authorized the state Department of Ecology to use the Trust Water Rights Program for water banking in the Yakima River Basin. This authority was extended statewide in 2009 when a section of the bill, Chapter 90.42 RCW, was amended. The statute allows the Department of Ecology to use water banking to accomplish the goals listed below:

- "Mitigate for new water uses.
- Hold water for beneficial uses consistent with terms established by the transferor.
- Meet future water supply needs.
- Provide a source of water to third parties on a temporary or permanent basis, for any allowed beneficial use."

The department is NOT allowed to use water banking to do the following:

- "Cause detriment or injury to existing rights.
- Issue temporary rights for new potable uses, or
- Administer federal project water rights."[23]

The legislation is general enough to enable water banks to take on many forms in Washington state. For example, they may be established as a clearinghouse to bring together

claimed by the USBR to support its Yakima Basin agricultural irrigation project. Authorized by Congress, the USBR operates five reservoirs to supply water to irrigators in Kittitas, Yakima and Benton counties. Since 1990, post-1905 water rights within the irrigation project (junior rights holders) have been pro-rated to less than 50 percent of a full water supply. In addition to the USBR water irrigation project, Congress has mandated that stream flows be maintained to support fisheries available to the Yakama Nation, which holds time immemorial water rights related to these stream flows. See Washington Department of Ecology, "Pending Water Right Applications in Subbasins of the Yakima River Basin," undated; Bob Barwin, "Water Right/Mitigation: Banking in the Yakima Basin," Edmonds, Wash.: Water Banking/Exchange Workshop, October 6, 2008; Courtney Flatt, "Water Banks Help Washington Land Developers," Eugene, Ore.: KLCC Public Radio, April 4, 2012; Washington State Department of Ecology, "Water Banking and Trust Water Programs: Important Water Management Tools," Publication No. 09-11-035, December 2009.

[22] Clifford, Laundry, and Larsen-Hayden, 2004; Laura Ziemer and Ada Montague, "Can Mitigation Water Banking Play a Role in Montana's Exempt Well Management?" *Trout Unlimited*, September 2011.

[23] Washington State Department of Ecology, "2010 Report to the Legislature: Water Banking in Washington State," Olympia, Wash:, Publication No. 11-11-072, June 2011.

sellers and buyers; as a process to mitigate the effects of new water use on existing uses; as a platform for rotational agricultural pools among farmers; or as a means of keeping water rights in trust to maintain in-stream flows.[24] As envisioned, the water banks in Kittitas County will benefit developers since the bank is a way of providing water sources for new needs. The banks also are a management tool to help the state meet minimum stream flow requirements resulting from the Endangered Species Act and fulfill Yakima Nation immemorial water rights. In addition, senior rights holders who face pro-rationing and junior rights holders (post-1905) who are at risk of curtailment when total water supply is below a given threshold benefit as well.

In 2012, the Kittitas Water Exchange had seven water banks: the Lamb and Anderson bank (within the Upper Kittitas Basin), the Roan bank (Swauk Basin), the SwiftWater Ranch bank (Teanaway Basin), the Masterson Ranch bank, the Reecer Creek Golf Course bank (Lower Kittitas Basin), the Williams and Amerivest bank (Lower Kittitas Basin) and the Roth-Clennon bank (Upper and Lower Kittitas Basin). The Upper and Lower Kittatas Exchange banks were the first two to become operational. These banks may have either groundwater or surface water that is "deposited" into a bank either temporarily or permanently for either in-stream or out-of-stream use.

The Department of Ecology manages a clearinghouse of available mitigation credits and determines whether the transaction is water neutral. Mitigation water rights that sellers offer are held by the trust (rights are not forfeited and maintain their priority date while in the trust). Each bank operates for a defined area to ensure that the mitigation water is applied within the same area affected by the new demand. The market determines prices.[25] Prices in the Lamb and Anderson bank located in the Upper Kittatas Exchange pilot were distorted by the first, large transaction with Suncadia, a development company. Suncadia purchased several pre-1905 irrigation water rights at approximately $1,300 per af and has sold over 30 af of mitigation credits for nearly $40,000 per af.[26]

The Kittitas Exchange water banking is one element of a broader effort to maintain surface water flows under the Washington Water Trust that was established by legislation. Among other things, the Water Trust legislation facilitates water banking statewide by preventing forfeiture of water rights and, while held in trust, the original priority date is retained. The legislation establishes a water-acquisition program to maintain stream flows where trout and salmon are at risk. The Yakima Basin Water Transfer program, begun in 2001 and administrated by the Yakima River Basin Water Enhancement Project, targets water rights to increase stream flows to

---

[24] Washington State Department of Ecology, 2009.

[25] Washington State Department of Ecology, "Suncadia Water-rights OK'd for Water-banking Program," news release, February 11, 2010.

[26] Landry, 2010.

benefit fish populations during critical periods. This program employs temporary leases and bilateral trades using market-based prices.[27]

While the Kittitas Exchange makes it easier for small landowners and developers to obtain water rights, the local community initially expressed distrust over the Upper Kittitas Exchange. The suspension of well permits was viewed as anti-growth. Some of the more recent issues identified with water banking in Washington state include the following:

- USBR limitations in pricing water because of Federal Acquisition Regulations (FAR). USBR leases and procures water in the state and partners with the Department of Ecology on water-trading activity. However, the FAR does not allow for water prices to exceed the value of the land combined with the water, which tends to lead to undervaluing the water rights and puts the Bureau at a disadvantage in the market.
- Managing localized economic effects with out-of-basin water rights transfers. While an analysis confirmed that water transfers benefit the overall state economy, more can be done to reduce or alleviate economic effects on communities that are selling water rights.
- The Department of Ecology is limited to water purchases. In some instances, sellers are reluctant to sell water separately from the land because they are concerned the remaining land would have limited uses and value. The department would have more options and could benefit multiple environmental concerns if this limitation was lifted.
- A lack of marketing and brokerage assistance such as technical assistance, purchasing expertise and managing and monitoring.[28]

Arkansas River Water Bank Pilot

The first experiment with water banking in Colorado was the Arkansas River Water Bank pilot program. The Arkansas River Basin in southeastern Colorado is Colorado's largest river basin, draining nearly 25,000 square miles. The Arkansas River supports recreation, several cities, and a large agricultural community of farmlands and rangelands that use a large web of ditches to transport water. Within the Arkansas River Basin, where approximately 20 percent of the state's population lives, virtually no water is available for new uses. Permanent sales of water to urban centers (Aurora and the Rocky Flats) as well as to investment groups (High Plains A&M[29] and the Fort Lyon Canal Company) affected agriculture in the basin and led to public concern over threats to agriculture, local economic conditions, and permanent access to water rights. The purpose of the bank was to allow users of Arkansas River water more flexibility in meeting water

[27] Ziemer and Montague, 2011.

[28] Peggy Clifford, "2010 Report to the Legislature: Water Banking in Washington State," Washington State Department of Ecology Water Resource Program, June 2010.

[29] Southwestern, affiliated with High Plains A&M, LLC, began acquiring water rights in 2001. High Plains purchased more than 18,280 acres of land and more than 21,300 shares (equivalent to more than 60,000 acre feet) of senior 1883 water rights in the Arkansas River Valley. See Chapter Five for more details. See also Southwestern Investor Group, "High Plains LLC Water Rights Project," undated; Dan Gordon, "Water Rights: Power Struggle in Southeastern Colorado," *Denver Post*, September 9, 2012.

needs while still meeting interstate compact obligations to Kansas.[30] In effect, the water bank was a way to help agricultural senior rights holders in the Arkansas River Basin maintain their water rights and financially benefit from their water rights, while providing water to high-growth urban areas that were junior rights holders. A map of the Arkansas River Basin within Colorado is shown in Figure A.3 below.

Figure A.3. The Arkansas River Basin in Colorado

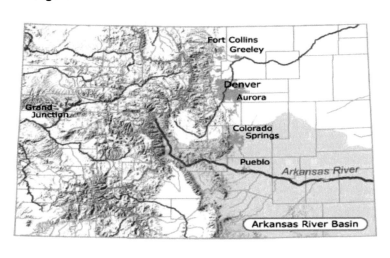

SOURCE: John D. Wiener, "Water Banking as Institutional Adaptation to Climate Variability: The Colorado Experiment," presentation at the AMS 2003 Annual Meeting, Symposium on Impacts of Water Variability: Benefits and Challenges, University of Colorado, Boulder, Colo., 2003.

The bank became active in 2003 after being established in 2001 by Colorado House Bills 1354 (allowing the State Engineer to promulgate banking rules) and in 2003 by House Bill 1381 (which made the bank permanent and restricted out-of-basin transfers). The State Engineer provides regulatory oversight, and the Southeastern Colorado Water Conservancy District (SCWCD) administers the bank. The Arkansas River Water Bank is a clearinghouse that facilitates water transfers through an online bulletin board. Water transfers are short-term (one year) bilateral trades between willing demanders (urban users) and willing providers (agricultural users), and prices are determined by market-based negotiations between parties. Dry-year options, or interruptible supply agreements, based on specific weather conditions and allowing for a schedule of prices also are possible.

No transactions were made in the first nine years since its establishment, yet the lessons from this bank are valuable. Suggested reasons why the bank did not complete any transactions include community support, administrative procedures, geography, pricing, and timing.

---

[30] Kansas has been claiming since 1902 that Colorado takes too much of the river's water. There have been several lawsuits before the U.S. Supreme Court, despite the interstate compact agreement between Colorado and Kansas in 1949. The most recent lawsuit, initiated in 1985, was largely resolved by 2009.

Originally, the bank was allowed to approve out-of-basin transfers without mitigation requirements. As a result of this possibility, the community did not support the bank for the fear of economic loss in the present and future, as well as significantly altering the community socially. Nor did the community have knowledge of a working water market to the same extent as agriculture within the Colorado–Big Thompson project area. This may have contributed to general distrust of water banks. Additionally, participants in the bank were asking for prices that were higher than the going market price of short-term leases, so demanders were able to acquire water in the lease market at a lower price. The administrative processes were expected to average three months, which is a significant amount of time to invest for a one-year lease.

Despite these failings, many in Colorado believe that water banks hold promise for dealing with water-management issues such as the vulnerability of municipalities to drought, permanent loss of agricultural land to development, and a potential compact call on Colorado River water. In fact, several other water bank proposals have been put forward. For example, the Arkansas Basin and Gunnison Basin roundtables[31] are considering a joint project to see if pre-1922 water rights in the Gunnison River basin could be banked in the Blue Mesa Reservoir as a hedge against a Colorado River compact call. While there has never been a call on the Colorado River, recent droughts and possible consequences of climate change suggest the upstream states (Colorado, New Mexico, Utah and Wyoming) could be vulnerable to a compact call from downstream states (Arizona, California and Nevada). Banking water in the Blue Mesa Reservoir potentially could protect those who rely on post-1922 water rights such as the Colorado–Big Thompson Project water, Denver and other municipalities in the region, Twin Lakes, and the Fryingpan Project should a compact call or shortages occur in the future (after the agreement for water shortage sharing signed by the states in 2007 expires in 2026).[32]

---

[31] The Colorado Water for the 21st Century Act established nine roundtables representing the eight major river basins and the Denver metropolitan area in order to address water management in a collaborative and inclusive manner and to encourage locally derived solutions. The roundtables include designated members that represent each county, each municipality and each water conservation district within the river basin. Other members represent the various interests such as agricultural, recreational, local water providers, industrial environmental interest within the river basin, half of which must own water rights.

[32] Chris Woodka, "Blue Mesa Seen as a Water Bank," *Pueblo Chieftain*, February 12, 2012; Colorado Water Conservation District, "Basin Roundtables," undated.

Bibliography

"Across the USA: Colorado," *USA Today*, July 11, 2012.

Adler, Jonathan H., "Warming Up to Water Markets," *Regulation*, Vol. 31, No. 4, Winter 2008–2009. As of October 29, 2015:
http://papers.ssrn.com/sol3/papers.cfm?abstract_id=1341061

Ahern, Nancy, "Green and 'Living' Buildings in Seattle: Seattle Public Utilities' Perspective," American Water Works Association Annual Conference and Exposition 2012, June 12, 2012. As of October 29, 2015:
http://toc.proceedings.com/15850webtoc.pdf

Allen, Greg, "Now Endangered, Florida's Silver Springs Once Lured Tourists," National Public Radio, April 13, 2013.

American Water Works Association, *Buried No Longer: Confronting America's Water Infrastructure Challenge*, 2012. As of October 29, 2015:
http://www.awwa.org/Portals/0/files/legreg/documents/BuriedNoLonger.pdf

———, *Principles of Water Rates, Fees, and Charges* (Sixth Edition), Denver, Colo.: American Water Works Association Manual of Practice, 2012.

———, "Tiered Water Rates," 2012.

———, "Water Reuse Rates and Charges: Survey Results," 2008.

———, "White Paper: Water Rights and Allocations for Sound Resource Management," June 28, 1995. As of October 29, 2015:
http://www.isws.illinois.edu/iswsdocs/wsp/AWWAWhitePaper.pdf

Amero, Richard W., "Lessons From Hetch Hetchy," BalboaParkHistory.net, June 30, 2000. As of October 29, 2015:
http://www.balboaparkhistory.net/glimpses/hetchy.htm

Anderson, Terry L., Brandon Scarborough, and Lawrence Watson, *Tapping Water Markets*, Washington, D.C.: Resources for the Future Press, 2012.

Arizona Corporation Commission, "Utilities Division," undated. As of October 29, 2015:
http://www.azcc.gov/divisions/utilities/water.asp

Arizona Department of Administration, "Arizona Population Projections 2006–2055," 2006. As of October 29, 2015:
http://www.workforce.az.gov/pubs/demography/2006-2055ArizonaProjections.xls

Arizona Department of Water Resources, "Active Management Areas and Irrigation Nonexpansion Areas," undated. As of October 29, 2015:
http://www.azwater.gov/AzDWR/WaterManagement/AMAs/

———, "Overview of the Arizona Groundwater Management Code," undated. As of October 29, 2015:
http://www.azwater.gov/AzDWR/WaterManagement/documents/Groundwater_Code.pdf

———, "Rural Water Resources Study/Rural Water Resources 2003 Questionnaire Report," October 2004. As of October 29, 2015:
http://azmemory.azlibrary.gov/cdm/singleitem/collection/statepubs/id/7040/rec/4

———, "Securing Arizona's Water Future," undated. As of October 29, 2015:
http://www.azwater.gov/azdwr/PublicInformationOfficer/documents/supplydemand.pdf

———, "Tucson AMA: Water Management," updated March 27, 2014. As of March 19, 2016:
http://www.azwater.gov/azdwr/WaterManagement/AMAs/TucsonAMA/images/AMAs.jpg

———, "Upper San Pedro," undated. As of October 29, 2015:
http://www.azwater.gov/azdwr/StatewidePlanning/RuralPrograms/map/UppSanPedPar.htm

———, "Active Management Areas (AMAs) and Irrigation Non-expansion Areas (INAs)," undated. As of October 29, 2015:
http://www.azwater.gov/AzDWR/WaterManagement/AMAs/

———, "Statewide Planning," undated. As of October 29, 2015:
http://www.azwater.gov/AzDWR/StatewidePlanning/RuralPrograms/default.htm

Arizona Senate Research Staff, "Arizona State Senate Background Brief: Arizona Water Banking Authority," Phoenix, Ariz., November 15, 2010.

Arizona State Senate, "Fact Sheet For H.B. 2835: Arizona Water Settlements Act Implementation," March 6, 2006. As of October 29, 2015:
http://www.azleg.gov/legtext/47leg/2r/summary/s.2835nrra-approp.doc.htm

"Arizona Town Holds Successful Effluent Water Auction," *Underground Infrastructure Management*, January/February 2008. As of October 29, 2015:
http://www.uimonline.com/inc/data/archives/2008-02-01.pdf

Arizona Water Banking Authority, homepage, undated. As of October 29, 2015:
http://www.azwaterbank.gov/

———, "Annual Report 2011," July 1, 2012. As of October 29, 2015:
http://www.azwaterbank.gov/Plans_and_Reports_Documents/documents/FinalAnnualReport2011.pdf

———, "Executive Summary," undated. As of October 29, 2015:
http://www.azwaterbank.gov/

"Arizona Water Rights Auction Tops $20m," *Global Water Intelligence*, Vol. 8, No. 11, November 2007. As of October 29, 2015:
http://www.globalwaterintel.com/archive/8/11/general/arizona-water-rights-auction-tops-20msup3sup.html

Associated Press, "Colorado Releases Draft Drought Mitigation Plan," *Denver Post*, July 23, 2013. As of October 29, 2015:
http://www.denverpost.com/breakingnews/ci_23712800/colorado-releases-draft-drought-mitigation-plan?source=rss

Atlanta Regional Commission, "Atlanta Region Plan 2040," 2011. As of October 29, 2015:
http://documents.atlantaregional.com/plan2040/docs/tp_PLAN2040RTP_ch2_072711.pdf

Autobee, Robert, "Colorado–Big Thompson Project," U.S. Bureau of Reclamation, 1996. As of October 29, 2015:
http://www.usbr.gov/projects//ImageServer?imgName=Doc_1303159857902.pdf

Ayotte, Joseph D., Jo Ann M. Gronberg, and Lori E. Apodaca, "Trace Elements and Radon in Groundwater Across the United States, 1992–2003," U.S. Geological Survey, Scientific Investigations Report 2011–5059, 2011. As of October 29, 2015:
http://pubs.usgs.gov/sir/2011/5059/pdf/sir2011-5059_report-covers_508.pdf

Badawy, Manuela, "Looking for Gold in Water Investments," Reuters, December 8, 2011. As of October 29, 2015:
http://www.reuters.com/article/2011/12/12/us-waterfunds-investments-idUSTRE7BB0VH20111212

Ballinger, Edward, Jr., "In Re the General Adjudication of all Rights to Use Water in the Gila River System and Source, Case Regarding Fort Huachuca, Order Granting the Special Master's Motion for Adoption of the April 4, 2008 Report Regarding Fort Huachuca," Civil No. w1-11-605, Superior Court of Arizona, Maricopa County, September 7, 2011. As of October 29, 2015:
http://www.superiorcourt.maricopa.gov/SuperiorCourt/GeneralStreamAdjudication/_ballinger/sp605ord090711.pdf

Barnett, Cynthia, "Water Works: Communities Imagine Ways of Making Every Drop Count," *Orion Magazine*, July/August 2013. As of October 29, 2015:
https://orionmagazine.org/article/water-works/

Baroni, Megan, "Pricing Strategies to Manage Water Demand," from *Whose Drop Is It Anyway? Legal Issues Surrounding Our Nation's Water Resources*, Washington, D.C.: American Bar Association, April 2011.

———, "Lessons from the "Tri-State" Water War," *American Bar Association*, Vol. 35, No. 2, Winter 2012. As of October 29, 2015:
http://www.americanbar.org/publications/state_local_law_news/2011_12/winter_2012/tri-state_water_war.html

Barringer, Felicity, "As 'Yuck Factor' Subsides, Treated Wastewater Flows from Taps," *New York Times*, February 9, 2012. As of October 29, 2015:
http://www.nytimes.com/2012/02/10/science/earth/despite-yuck-factor-treated-wastewater-used-for-drinking.html?pagewanted=all

———, "In California, What Price Water?" *New York Times*, February 28, 2013. As of October 29, 2015:
http://www.nytimes.com/2013/03/01/business/energy-environment/a-costly-california-desalination-plant-bets-on-future-affordability.html

———, "Storing Water for a Dry Day Leads to Suits," *New York Times*, July 26, 2011. As of October 29, 2015:
http://www.nytimes.com/2011/07/27/science/earth/27waterbank.html

Barwin, Bob, "Water Right/Mitigation: Banking in the Yakima Basin," Edmonds, Wash.:, Water Banking/Exchange Workshop, October 6, 2008. As of October 29, 2015:
http://www.ecy.wa.gov/programs/wr/instream-flows/Images/trust/bbarwin_mitbank_112108.pdf

Basta, Elizabeth, and Bonnie Colby, "Water Market Trends: Transactions, Quantities, and Prices," *Appraisal Journal*, Vol. 78, No. 1, January 1, 2010.

Beaujon, David, "Rainwater Harvesting in Colorado," Issue Brief 09-02, Colorado Legislative Council Staff, August 2009. As of October 29, 2015:
http://www.colorado.gov/cs/Satellite?blobcol=urldata&blobheader=application%2Fpdf&blobkey=id&blobtable=MungoBlobs&blobwhere=1251597990496&ssbinary=true

Beecher, Janice A., "Primer on Water Pricing," Lansing, Mich.: Michigan State University, Institute of Public Utilities Regulatory Research and Education, November 1, 2011. As of October 29, 2015:
http://ipu.msu.edu/research/pdfs/IPU%20Primer%20on%20Water%20Pricing%20Beecher%20(2011).pdf

Berwyn, Bob, "West Slope Eyes 'Water Bank' Pool of Senior Water Rights Could Avert Drought Cuts," *Summit Daily News*, August 31, 2008.

Billington, David P., Donald C. Jackson, and Martin V. Melosi, *The History of Large Federal Dams: Planning, Design, and Construction in the Era of Big Dams*, Denver, Colo.: U.S. Department of the Interior, U.S. Bureau of Reclamation, 2005.

Bobb, Nancy E., and Greg A. Kolle, "Bridging the Bay," *Public Roads*, FHWA-HRT-12-006, Vol. 76, No. 2, September/October 2012. As of October 29, 2015: http://www.fhwa.dot.gov/publications/publicroads/12septoct/01.cfm

Boxall, Bettina, "Seawater Desalination Plant Might be Just a Drop in the Bucket," *Los Angeles Times*, February 17, 2013. As of October 29, 2015: http://articles.latimes.com/2013/feb/17/local/la-me-carlsbad-desalination-20130218

Bracken, Nathan, and the Western States Water Council, "Water Reuse in the West: State Programs and Institutional Issues," Salt Lake City, Utah: Western States Water Council, July 2011. As of October 29, 2015: http://www.westernstateswater.org/wp-content/uploads/2012/10/Water-Reuse-Report-Final-Published-Version-Summer-2012-1.pdf

Brandt, Peiffer, and B. Ramaley, "Balancing Fixed Costs and Revenues: Newport News Waterworks' Shifting Revenues Recovery to Better Match Fixed vs. Variable Costs," American Water Works Association Annual Conference and Exposition 2012, June 13, 2012. As of October 29, 2015: http://toc.proceedings.com/15850webtoc.pdf

Brean, Henry, "Water Authority Puts Water Banking Agreement with Arizona on Hold," *Las Vegas Review-Journal*, April 18, 2013. As of November 9, 2015: http://www.reviewjournal.com/news/water-environment/water-authority-puts-water-banking-agreement-arizona-hold

Brewer, Jedidiah, Robert Glennon, Alan Ker and Gary Libecap, "Water Markets in the West: Prices, Trading and Contractual Forms," *Economic Inquiry*, Vol. 46, No. 2, April 2008. As of October 29, 2015: http://www2.bren.ucsb.edu/~glibecap/Econ%20Inquiry.pdf

Brookshire, David S., Bonnie Colby, Mary Ewers, and Phillip T. Ganderton, "Market Prices for Water in the Semi-Arid West of the United States," *Water Resources Research*, Vol. 40, 2004. As of October 29, 2015: http://www.geo.oregonstate.edu/classes/ecosys_info/readings/2003WR002846.pdf

Brookshire, David, Philip Ganderton, Mary Ewers, Bonnie Colby, and Steve Stewart, "Water Markets in the Southwest: Why and Where?" *Southwest Hydrology*, Vol. 3, March/April 2004.

Burgarino, Paul, "Use of Recycled Water Trickles into Delta Region," *Contra Costa Times*, March 8, 2013. As of November 2, 2015: http://www.contracostatimes.com/ci_22751013/use-recycled-water-trickles-into-delta-region

Burke, Garance, "Fracking Fuels Water Fights in Nation's Dry Spots," Associated Press, June 17, 2013. As of November 2, 2015:
http://news.yahoo.com/fracking-fuels-water-fights-nations-133753148.html?goback=.gde_733277_member_250606606

California Department of Water Resources, "California Water Plan Update 2009: Chapter 15, Federal Interests," 2009. As of November 2, 2015:
http://www.waterplan.water.ca.gov/cwpu2009/

———, "Integrated Regional Water Management Grants." As of November 2, 2015:
http://www.water.ca.gov/irwm/grants/index.cfm

———, "Strategic Plan for the Future of Integrated Regional Water Management in California," September 2012. As of November 2, 2015:
http://www.water.ca.gov/irwm/stratplan/documents/DWR_BrochureV9B.pdf

California Environmental Protection Agency, State Water Resources Control Board, "The Water Rights Process," undated. As of November 2, 2015:
http://www.waterboards.ca.gov/waterrights/board_info/water_rights_process.shtml#law

California Water Impact Network, "The Kern Water Bank," undated. As of November 2, 2015:
http://www.c-win.org/kern-water-bank.html

Caprioli, Jennifer M., "Water Resources Management at Fort Huachuca Continues 15 Years Later," U.S. Army news article, July 26, 2010. As of November 2, 2015:
http://www.army.mil/-news/2010/07/26/42850-water-resources-management-at-fort-huachuca-continues-15-years-later/

Carey, Janis M., and David L. Sunding, "Emerging Markets in Water: A Comparative Institutional Analysis of the Central Valley and the Colorado–Big Thompson Projects," *Natural Resources Journal*, Vol. 41, 2001, pp. 283–328. As of October 29, 2015:
http://lawschool.unm.edu/nrj/volumes/41/2/03_carey_emerging.pdf

Caroom, Doug, and Paul Elliott, "Water Rights Adjudication—Texas Style," *Texas Bar Journal*, Vol. 44, No. 10, November 1981.

Central Kansas Water Bank Association, *Charter of the Central Kansas Water Bank Association,* June 2005. As of November 2, 2015:
http://www.gmd5.org/Water_Bank/Archive/2013-03-12_CKWBA_Charter.pdf

———, "Kansas Water Banking Act 2001," undated. As of November 2, 2015:
Archive/2001WaterBankingAct.pdf

Central Platte Natural Resources Division, "Water Banking Program," undated. As of November 2, 2015:
http://www.cpnrd.org/Water_Bank.html

Chacon, Daniel, "SDS Water Rate Hikes May be Lower than Planned," *The Gazette*, April 18, 2012. As of November 2, 2015:
http://www.gazette.com/articles/rate-137124-sds-water.html

Chan, Chris, Patrick Laplagne, and David Appels, "The Role of Auctions in Allocating Public Resources," Productivity Commission Staff Research Paper, Productivity Commission, Melbourne, Australia, 2003. As of November 2, 2015:
https://ideas.repec.org/p/wpa/wuwpmi/0304007.html

Chang, Ike, Steven Galing, Carolyn Wong, Howell Yee, Elliot Axelband, Mark Onesi, and Kenneth P. Horn, *Use of Public-Private Partnerships to Meet Future Army Needs*, Santa Monica, Calif.: RAND Corporation, MG-997-A, 1999. As of November 2, 2015:
http://www.rand.org/pubs/monograph_reports/MR997.html

Characklis, Gregory W., Brian R. Kirsch, Jocelyn Ramsey, Karen E. M. Dillard, and C. T. Kelley, "Developing Portfolios of Water Supply Transfers," *Water Resources Research*, July 11, 2005. As of November 2, 2015:
http://www4.ncsu.edu/~ctk/PAPERS/Portfolios_WRR.pdf

Chattahoochee Riverkeeper, "Charting a New Course for Georgia's Water Security," July 2010.

Chesapeake Bay Commission, "History of the Commission," undated. As of November 2, 2015:
http://www.chesbay.us/history.html

Chesapeake Bay Program, "1987 Chesapeake Bay Agreement," undated. As of October 29, 2015:
http://www.chesapeakebay.net/content/publications/cbp_12510.pdf

———, "Programs & Projects," undated. As of November 2, 2015:
http://www.chesapeakebay.net/about/programs

———, "Watershed Implementation Plans," undated. As of November 2, 2015:
http://www.chesapeakebay.net/about/programs/watershed

"Chino Basin Auction Postponed as Bidders Take Fright," *Global Water Market Intelligence*, November 5, 2009. As of October 29, 2015:
http://www.globalwaterintel.com/news/2009/45/chino-basin-auction-postponed-bidders-take-fright.html

Chino Basin Watermaster, "Water Auction Legal Summary," undated. As of November 2, 2015:
http://www.cbwm.org/waterauction/docs/54001.CBWM.Summary.pdf

———, "Working Collaboratively to Maximize the Benefits from the Chino Groundwater Basin," undated. As of November 2, 2015:
http://www.cbwm.org/waterauction/docs/9_2_09.FINAL.pdf

Cianci, Michael J., Jr., James F. Williams, and Eric S. Binkley, "The New National Defense Water Right - An Alternative to Federal Reserved Water Rights For Military Installations," *Air Force Law Review*, Air Force Judge Advocate General School, Vol. 48=, 2000. As of November 2, 2015:
https://litigation-essentials.lexisnexis.com/webcd/app?action=DocumentDisplay&crawlid=1&srctype=smi&srcid=3B15&doctype=cite&docid=48+A.F.+L.+Rev.+159&key=bf20a1cf97593db0db1d557b1decf539

City of Austin and Lower Colorado River Authority, "Supplemental Water Supply Agreement," November 14, 2007. As of November 2, 2015:
http://www.lcra.org/water/water-supply/water-supply-contracts/Documents/lcra_austin_water_agreement07.pdf

City of Lake Oswego and the City of Tigard, "Intergovernmental Agreement Regarding Water Supply Facilities, Design, Construction, and Operation," August 2008.

City of Palo Alto, "Palo Alto Recycled Water Pipeline Project," fact sheet, undated.

City of Phoenix, "Historical Population and Water Use," undated. As of November 2, 2015:
https://www.phoenix.gov/waterservices/resourcesconservation/yourwater/historicaluse

City of Redwood City, "Redwood City Recycled Water Project: Seaport," fact sheet, undated.

Clements, Janet, and Robert Raucher, "Expanding Utility Services Beyond Water Supply to Improve Customer Satisfaction and Utility Effectiveness," presentation at the American Water Works Association 2012 Annual Conference, June 13, 2012.

Clifford, Peggy, "2010 Report to the Legislature: Water Banking in Washington State," Water Resource Program, Washington State Department of Ecology, June 2010. As of November 2, 2015:
https://fortress.wa.gov/ecy/publications/SummaryPages/1111072.html

Clifford, Peggy, Clay Laundry, and Andrea Larsen-Hayden, "Analysis of Water Banks in the Western States," Olympia, Wash.: Washington Department of Ecology Water Resources Program, Report 04-11-011, July 2004. As of November 2, 2015:
http://www.ecy.wa.gov/biblio/0411011.html

Colby, Bonnie, "Reallocating Water: Evolving Markets, Values and Prices in the Western United States," *Journal of Contemporary Water Research and Education*, Vol. 92, 1993. As of November 2, 2015:
http://opensiuc.lib.siu.edu/jcwre/vol92/iss1/5/

Colby, Bonnie, et al., "Arizona's Water Future: Challenges and Opportunities," Tucson, Ariz.: University of Arizona Office of Economic Development and Water Resources Research Center, 2004. As of November 2, 2015:
https://portal.azoah.com/08A-AWS001-DWR/Omnia/Exhibit%20404%20Arizona's%20Water%20Future.pdf

Colby, Bonnie, and Katharine Jacobs (eds.), *Arizona Water Policy: Management Innovations in an Urbanizing, Arid Region*, Washington, D.C.: Resources for the Future, 2006.

Colorado Department of Natural Resources, "The Colorado Drought Mitigation and Response Plan," September 2010. As of November 2, 2015:
http://cwcbweblink.state.co.us/WebLink/0/doc/173111/Electronic.aspx?searchid=45a1d11c-9ccf-474b-bed4-2bccb2988870

Colorado Department of Public Health And Environment, "Regulation No. 84 Reclaimed Water Control Regulation," Denver, Colo.: Water Quality Control Commission, 5 CCR 1002-84, August 2010. As of November 2, 2015:
https://www.colorado.gov/pacific/sites/default/files/Regulation-84.pdf

Colorado Division of Water Resources, "A Summary of Compacts and Litigation Governing Colorado's Use of Interstate Streams," 2006. As of November 2, 2015:
http://water.state.co.us/DWRIPub/Documents/compactsreport.pdf

Colorado Governor's Office, "Directing the Colorado Water Conservation Board to Commence Work on the Colorado Water Plan," Executive Order D2013-005, May 14, 2013. As of November 5, 2015:
http://cwcbweblink.state.co.us/WebLink/ElectronicFile.aspx?docid=171100&searchid=c428f27e-6b83-4a97-908c-31bb6996cf74&dbid=0

Colorado River Governance Initiative, "Colorado River Law and Policy: Frequently Asked Questions (FAQs)," Boulder, Colo.: University of Colorado-Boulder Natural Resources Law Center, Western Water Policy Program, March 2011. As of November 5, 2015:
http://scholar.law.colorado.edu/cgi/viewcontent.cgi?article=1005&context=books_reports_studies"

Colorado Springs Utilities, "2008–2012 Water Conservation Plan," January 30, 2008. As of November 2, 2015:
http://www.csu.org/residential/water/Documents/item14309.pdf

———, "Business," undated. As of November 2, 2015:
https://www.csu.org/Pages/business.aspx.

———, "Residential," undated. As of November 2, 2015:
https://www.csu.org/Pages/residential.aspx

———, "Water Rate Schedules," March 26, 2013. As of October 29, 2015:
https://www.csu.org/CSUDocuments/tariffwater.pdf

———, "Water Shortage Rate Changes Effective August 1, 2013," undated.

Colorado Water Conservation Board, "Basin Roundtables," undated. As of November 2, 2015:
http://cwcb.state.co.us/water-management/basin-roundtables/Pages/main.aspx

———, "Colorado Water Plan Interim Website," undated.

———, "Colorado's Water Needs," undated. As of October 29, 2015:
http://cwcb.state.co.us/water-management/water-supply-
planning/Pages/ColoradosWaterSupplyNeeds.aspx

———, "Colorado's Water Supply Future: State of Colorado 2050 Municipal and Industrial
Water Use Projections," July 2010. As of November 2, 2015:
http://cwcb.state.co.us/water-management/water-supply-
planning/documents/swsi2010/appendix%20h_state%20of%20colorado%202050%20munici
pal%20and%20industrial%20water%20use%20projections.pdf

———, "Drought," undated. As of November 2, 2015:
http://cwcb.state.co.us/water-management/drought/Pages/main.aspx

———, "The Municipal & Industrial Water Supply and Demand Gap," undated. As of October
29, 2015:
http://cwcb.state.co.us/water-management/water-supply-
planning/Pages/TheWaterSupplyGap.aspx

———, "Water Supply Planning," undated. As of November 2, 2015:
http://cwcb.state.co.us/water-management/water-supply-planning/Pages/main.aspx

Cooley, Heather, Peter H. Gleick, and Gary Wolff, "Desalination, with a Grain of Salt: A
California Perspective," Pacific Institute, June 2006. As of November 2, 2015:
http://pacinst.org/publication/desalination-with-a-grain-of-salt-a-california-perspective-2/

Council of Great Lakes Governors, "Overview," undated. As of November 2, 2015:
http://www.cglslgp.org/about-us/

———, "Council of Great Lakes Governors Projects," undated. As of November 2, 2015:
http://www.cglslgp.org/about-us/projects/

Criddle, Craig, "Wastewater as a Valuable Resource," Palo Alto, Calif.: Stanford University
Woods Institute for the Environment Solution Brief, May 2010. As of November 2, 2015:
https://woods.stanford.edu/sites/default/files/files/Waster-Salon-I-Solution-Brief-Craig-
Criddle-20100316.pdf

Cummings, Ronald G., Charles A. Holt, and Susan K. Laury, "Using Laboratory Experiments for Policy Making: An Example from the Georgia Irrigation Reduction Auction," Atlanta, Ga.: Georgia State University Andrew Young School of Policy Studies Research Paper Series, Working Paper 06-14, September 2003. As of November 2, 2015: http://papers.ssrn.com/sol3/papers.cfm?abstract_id=893800

Davis, Tony, "Arizona's Drinking-Water Needs Will Force Trade-Offs," *Arizona Daily Star*, February 23, 2014. As of November 9, 2015: http://tucson.com/news/science/environment/arizona-s-drinking-water-needs-will-force-trade-offs/article_abaab24a-954b-5825-9939-e694194507b7.html

Dellapenna, Joseph W., "Special Challenges to Water Markets in Riparian States," *Georgia State University Law Review*, Vol. 21, No., 2, Winter 2004. As of November 2, 2015: http://readingroom.law.gsu.edu/cgi/viewcontent.cgi?article=2070&context=gsulr

Denver Water, "2012 Billing Rates," undated.

DiNatale, Kelly, "Purple Mountain Majesties—Water Reuse in the Rockies," presentation at the 2009 Water Reuse Workshop, August 13, 2009. As of November 2, 2015: http://dinatalewater.com/presentations/opportunities_constraints_reuse/Slide1.jpg

Dixon, Lloyd, Nancy Moore, and Susan Schechter, *California's 1991 Drought Water Bank*, Santa Monica, Calif.: RAND Corporation, MR-301-CDWR, 1993. As of November 2, 2015: http://www.rand.org/pubs/monograph_reports/MR301.html

Donohew, Zack, "Water Transfer Level Dataset," undated. Santa Barbara, Calif.: University of California, Santa Barbara. As of November 2, 2015: http://www.bren.ucsb.edu/news/water_trans_10_intro.doc

Dove, Justin, "Three Easy Ways to Invest in Water," *InvestmentU,* April 2, 2012. As of November 2, 2015: www.investmentu.com/2012/April/invest-in-water.html

Dziegielewski, Benedykt and Jack Kiefer, "U.S. Water Demand, Supply and Allocation: Trends and Outlook," Washington, D.C.: U.S. Army Corps of Engineers Institute for Water Resources, Report 2007-R-3, December 2008. As of November 2, 2015: http://planning.usace.army.mil/toolbox/library/IWRServer/2007-R-03.pdf

Ecosystem Marketplace Team, "Water Trading: The Basics," April 16, 2008. As of November 2, 2015: http://www.ecosystemmarketplace.com/pages/dynamic/article.page.php?page_id=5788&section=home&eod=1

Eden, Susanna, Robert Glennon, Alan Ker, et al., "Agricultural Water to Municipal Use: The Legal and Institutional Context for Voluntary Transactions in Arizona," *The Water Report*, December 15, 2008. As of November 2, 2015:
http://wrrc.arizona.edu/sites/wrrc.arizona.edu/files/ag_to_muni_article.pdf

El Paso Water Utilities, "DESALINATION: Setting the Stage for the Future," 2007. As of November 2, 2015:
http://epwu.org/water/desal_info.html

———, "Reclaimed Water," 2007. As of November 2, 2015:
http://www.epwu.org/reclaimed_water/

———, "Wastewater Treatment: Fred Hervey Water Reclamation Plant," 2007. As of November 2, 2015:
http://www.epwu.org/wastewater/fred_hervey_reclaimation.html

ENR Mountain States, "Fort Carson Brigade Battalion Headquarters Earns LEED Platinum," July 5, 2012. As of November 2, 2015:
http://mountainstates.construction.com/mountainstates_construction_projects/2012/0705-fort-carson-brigade-battalion-headquarters-earns-leed-platinum.asp.

Equinox Center, "Water Pricing Primer," October 2009.

Fairfax County, Virginia, "What Is Stormwater Management?" 2013. As of November 2, 2015:
http://www.fairfaxcounty.gov/dpwes/stormwater/whatis_swm.htm

Faulkner, Timothy, "Sustainable Water: Reaching Net Zero*," U.S. Army Journal of Installation Management*, Spring 2011. As of November 2, 2015:
http://www.imcom.army.mil/portals/0/hq/about/publications/journal/562DBF6B-423D-452D-4F468DA7B4C7AF46.pdf

Feller, Joseph M., "The Adjudication that Ate Arizona Water Law," *Arizona Law Review*, Vol. 49, 2007. As of November 2, 2015:
http://www.law.asu.edu/files/Faculty/Publications/915.pdf

Finley, Bruce, "Fracking Bidders Top Farmers at Water Auction," *The Denver Post*, April 2, 2012. As of November 2, 2015:
http://www.denverpost.com/entertainmentcolumnists/ci_20306480/fracking-bidders-top-farmers-at-water-auction

Flatt, Courtney, "Water Banks Help Washington Land Developers," Eugene, Ore: KLCC Public Radio, April 4, 2012. As of November 2, 2015:
http://klcc.org/Feature.asp?FeatureID=3296

Florida Department of Environmental Protection, "2011 Reuse Inventory," Tallahassee, Fla.: Water Reuse Program, May 2012.

Foos, Annabelle, "Geology of the Colorado Plateau," Akron, Ohio: University of Akron, 1999. As of November 2, 2015:
http://nature.nps.gov/geology/education/foos/plateau.pdf

Fort Benning, "Joint Land Use Study," 2008. As of November 2, 2015:
http://www.mrrpc.com/Misc_pdfs/Ft_Benning_JLUS_Report.pdf

Fort Bragg, "Sustainable Fort Bragg: Water," undated. As of November 2, 2015:
http://sustainablefortbragg.com/water/

Fort Carson, "Fort Carson 2006 Sustainability Report," 2006.

———, "Fort Carson Sustainability Baseline," August 2002.

———, "Fort Carson Sustainability Guide 2011–2012: Knowing Your Piece of the Sustainability Puzzle," 2011. As of November 2, 2015:
http://www.carson.army.mil/DPW/Documents/sustainability-guide.pdf

Fort Huachuca Directorate of Public Works, "Water Resource Management Plan: Fort Huachuca, AZ," April 3, 2009.

"Fracking Companies Make Top Bids for Water Alongside Colorado Farmer," *Huffington Post*, April 4, 2012. As of October 29, 2015:
http://www.huffingtonpost.com/2012/04/02/fracking-bidders_n_1398786.html?ref=denver

Friederici, Peter, "Making an Effluent Market," *High Country News*, September 17, 2007. As of November 2, 2015:
http://www.hcn.org/issues/354/17235/print_view

Galbraith, Kate, "Texas' Water Woes Spark Interest in Desalination," *The Texas Tribune*, June 10, 2012. As of November 2, 2015:
http://www.texastribune.org/2012/06/10/texas-water-woes-spark-interest-desalination/

Ganzel, Bill, "State to State Water Agreements," Ganzel Group, undated. As of November 2, 2015:
http://www.livinghistoryfarm.org/farminginthe50s/water_13.html

Garrity, Michael, "Big Opportunity (and Big Misperceptions) in the Yakima River Basin," American Rivers River Blog, November 23, 2010.

Gelt, Joe, "Land Subsidence, Earth Fissures Change Arizona's Landscape," *Arroyo*, Vol. 6, No. 2, Summer 1992. As of November 2, 2015:
http://worldcat.org/arcviewer/2/WCA/2009/12/14/H1260834861347/viewer/file2.html

Georgia Department of Natural Resources, "Middle Chattahoochee Regional Plan," September 2011. As of November 2, 2015:
http://www.middlechattahoochee.org/pages/our_plan/Middle_Chattahoochee_Regional_Water_Plan.php

Georgia Environmental Protection Division, "Georgia's Water Future in Focus: Highlights of Regional Water Planning 2009–2011," December 2011. As of November 2, 2015:
http://www.georgiawaterplanning.org/documents/Highlights_of_Regional_Water_Planning.pdf

Gerlak, Andrea K., and John E. Thorson, "General Stream Adjudications Today: An Introduction," *Journal of Contemporary Water Research & Education*, No. 133, May 2006. As of November 2, 2015:
http://opensiuc.lib.siu.edu/jcwre/vol133/iss1/1/

Getches, David H., "The Metamorphosis of Western Water Policy: Have Federal Laws and Local Decisions Eclipsed the States' Role?" *Stanford Environmental Law Journal,* Vol. 20, No. 3, January 2001. As of November 2, 2015:
https://lawweb.colorado.edu/profiles/pubpdfs/getches/GetchesSELJ.pdf

Gibbons, Christia, "Prescott Valley Water Auction Could Impact Phoenix," *Phoenix Business Journal,* October 29, 2006. As of November 2, 2015:
http://www.bizjournals.com/phoenix/stories/2006/10/30/story18.html?page=all

Gibson, Campbell, "Population of the 100 Largest Cities and Other Urban Places in the United States: 1790 to 1990," Washington, D.C.: U.S. Census Bureau Population Division Working Paper No. 27, June 1998. As of November 2, 2015:
http://www.census.gov/population/www/documentation/twps0027/twps0027.html

Glennon, Robert, *Unquenchable: America's Water Crisis and What to Do About It*, Washington, D.C.: Island Press, 2009.

Goodnough, Abby, Monica Davey, and Mitch Smith, "When the Water Turned Brown," *New York Times*, January 23, 2016.

Gordon, Dan, "Water Rights: Power Struggle in Southeastern Colorado," *Denver Post*, September 9, 2012. As of November 2, 2015:
http://www.denverpost.com/ci_21485072/where-money-flows

Graf, William L., "Dam Nation: A Geographic Census of American Dams and Their Large-Scale Hydrologic Impacts," *Water Resources Research*, Vol. 35, No. 4, 1999. As of November 02, 2015:
http://scholarcommons.sc.edu/cgi/viewcontent.cgi?article=1052&context=geog_facpub

Greater Monterey County Integrated Regional Water Management Program, website, undated. As of October 29, 2015:
http://www.greatermontereyirwmp.org/

Groves, David G., Jordan R. Fischbach, Evan Bloom, Debra S. Knopman, and Ryan Keefe, *Adapting to a Changing Colorado River: Making Future Water Deliveries More Reliable Through Robust Management Strategies*, Santa Monica, Calif.: RAND Corporation, RR-242-BOR, 2013. As of November 2, 2015:
http://www.rand.org/pubs/research_reports/RR242.html

Hagy, James D., Walter R. Boynton, Carolyn W. Keefe, and Kathryn V. Wood, "Hypoxia in Chesapeake Bay, 1950-2001: Long-term Change in Relation to Nutrient Loading and River Flow," *Estuaries*, Vol. 27, No. 4, August 2004. As of November 2, 2015:
http://link.springer.com/article/10.1007%2FBF02907650

Hammack, Katherine, "Energy and Sustainability Priorities and Opportunities," *U.S. Army Journal of Installation Management*, Spring 2011. As of November 2, 2015:
http://www.imcom.army.mil/portals/0/hq/about/publications/journal/562DBF6B-423D-452D-4F468DA7B4C7AF46.pdf

Hanak, Ellen, "Who Should be Allowed to Sell Water in California? Third Party Issues and the Water Market," San Francisco, Calif.: Public Policy Institute of California, July 2003. As of November 2, 2015:
http://www.ppic.org/main/publication.asp?i=337

Hanak, Ellen, and Elizabeth Stryjewski, "California's Water Market, by the Numbers: Update 2012," San Francisco, Calif.: Public Policy Institute of California, Report Number R-1114EHR, November 2012. As of November 2, 2015:
http://www.ppic.org/main/publication.asp?i=1041

Hardner, Jared, and R. E. Gullison, "Independent External Evaluation of the Columbia Basin Water Transactions Program (2003–2006)," Hardner and Gullison Associates LLC, October 7, 2007. As of November 2, 2015:
http://www.nfwf.org/cbwtp/Documents/CBWTP_Eval_Report_10-7_FINAL.pdf

Hartwell, Ray, and Bruce Aylward, "Auctions and the Reallocation of Water Rights in Central Oregon," Deschutes River Conservancy, River Paper series No. 1, April 2007. As of November 2, 2015:
http://www.ecosystemeconomics.com/Resources_files/Hartwell%20%26%20Aylward%20(2007)%20Auctions.pdf

Healy, Jack, "For Farms in the West, Oil Wells are Thirsty Rivals," *New York Times*, September 5, 2012. As of November 2, 2015:
http://www.sott.net/article/251187-For-Farms-in-the-West-Oil-Wells-Are-Thirsty-Rivals

Heimbuch, Jaymi, "Desalination Plant Helps Save a California Coastal Community," The Learning Channel, 2013.

Henry, Terrence, "When Wells Run Dry: Spicewood Beach, Texas is Out of Water," NPR, January 31, 2012. As of November 2, 2015: http://stateimpact.npr.org/texas/2012/01/31/when-wells-run-dry-spicewood-beach-is-out-of-water/

"Hetch Hetchy Reservoir," City of Mountain View, undated.

"Hetch-Hetchy Valley, Sierra Nevada Mountains, CA," Washington, D.C.: Library of Congress Prints and Photographs Division, 1911. As of October 29, 2015: http://lcweb2.loc.gov/service/pnp/pan/6a19000/6a19500/6a19572r.jpg

Hollis, Paul L., "Farmer's Irrigation Lawsuit Dismissed," *Southeast Farm Press*, January 16, 2002. As of November 2, 2015: http://southeastfarmpress.com/farmers-irrigation-lawsuit-dismissed

Holmquist, Abigail, Todd Cristiano, Marc Waage, Steve Price, and Mary Stahl, "Denver Water's Plan for a Sustainable Reuse System," presentation at the 2011 WateReuse Symposium, September 12–14, 2011.

Howe, Charles W., and John D. Weiner, "The Arkansas Water Bank Pilot Program—Progress So Far," Boulder, Colo.: National Center of Atmospheric Research, Water Conference, Institute for the Study of Society and Environment, 2002. As of November 2, 2015: http://www.isse.ucar.edu/water_conference/CD_files/Posters/Howe%20and%20Weiner,%20 Arkansas%20Water%20Bank.ppt

Howe, Charles W., and Christopher Goemans, "Water Transfers and Their Impacts: Lessons From Three Colorado Water Markets," *Journal of The American Water Resources Association*, Vol. 39, No. 5, October 2003. As of November 2, 2015: http://onlinelibrary.wiley.com/doi/10.1111/j.1752-1688.2003.tb03692.x/abstract

"Innovations in American Government Awards: California Drought Water Bank," Cambridge, Mass.: Kennedy School of Public Policy, Harvard University, 1995. As of October 29, 2015: http://www.innovations.harvard.edu/awards.html?id=563498

International Boundary and Water Commission of United States and Mexico, "Minute No. 242, Permanent and Definitive Solution to the International Problem of Salinity of the Colorado River," August 30, 1973. As of November 2, 2015: http://www.ibwc.gov/Files/Minutes/Min242.pdf

Ironhouse Sanitary District, "Ironhouse Sanitary District Recycled Water Project," fact sheet, Oakley, Calif., undated.

Iseman, Tom, "Banking on Colorado Water," *PERC Reports*, Vol. 28, No. 1, Spring 2010. As of November 2, 2015:
http://www.perc.org/articles/article1227.php

———, "Innovative Water Transfers: West-Wide Practice and Potential," American Water Resources Association, April 27, 2011.

Ivahnenko, Tamara and Jennifer L. Flynn, "Estimated Withdrawals and Use of Water in Colorado, 2005," U.S. Geological Survey, April 2010. As of November 2, 2015:
http://pubs.usgs.gov/sir/2010/5002/pdf/SIR10-5002.pdf

Jarvis, Glenn, "Historical Development of Texas Surface Water Law: Background of the Appropriation and Permitting System," McAllen, Tex., April 2008. As of November 2, 2015:
http://www.glennjarvis.com/water-rights-adjudication/HistoricalDevelopment_TexasWaterLaw.pdf

Jeanes, Michael, "The Water Case Turns 30," *Maricopa Lawyer*, Vol. 28, No. 5, May 2009. As of November 2, 2015:
http://www.clerkofcourt.maricopa.gov/news/ClerksCorner05-09.pdf

Jenicek, Elisabeth M., Rebecca A. Carroll, Laura E. Curvey, MeLena S. Hessel, Ryan M. Holmes, and Elizabeth Pearson, "Water Sustainability Assessment for Ten Army Installations," U.S. Army Corps of Engineers Engineer Research and Development Center, ERDC/CERL TR-11-5, March 2011. As of November 2, 2015:
http://www.aepi.army.mil/docs/whatsnew/ERDC-CERL_TR-11-5%20Water%20Sustainability%20Assessment%20for%20Ten%20Army%20Installations.pdf

Jensen, Ric, "Why Droughts Plague Texas," *Texas Water Resources*, Vol. 22, No. 2, Summer 1996. As of November 2, 2015:
http://twri.tamu.edu/newsletters/texaswaterresources/twr-v22n2.pdf

Job, Charles A., *Groundwater Economics*, Boca Raton, Fla: CRC Press, 2010.

Johnson Controls, "Energy Savings Performance Contract Feasibility Study Report: Fort Bliss, TX, Project 7," June 6, 2013.

Johnson, Nathan C., "Protecting Our Water Compacts: The Looming Threat of Unilateral Congressional Interaction," *Wisconsin Law Review*, Vol. 875, 2010. As of November 2, 2015:
http://papers.ssrn.com/sol3/papers.cfm?abstract_id=1803583

Johnson, Tad, "Battling Seawater Intrusion in the Central and West Coast Basins," Lakewood, Calif.: Water Replenishment District of Southern California, Technical Bulletin, Vol. 13, Fall 2007. As of November 2, 2015:
http://www.wrd.org/engineering/seawater-intrusion-los-angeles.php

Kansas Department of Agriculture, "Kansas-Colorado Arkansas River Compact Fact Sheet," August 2009. As of November 2, 2015:
http://www.ksda.gov/includes/document_center/dwr/Publications/ArkCompactFactSheetAug 2009.pdf

Kansas Department of Agriculture, "Voluntary Incentive Program," undated.

Karl, Thomas R., Jerry M. Melillo, and Thomas C. Peterson (eds.), *Global Climate Change Impacts in the United States,* New York: United States Global Change Research Program. Cambridge University Press, 2009. As of November 2, 2015:
http://downloads.globalchange.gov/usimpacts/pdfs/climate-impacts-report.pdf

Kats, Greg, Leon Alevantis, Adam Berman, Evan Mills and Jeff Perlman, "The Costs and Financial Benefits of Green Buildings: A Report to California's Sustainabile Building Task Force," U.S. Green Building Council, October 2003. As of November 2, 2015:
http://www.usgbc.org/Docs/News/News477.pdf

Kenney, Douglas S., Michael Mazzone, and Jacob Bedingfield, "Relative Costs of New Water Supply Options for Front Range Cities," *Colorado Water*, September/October 2010. As of November 2, 2015:
http://www.cwi.colostate.edu/ThePoudreRunsThroughIt/files/Relative_costs_of_new_water_ supply_options.pdf

Kenny, Alice, "Pennsylvania Water Deal: Blip or Boom?" *Ecosystem Marketplace*, June 3, 2008. As of November 2, 2015:
http://www.ecosystemmarketplace.com/pages/dynamic/article.page.php?page_id=5905&sect ion=home&eod=1

Kidd, Richard, "Comments at the Federal Utility Partnership Working Group (FUPWG) Meeting, November 19–20, 2008, Williamsburg Virginia," *FUPWG Fall 2008 Report*, Fall 2008.

Kieser, Mark, and Andrew Feng Fang, "U.S. Water Trading: The Infrastructure," Ecosystem Marketplace, May 6, 2008. As of November 2, 2015:
www.ecosystemmarketplace.com/pages/dynamic/article.page.php?page_id=5793&section=h ome&eod=1

Kosloff, Tracy, "A WISE Project for the Denver Metro Area," American Water Resources Association—Colorado Section, March 30, 2010. As of November 2, 2015: http://www.awracolorado.org/water-infrastructure-and-supply-efficiency-wise-project-mar-30-2010/

Kumar, Supriya, "The Looming Threat of Water Scarcity," World Watch Institute, March 19, 2013. As of November 2, 2015: http://vitalsigns.worldwatch.org/node/180

Lacan, Igor, "SB 918 Senate Bill - Bill Analysis," Sacramento, Calif.: California Assembly Committee on Water, Parks and Wildlife, June 1, 2010. As of November 2, 2015: http://www.leginfo.ca.gov/pub/09-10/bill/sen/sb_0901-0950/sb_918_cfa_20100628_114639_asm_comm.html

Lachman, Beth E., Kimberly Curry Hall, Aimee E. Curtright, and Kimberly M. Colloton, *Making the Connection: Beneficial Collaboration Between Army Installations and Energy Utility Companies*, Santa Monica, Calif.: RAND Corporation, MG-1126-A, 2011. As of November 2, 2015: http://www.rand.org/pubs/monographs/MG1126.html

Lachman, Beth E., Ellen M. Pint, Gary Cecchine, and Kimberly Colloton, *Developing Headquarters Guidance for Army Installations Sustainability Plans in 2007*, Santa Monica, Calif.: RAND Corporation, MG-837-A, 2009. As of November 2, 2015: http://www.rand.org/pubs/monographs/MG837.html

Lachman, Beth E., Susan A. Resetar, Geoffrey McGovern, Katherine Pfrommer, and Jerry M. Sollinger, "A Guide to Army Installation Water Rights: How to Establish, Maintain, Document, and Enhance Your Water Rights," unpublished RAND Corporation research, March 14, 2015.

Lacombe, Pierre, and Glen Carleton, "Hydrogeologic Framework, Availability of Water Supplies, and Saltwater Intrusion, Cape May County, New Jersey," U.S. Geological Survey, Water Resources Investigations Report 01-4246, 2002. As of November 2, 2015: http://pubs.usgs.gov/wri/wri014246/

Lake Oswego Tigard Water Partnership, "About the Partnership," undated. As of November 2, 2015: http://www.lotigardwater.org/?p=about-the-partnership

———, "Lake Oswego Tigard Water Partnership," fact sheet, undated.

Landry, Clay, "The Wet Water Market - The Case for Water Rights," *Global Water Intelligence*, Vol. 1, No. 1, October 2010.

Lee, Terence R. and Andrei S. Jouravlev, "Prices, Property and Markets in Water Allocation," Santiago, Chile: United Nations Economic Commission for Latin America and The Caribbean, 1998. As of November 2, 2015:
http://archivo.cepal.org/pdfs/1998/S9800052.pdf

Leonard Rice Engineers, Inc., et al., "Holistic Approach to Sustainable Water Management in Northwest Douglas County," January 2007. As of November 2, 2015:
http://www.tribesandclimatechange.org/docs/tribes_244.pdf

Leonard, Abby, "Architect Frank Gehry Talks LEED and the Future of Green Building," PBS, June 14, 2010. As of November 2, 2015:
http://www.pbs.org/wnet/need-to-know/culture/architect-frank-gehry-talks-leed-and-the-future-of-green-building/1458/

Lepper, Troy, "Water Banking in the West: Using Water Banks to Create Markets in the Arkansas Valley," Fort Collins, Colo.: Colorado State University Sociology Lab, 2003.

Leurig, Sharlene, "Water Ripples: Expanding Risks for U.S. Water Suppliers," *Ceres*, December 2012. As of November 2, 2015:
https://www.ceres.org/resources/reports/water-ripples-expanding-risks-for-u.s.-water-providers/view

Libecap, Gary, "Chinatown: Owens Valley and Western Water Reallocation—Getting the Record Straight and What it Means for Water Markets," *Texas Law Review*, Vol. 83, No. 7, June 2005. As of November 2, 2015:
http://papers.ssrn.com/sol3/papers.cfm?abstract_id=874818

———, "Water Woes: Using Markets to Quench the Thirst of the American West," *The Milken Institute Review*, Fourth Quarter 2010. As of November 2, 2015:
http://www2.bren.ucsb.edu/~glibecap/milkenInstitutequarterlyreport.pdf

Libecap, Gary, and Jane Shaw, "Rescuing Water Markets: Lessons from Owens Valley," *PERC Policy Series,* Issue Number PS-33, July 2005. As of November 2, 2015:
http://www.perc.org/sites/default/files/ps33.pdf

Little, Amanda, "Can Desalination Counter the Drought?" *The New Yorker*, July 22, 2015.

Los Angeles Aqueduct, "The Story of the Los Angeles Aqueduct," undated. As of November 2, 2015:
http://wsoweb.ladwp.com/Aqueduct/historyoflaa/

Los Angeles Department of Water and Power, "Mayor Garcetti Announces Increased LADWP Rebate for Residential Turf Removal to Highest Level in Southern California—$3.75," November 3, 2014. As of February 23, 2016:
http://www.ladwpnews.com/go/doc/1475/2404414/Residential-Cash-In-Your-Lawn-Rebate-now-3-75

Lovett, Ian, "Arid Southwest Cities' Plea: Lose the Lawn," *New York Times*, August 11, 2013. As of November 2, 2015:
http://www.nytimes.com/2013/08/12/us/to-save-water-parched-southwest-cities-ask-homeowners-to-lose-their-lawns.html?_r=0

Lower Colorado River Authority, "Water Partnership with City of Austin," 2013. As of November 2, 2015:
http://www.lcra.org/water/supply/contracts/austinagreement.html

Luke, Rob, "Voluntary Water Markets: The Demand Dilemma," *Ecosystem Marketplace,* May 26, 2008. As of November 2, 2015:
http://www.ecosystemmarketplace.com/pages/dynamic/article.page.php?page_id=5901&section=home&eod=1

Maricopa County Superior Court, "Arizona's General Stream Adjudications," Maricopa County, Ariz., undated. As of November 5, 2015:
http://www.superiorcourt.maricopa.gov/SuperiorCourt/GeneralStreamAdjudication/Index.asp

Massera, Paul, "Integrated Water Management in the California Water Plan," Integrated Regional Water Management (IRWM) Conference, May 24, 2011.

Melton, Paula, "International Green Construction Code Passes," *Environmental Building News,* November 8, 2011. As of November 2, 2015:
http://www.buildinggreen.com/auth/article.cfm/2011/11/8/International-Green-Construction-Code-Passes/

Mitchell, Becky, "Colorado Water Plan," American Water Resources Association—Colorado Section, June 2013. As of November 2, 2015:
http://www.awracolorado.org/wp-content/uploads/2013/08/Colorado-Water-Plan-template-Powerpoint-7-8-2013_Becky-Mitchell.pdf

Monterey County Water Resources Agency Board Of Directors, "Greater Monterey County Regional Full Planning Packet," December 17, 2009.

Moore, Paula, "Southern Delivery System Pipeline Working its Way toward Colorado Springs," ENR Mountain States, April 23, 2012. As of November 2, 2015:
http://mountainstates.construction.com/mountainstates_construction_projects/2012/0423-water-supply-boostfor-southern-colorado.asp

Moore, Rhonda, "Castle Rock Still Wants WISE Partnership Water but There are Worries about Rates," *Castle Rock News-Press*, February 28, 2013. As of November 2, 2015: https://coyotegulch.wordpress.com/2013/02/28/castle-rock-still-wants-wise-partnership-water-but-there-are-worries-about-rates/

Naeser, Robert Benjamin, and Mark Griffin Smith, "Playing With Borrowed Water: Conflicts Over Instream Flows on the Upper Arkansas River," *Natural Resources Journal,* Vol. 35, Winter 1995. As of November 2, 2015: http://lawlibrary.unm.edu/nrj/35/1/04_naeser_playing.pdf

Nataraj, Shanthi, and W. Michael Hanemann, "Does marginal price matter? A regression discontinuity approach to estimating water demand," *Journal of Environmental Economics and Management*, Vol. 61, No. 2, March 2011, pp. 198–212. As of November 2, 2015: http://www.sciencedirect.com/science/article/pii/S0095069610000823

National Aeronautics and Space Administration, "Owens Lake, California," September 12, 2011. As of November 2, 2015: http://earthobservatory.nasa.gov/IOTD/view.php?id=52072

National Agricultural Law Center, "Water Law: An Overview," Fayetteville, Ark.: University of Arkansas School of Law, undated. As of October 29, 2015: http://nationalaglawcenter.org/overview/water-law/

National Association of Water Companies, "Quick Facts of U.S. Water Public-Private Partnerships," undated. As of November 2, 2015: http://www.nawc.org/uploads/documents-and-publications/documents/document_54e30a7f-65fe-465a-ab91-1ce21d0d44ad.pdf

National Drought Mitigation Center, "State Drought Planning," 2008. As of November 2, 2015: http://www.rrbdp.org/planning_state.html

National Judicial College, "Dividing the Waters," undated. As of November 2, 2015: http://www.judges.org/dtw/

National Park Service, "Withdrawal of National Historic Landmark Designation: Roosevelt Dam," undated. As of November 2, 2015: http://www.nps.gov/history/nhl/DOE_dedesignations/Roosevelt.htm

National Research Council, "Central Arizona: The Endless Search for New Supplies to Water the Desert," *Water Transfers in the West: Efficiency, Equity, and the Environment*, Washington, D.C.: National Academies Press, 1992.

———, "Water Reuse: Potential for Expanding the Nation's Water Supply Through Reuse of Municipal Wastewater," Washington, D.C.: National Academies Press, 2012. As of November 5, 2015:
http://www.nap.edu/catalog/13303/water-reuse-potential-for-expanding-the-nations-water-supply-through

Nature Conservancy, "Arizona San Pedro River," last updated January 8, 2013. As of November 5, 2015:
http://www.nature.org/ourinitiatives/regions/northamerica/unitedstates/arizona/placesweprotect/san-pedro-river.xml

Neely, Betsy, et al., *Central Shortgrass Prairie Ecoregional Assessment and Partnership Initiative: Final Report*, Nature Conservancy of Colorado and the Shortgrass Prairie Partnership, November 2006.

Nelson, Barry, "Southern California's New Wave of Local Water Supplies," Natural Resources Defense Council Staff Blog, February 3, 2012. As of November 2, 2015:
http://switchboard.nrdc.org/blogs/bnelson/southern_californias_new_wave_1.html

Neuman, Janet C., "Have We Got a Deal for You: Can the East Borrow from the Western Water Marketing Experience?" *Georgia State University Law Review*, Vol. 21, Winter 2004. As of November 2, 2015:
http://papers.ssrn.com/sol3/papers.cfm?abstract_id=969742

Nichols, Peter D., Megan K. Murphy, and Douglas S. Kenney, "Water and Growth in Colorado: A Review of Legal and Policy Issues," Denver, Colo.: University of Colorado School of Law Natural Resources Law Center, 2001. As of November 2, 2015:
http://scholar.law.colorado.edu/books_reports_studies/29/

Norment, Richard, "The Framework for Public-Private Partnerships vs. Privatization," Washington, D.C.: National Council for Public-Private Partnerships, 2013.

Northern Colorado Water Conservancy District, "The Colorado Big Thompson Project: Historical, Logistical, and Political Aspects of This Pioneering Water Delivery System," undated. As of November 5, 2015:
http://www.northernwater.org/docs/MediaAndNews/CBT_NwBroch.pdf

Northern Water, "Colorado–Big Thompson Project," undated. As of October 29, 2015:
http://www.northernwater.org/WaterProjects/C-BTProject.aspx

O'Donnell, Michael, and Bonnie Colby, "Dry-Year Supply Reliability Contracts: A Tool for Water Managers," Tucson, Ariz.: University of Arizona Department of Agricultural and Resource Economics, October 2009. As of November 5, 2015:
https://ag.arizona.edu/arec/pubs/facultypubs/ewsr-dyo-Final-5-12-10.pdf

———, "Water Auction Design for Reliability: Design, Implementation and Evaluation," Tucson, Ariz.: University of Arizona Department of Agricultural and Resource Economics, May 27, 2009. As of November 5, 2015: https://ag.arizona.edu/arec/pubs/facultypubs/ewsr-AUCTION-final-5-12-10.pdf

———, "Water Banks: A Tool for Enhancing Water Supply Reliability," Tucson, Ariz.: University of Arizona Department of Agricultural and Resource Economics, January 2010. As of November 5, 2015: https://ag.arizona.edu/arec/pubs/facultypubs/ewsr-Banks-final-5-12-10.pdf

Official Code of Georgia Annotated (O.C.G.A.), "Policy statement for comprehensive state-wide water management plan; guiding principles; requirements of plan; regional plans; compliance with plans," § 12-5-522, 2010. As of November 5, 2015: http://law.justia.com/codes/georgia/2010/title-12/chapter-5/article-8/12-5-522

Olmstead, Sheila M., "Managing Water Demand: Price vs. Non-Price Conservation Programs," Pioneer Institute, July 2007. As of November 5, 2015: http://www.hks.harvard.edu/fs/rstavins/Monographs_&_Reports/Pioneer_Olmstead_Stavins_Water.pdf

———, "The Economics of Managing Scarce Water Resources," *Review of Environmental Economics and Policy*, Vol. 4, No. 2, 2010. As of November 5, 2015: http://reep.oxfordjournals.org/content/4/2/179.abstract

Olmstead, Sheila M., and Robert N. Stavins, "Comparing Price and Non-price Approaches to Urban Water Conservation," Cambridge, Mass.: National Bureau of Economic Research, Working Paper 14147, June 2008. As of November 5, 2015, 2012: http://www.nber.org/papers/w14147

Oney, Dan, "Recycled Water Coalition Expands Membership to Delta, Central Valley; Plans 20 New Projects," PublicCEO.com, February 22, 2013. As of November 5, 2015: http://www.publicceo.com/2013/02/14998/

Passel, Jeffrey, and D'Vera Cohn, "U.S. Population Projections: 2005–2050," Pew Research Center, February 11, 2008. As of November 5, 2015: http://www.pewsocialtrends.org/2008/02/11/us-population-projections-2005-2050/

Petrie, Ragan, Susan K. Laury, and Stephanie Hill, "Crops, Water Usage, and Auction Experience in the 2002 Irrigation Reduction Auction," Atlanta, Ga.: Georgia State University Water Policy Working Paper #2004-014, December 2004. As of November 5, 2015: http://www.edc.uri.edu/temp/ci/ciip/SummerPracticum/Docs2006/Irrigation%20Auction.pdf

Phillips, Caitrin, "Imported vs. Local Water Supplies: The Planning Decisions Facing Southern California Water Agencies," paper prepared for the Natural Resources Defense Council, August 3, 2011. As of November 5, 2015:
http://switchboard.nrdc.org/blogs/bnelson/Local%20vs%20Imported_Final%208-4-11.pdf

Pikes Peak Area Council of Governments, "Pikes Peak Region Sustainability Project, Regional Sustainability Stretch Goals for 2030," December 6, 2010. As of November 5, 2015:
http://www.ppacg.org/files/SUSTAIN/DOCUMENTS/PPR2030_final.pdf

Pima County Geographic Information Systems, "Arizona's Assured Water Supply Rules," undated. As of November 5, 2015:
http://gis.pima.gov/data/layers/cagrdsub/aws.htm

Pittman, Craig, "Florida's Aquifer Models Full of Holes, Allowing More Water Permits and Pollution," *Tampa Bay Times,* January 27, 2013. As of November 5, 2015:
http://www.tampabay.com/news/environment/water/floridas-aquifer-models-full-of-holes-allowing-more-water-permits-and/1272555

Pitzer, Gary, Susanna Eden, and Joe Gelt, "Layperson's Guide to Arizona's Water," Tucson, Ariz.: Water Resources Research Center and Water Education Foundation, 2007. As of November 5, 2015:
https://wrrc.arizona.edu/publications/laypersons-guide-arizona-water/laypersons-guide-arizona-water

PRISM Group and Oregon Climate Service, "Average Annual Precipitation 1971–2000 Colorado," 2006.

Pritchett, James, et al., "Water as a Crop: Limited Irrigation and Water Leasing in Colorado," *Applied Economic Perspectives Policy,* Vol. 30, No. 3, 2008. As of November 5, 2015:
http://econpapers.repec.org/article/ouprevage/v_3a30_3ay_3a2008_3ai_3a3_3ap_3a435-444.htm

Public Law 91-604, Clean Air Act Amendments of 1970, December 31, 1970.

Public Law 92-500, Clean Water Act, October 18, 1972.

Public Law 93-205, Endangered Species Act, December 28, 1973.

Public Law 94–579, Federal Land Policy and Management Act of 1976, as amended through May 7, 2001.

Public Law 94-588, National Forest Management Act, October 22, 1976.

Public Law 109–58, Energy Policy Act, August 8, 2005.

Puget Sound Partnership, "About the Partnership," undated. As of November 5, 2015:
http://www.psp.wa.gov/

Quinlan, Paul, "Chesapeake Bay: Cap and Trade for Water Pollution—Trendy, Hip, Glitzy and Controversial," *Environment and Energy News*, May 8, 2012. As of November 5, 2015: http://www.eenews.net/stories/1059964052

Rascon, Matt, and R. Stickney, "Carlsbad Desalination Plant Opens," San Diego 7 NBC News, December 14, 2015. As of February 23, 2016: http://www.nbcsandiego.com/news/local/Carlsbad-Desalination-Plant-Opens-361844511.html

Ray, Tina, "Fort Bragg Focuses on Energy, Water Conservation," U.S. Army.mil, September 16, 2010. As of November 5, 2015: http://www.army.mil/article/45274/fort-bragg-focuses-on-energy-water-conservation/

Renshaw, Jeffrey A, "Utility Privatization in the Military Services: Issues, Problems, and Potential Solutions," *Air Force Law Review*, January 1, 2002.

"Reuse: The WISE Partnership Gets Approval from the Denver Water Board," *Denver Business Journal,* August 20, 2013. As of November 5, 2015: https://coyotegulch.wordpress.com/2013/08/20/reuse-the-wise-partnership-gets-approval-from-the-denver-water-board/

Robison, Jason, Katja Bratrschovsky, Jaime Latcham, Eliza Morris, Vanessa Palmer, and Arturo Villanueva, "Forging Ahead in the Era of Limits: The Evolution of Interstate Water Policy in the Colorado River Basin: Colorado River Basin Background Paper," Cambridge, Mass.: Harvard University Water Federalism Project, April 19–21, 2012.

Rogers, Jedediah S., "Fryingpan-Arkansas Project," U.S. Bureau of Reclamation, 2006. As of November 5, 2015: http://www.usbr.gov/projects//ImageServer?imgName=Doc_1305042036789.pdf

Ruble, John, "Successfully Reducing Your Groundwater Footprint," Fort Huachuca, Ariz.: PowerPoint presentation, Directorate of Public Works, 2011.

San Diego Coastkeeper, "Desalination in San Diego," undated. As of November 5, 2015: http://www.sdcoastkeeper.org/learn/san-diegos-water-supply/desalination.html

Sanburn, Josh, "The Toxic Tap: How a Disastrous Chain of Events Corroded Flint's Water System—and the Public Trust," *Time*, February 1, 2016.

Scanlon, Bridget, et al., "Groundwater Depletion and Sustainability of Irrigation in the U.S. High Plains and Central Valley," *Proceedings of the National Academy of Sciences*, Vol. 109, No. 24, May 29, 2012. As of November 5, 2015: http://www.pnas.org/content/109/24/9320.abstract

Scholze, Richard J., "Water Reuse and Wastewater Recycling at U.S. Army Installations: Policy Implications," U.S. Army Corps of Engineers, ERDC/CERL SR-11-7, June 2011. As of November 5, 2015:
http://www.aepi.army.mil/docs/whatsnew/ERDC-CERL_SR-11-7(AEPI).pdf

Scott, Christopher "Case Study: Effluent Auction in Prescott Valley, Arizona," Tucson, Ariz.: University of Arizona, 2012. As of November 5, 2015:
http://aquasec.org/wrpg/wp-content/uploads/2011/09/Scott-InPress-US-AZ-PrescottValley-EPA-Reuse.pdf

Selman, Mindy, Suzie Greenhalgh, Cy Jones, Evan Branosky, and Jenny Guiling, "U.S. Water Quality Trading: Growing Pains and Evolving Drivers: The State of the Market Today," *Ecosystem Marketplace,* May 14, 2008. As of November 5, 2015:
http://www.ecosystemmarketplace.com/pages/dynamic/article.page.php?page_id=5796&section=home&eod=1

Selman, Mindy, Suzie Greenhalgh, Evan Branosky, Cy Jones, and Jenny Guiling, *Water Quality Trading Programs: An International Overview*, Washington, D.C.: World Resources Institute, March 2009.

Shipman, Taylor, "Why Not Auction Water? Perspectives on Auctioning the Next Bucket in Arizona," presentation at AWRA Annual Meeting, Albuquerque, N.M., November 9, 2011. As of November 5, 2015:
http://www.kysq.org/docs/AWRA_Shipman.pdf

Shipman, Taylor and Sharon B. Megdal, "Arizona's Groundwater Savings Program," *Southwest Hydrology*, May/June 2008.

Shipp, Allison, and Gail E. Cordy, "The USGS Role in TMDL Assessments," Washington, D.C.: U.S. Geological Survey, Fact Sheet FS 130-01, undated. As of November 5, 2015:
http://pubs.usgs.gov/fs/FS-130-01/

Smith, Duane A., "The River of Sorrows: The History of the Lower Dolores River Valley," National Park Service, undated. As of November 5, 2015:
http://www.nps.gov/parkhistory/online_books/rmr/river_of_sorrows/chap1.htm

Smith, Mike, "City, Casino Partner on Water Services," *The Duncan Banner*, August 14, 2013. As of November 5, 2015:
http://duncanbanner.com/x1664884769/City-Casino-partner-on-water-services

Smith, William James, Jr. and Young-Doo Wang, "Conservation Rates: The Best 'New' Source of Urban Water," *Water and Environment Journal*, Chartered Institute of Water and Environmental Management, 2007. As of November 5, 2015:
http://onlinelibrary.wiley.com/doi/10.1111/j.1747-6593.2007.00085.x/abstract

Southern California Water Committee, "Stormwater Capture: Opportunities to Increase Water Supplies in Southern California," January 2012. As of November 5, 2015:
http://socalwater.org/images/SCWC_Stormwater_White_Paper__Case_Studies.Smaller.pdf

Southern Minnesota Beet Sugar Cooperative, "SMBSC Facts," undated. As of November 5, 2015:
http://www.smbsc.com/OurCooperative/overview.aspx

———, "Southern Minnesota Beet Sugar Cooperative Far Exceeds Annual Phosphorous Offset Requirements," undated. As of November 5, 2015:
http://smbsc.com/pdf/2013SustainabilityOverview.pdf

Southwest Florida Water Management District, "Reclaimed Water Guide," 1999. As of November 5, 2015:
http://www.swfwmd.state.fl.us/download/view/site_file_sets/118/reclaimed-water-guide.pdf

Southwestern Investor Group, "High Plains LLC Water Rights Project," undated. As of November 5, 2015:
http://swinvest.com/Project%20Pages/Highplains.html

Southwestern Water Conservation District, "The Water Information Program: Providing Water Information to the Communities of Southwest Colorado," undated. As of November 5, 2015:
http://www.waterinfo.org

Spotts, Pete, "Southern Great Plains Could Run Out of Groundwater in 30 Years, Study Finds," *Christian Science Monitor*, May 30, 2012. As of November 5, 2015:
http://www.csmonitor.com/Environment/2012/0530/Southern-Great-Plains-could-run-out-of-groundwater-in-30-years-study-finds

Stamper, Vicki, Cindy Copeland, and Megan Williams, "Poisoning the Great Lakes: Mercury Emissions from Coal-Fired Power Plants in the Great Lakes Region," National Resources Defense Council, 2012. As of November 5, 2015:
http://www.nrdc.org/air/files/poisoning-the-great-lakes.pdf

Staudenmaier, L. William, "Between a Rock and a Dry Place: The Rural Water Supply Challenge for Arizona, *Arizona Law Review,* Vol. 49, 2007. As of November 5, 2015:
http://www.arizonalawreview.org/pdf/49-2/49arizlrev321.pdf

Steinitz, Carl, et al, "Alternative Futures for Landscapes in the Upper San Pedro River Basin of Arizona and Sonora," U.S. Forest Service Gen. Tech. Rep. PSW-GTR-191, 2005.

Stelting, Neal, "An Overview of Water Markets in the West," Laramie, Wyo.: WestWater Research, October 27, 2004.

Steven Winter Associates, "GSA LEED Cost Study," October 2004. As of September 26, 2013:
http://www.wbdg.org/ccb/GSAMAN/gsaleed.pdf

Stoughton, Kate McMordie, "Fort Carson Net Zero Water Balance," Pacific Northwest National Laboratory, April 2012.

Stover, Susan, et al., "Central Kansas Water Bank Association Five Year Review and Recommendations," January 20, 2011.

Streeter, Tracy, "Testimony before the House Agriculture and Natural Resources Committee on HB 2516," Kansas Water Authority, January 31, 2012. As of November 5, 2015:
http://www.kwo.org/news_government/Testimony/testimony_HB2516_Wbank__Streeter_01 3112_ss.pdf

Strommer, Jane, "Successful Bay Area Recycled Water Coalition is Expanding Across Mid-Pacific Region and Open to New Member Agencies," Central Valley/Sierra Foothills WaterReuse California, Central Valley/Sierra Foothills Chapter Newsletter, Vol. 1, No. 2, December 2012.

Sutherland, Andrea, "Fort Carson Builds Toward Energy, Water, Waste Goals," *Public Works Digest*, Vol. 22, No. 5, September/October 2011.

Templer, Otis, "Water Law," Texas State Historical Association, undated. As of November 5, 2015:
http://www.tshaonline.org/handbook/online/articles/gyw01

Texas Water Code, Chapter 16.051, undated. As of November 5, 2015:
http://www.statutes.legis.state.tx.us/Docs/WA/htm/WA.16.htm

Texas Water Development Board, "Water for Texas: 2012 State Water Plan," January 2012.

"The Right Way to Invest in Water?" *Global Water Intelligence*, Vol. 10, No. 5, May 2009. As of October 29, 2015:
http://www.globalwaterintel.com/archive/10/5/market-insight/right-way-invest-water.html

TheWaterPage.com, "Water Scarcity," undated. As of November 5, 2015:
http://www.thewaterpage.com/drought_water_scarcity.htm

"The West's New Gold Rush for Water Rights," *Global Water Intelligence*, Vol. 8, No. 12, December 2007. As of October 29, 2015:
http://www.globalwaterintel.com/archive/8/12/general/the-wests-new-gold-rush-for-water-rights.html

Theobald, David M., "Defining and Mapping Rural Sprawl: Examples from the Northwest U.S.," Fort Collins, Colo.: Colorado State University, September 16, 2003.

Trapp, Ely, "TEAD Team Wins Federal Energy Conservation Award," U.S. Army, news article, November 9, 2009. As of November 15, 2015:
http://www.army.mil/article/30099/TEAD_team_wins_Federal_Energy_Conservation_Award/

Turner, Jim, *Arizona: A Celebration of the Grand Canyon State,* Layton, Utah: Gibbs Smith, 2011.

Umof, Allie Alexis, "An Analysis of the 1944 U.S.-Mexico Water Treaty: Its Past, Present and Future," *Environmental Law and Policy Journal*, Vol. 32, No. 1, 2008. As of November 5, 2015:
http://environs.law.ucdavis.edu/volumes/32/1/umoff.pdf

United Nations World Commission on Environment and Development, *Our Common Future,* 1987.

Upper San Pedro Partnership, homepage, undated. As of October 29, 2015:
http://www.usppartnership.com/

———, "About Us," undated. As of November 5, 2015:
http://www.usppartnership.com/press_mission.htm

URS Corporation, "Energy Development Water Needs Assessment," September 2008. As of November 5, 2015:
http://cwcbweblink.state.co.us/weblink/ElectronicFile.aspx?docid=127791

U.S. Army Corps of Engineers, "Building Strong Collaborative Relationships for a Sustainable Water Resources Future: State of Georgia Summary of State Water Planning," Washington, D.C., December 2009. As of November 5, 2015:
http://www.building-collaboration-for-water.org/Documents/StateSummaries/GA%201209.pdf

U.S. Army Environmental Command, "Fort Huachuca ACUB," fact sheet, undated. As of November 5, 2015: www.repi.mil/Documents/FactSheets/FortHuachuca.pdf

U.S. Army Environmental Policy Institute, *Army Water Security Strategy*, December 2011. As of October 29, 2015:
http://www.aepi.army.mil/docs/whatsnew/ArmyWaterStrategy.pdf

U.S. Bureau of Reclamation, "Salt River Project," last updated July 8, 2015. As of November 2, 2015:
http://www.usbr.gov/lc/phoenix/projects/saltriverproj.html

———, Colorado–Big Thompson Project, last updated July 18, 2013. As of December 15, 2012:
http://www.usbr.gov/projects/Project.jsp?proj_Name=Colorado–Big%20Thompson%20Project

———, "Title XVI—Water Reclamation & Reuse Program," June 17, 2013. As of November 2, 2015:
http://www.usbr.gov/WaterSMART/title/

———, "Fryingpan-Arkansas Project," last updated April 4, 2013. As of December 15, 2015:
http://www.usbr.gov/projects/Project.jsp?proj_Name=Fryingpan-Arkansas%20Project

———, "The Law of the River," March 2008. As of November 2, 2015:
http://www.usbr.gov/lc/region/g1000/lawofrvr.html

U.S. Census Bureau, "Annual Estimates of the Population for the United States, Regions, States, and Puerto Rico: April 1, 2010 to July 1, 2011," NST-EST2011-01, December 2009. As of November 5, 2015:
http://www.census.gov/popest/data/state/totals/2011/tables/NST-EST2011-01.xls

———, "Colorado 2010 Census Results," August 2012. As of November 5, 2015:
http://www.census.gov/prod/cen2010/cph-2-7.pdf

———, "Cumulative Estimates of Resident Population Change for the United States, Regions, States and Puerto Rico and Region and State Rankings: April 1, 2000 to July 1, 2009 (NST-EST2009-02)," December 22, 2009.

———, "Population Change and Distribution: 1990 to 2001," April 2001. As of November 5, 2015:
http://www.census.gov/prod/2001pubs/c2kbr01-2.pdf

———, "Resident Population Data," undated. As of November 5, 2015:
http://www.census.gov/2010census/data/apportionment-dens-text.php

———, "State & County QuickFacts," undated, As of November 5, 2015:
http://quickfacts.census.gov/qfd/states/04000.html

U.S. Conference of Mayors, "Executive Summary: Filling the Void," 2011. As of November 5, 2015: www.usmayors.org/bestpractices/pubpri11/Schenectady.pdf

———, "Four Cities Honored for Excellence & Innovation in Public-Private Partnerships," January 21, 2011. As of November 5, 2015:
http://usmayors.org/pressreleases/uploads/2013/0119-release-publicprivatepartnership.pdf

U.S. Department of the Army, "Alternative Financing," Washington, D.C.: Army Energy and Water Management Program, undated. As of October 29, 2015:
http://army-energy.hqda.pentagon.mil/funding/alternative.asp

———, "Appropriated Funds," Washington, D.C.: Army Energy and Water Management Program, undated. As of October 29, 2015:
http://army-energy.hqda.pentagon.mil/funding/appropriated.asp

349

—————, *Army Energy and Water Campaign Plan for Installations: FY2008–2013*, 2006.

—————, Army Regulation 420-1, Section 22-12, undated.

—————, Army Regulation 420-1, Section 23-20, undated.

—————, *Army Energy Security Implementation Strategy,* 2009.

—————, *Installation Management Community Campaign Plan: 2010–2017,* Annex F, March 2010.

—————, *Installation Management Community Campaign Plan: 2012–2017*, Annex F, November 2011.

—————, *Installation Management Water Portfolio: 2011–2017*, April 2011.

—————, "Sustainable Design and Development Policy Update," Washington, DC: Office of the Assistant Secretary of the Army, Installations, Energy and Environment (OASA IE&E), October 27, 2010.

—————, "Water Rights Policy for Army Installations in the United States," Washington, D.C.: Secretary of the Army, Army Directive 2014-8, May 12, 2014.

U.S. Department of Defense, *American Recovery and Reinvestment Act of 2009*, June 2010. As of November 2, 2015:
http://www.defense.gov/recovery/plans_reports/2010/pdfs/DoD%20ECIP%20Program%20Pl an%20Update_FINAL_062110.pdf

—————, *Department of Defense Strategic Sustainability Performance Plan: FY 2011*, Washington, D.C.: DoD Environment, Safety and Occupational Health Network and Information Exchange, October 2011.

—————, *Department of Defense Strategic Sustainability Performance Plan: FY 2012*, Washington, D.C.: DoD Environment, Safety and Occupational Health Network and Information Exchange, August 26, 2010. As of November 2, 2015:
http://www.denix.osd.mil/sustainability/upload/dod-sspp-public-26aug10.pdf

—————, "Energy Savings Performance Contracts and Utility Energy Service Contracts Memorandum," Washington, D.C.: Office of the Under Secretary of Defense, January 24, 2008.

—————, "FY 2008 Energy Conservation Investment Program (ECIP) Projects," Washington, D.C.: Office of the Undersecretary of Defense for Acquisition, Technology and Logistics, 2008.

—————, "FY 2012 Energy Conservation Investment Program (ECIP)," Washington, D.C.: Office of the Under Secretary of Defense (Comptroller), 2012.

———, "FY 2013 Energy Conservation Investment Program (ECIP)," Washington, D.C.: Office of the Under Secretary of Defense (Comptroller), 2013.

———, "REPI (Readiness and Environmental Protection Program)," undated. As of November 5, 2015:
http://www.repi.mil/

———, "U.S. Army: Fort Huachuca: Arizona," Washington, D.C.: Readiness and Environmental Protection Integration (REPI) Program fact sheet, undated.

———, "Water Rights and Water Resources Management on Department of Defense Installations and Ranges in the United States and Territories," Washington, D.C.: Office of the Under Secretary of Defense, May 23, 2014.

U.S. Department of Energy, "2012 Federal Energy and Water Management Award Winners," Washington, D.C.: Federal Energy Management Program, February 7, 2013. As of November 2, 2015:
http://energy.gov/eere/femp/2012-federal-energy-and-water-management-award-winners

———, "DOE IDIQ ESPC Awarded Projects," January 18, 2013. As of November 5, 2015:
www1.eere.energy.gov/femp/pdfs/do_awardedcontracts.pdf

———, "Energy Savings Performance Contracts (ESPCs)," September 2012.

———, "Federal Energy and Water Efficiency Project Financing," Washington, D.C.: Federal Energy Management Program, undated. As of November 2, 2015:
http://www1.eere.energy.gov/femp/financing/espcs.html

———, "Utility Partnerships Program Overview," Washington, D.C.: Federal Energy Management Program, undated. As of November 2, 2015:
http://energy.gov/eere/femp/downloads/utility-partnerships-program-overview

U.S. Department of the Interior and U.S. Geological Survey, "Arizona," National Atlas of the United States of America, undated. As of November 2, 2015:
http://www.nationalatlas.gov/printable/images/pdf/precip/pageprecip_az3.pdf

———, "Colorado," National Atlas of the United States of America, undated. As of October 29, 2015:
http://nationalmap.gov/small_scale/printable/images/pdf/fedlands/CO.pdf

U.S. Department of Justice, "Federal Reserved Water Rights and State Law Claims," Washington, D.C.: Environment and Natural Resource Division, undated. As of November 5, 2015:
http://www.justice.gov/enrd/3245.htm

———, "The McCarran Amendment," Washington, D.C.: Environment and Natural Resource Division, undated. As of November 5, 2015:
http://www.justice.gov/enrd/3248.htm

U.S. Department of State, "Utilization of the Waters of the Colorado and Tijuana Rivers and of the River Grande," U.S. Government Printing Office, 1946. As of October 29, 2015:
http://www.ibwc.state.gov/Files/1944Treaty.pdf

U.S. Environmental Protection Agency, "2010 Biennial National Listing of Fish Advisories," EPA-820-F-11-009, September 2011. As of November 5, 2015:
http://water.epa.gov/scitech/swguidance/fishshellfish/fishadvisories/upload/National-Listing-of-Fish-Advisories-Technical-Fact-sheet-2010.pdf

———, "Cases in Water Conservation: How Efficiency Programs Help Water Utilities Save Water and Avoid Costs," 2002. As of November 5, 2015:
http://www.epa.gov/WaterSense/docs/utilityconservation_508.pdf

———, "Chesapeake Bay Watershed, Virginia: Watershed-Based General Permit for Nutrient Discharges and Nutrient Trading," undated.

———, "Climate Impacts on Water Resources," undated.

———, "Combined Sewer Overflows," February 16, 2012. As of November 5, 2015:
http://cfpub.epa.gov/npdes/home.cfm?program_id=5

———, "EPA's 2007 Drinking Water Infrastructure Needs Survey and Assessment Fact Sheet," February 2009. As of November 5, 2015:
http://water.epa.gov/infrastructure/drinkingwater/dwns/upload/2009_03_26_needssurvey_2007_fs_needssurvey_2007.pdf

———, "Final Water Quality Trading Policy," January 13, 2003. As of November 5, 2015:
http://water.epa.gov/type/watersheds/trading/finalpolicy2003.cfm

———, "Flint Drinking Water Response," last accessed February 19, 2016. As of February 23, 2016:
http://www.epa.gov/flint

———, "Green Infrastructure Case Studies: Municipal Policies for Managing Stormwater With Green Infrastructure," EPA-841-F-10-004, August 2010. As of November 5, 2015:
http://nepis.epa.gov/Exe/ZyNET.exe/P100FTEM.TXT?ZyActionD=ZyDocument&Client=EPA&Index=2006+Thru+2010&Docs=&Query=&Time=&EndTime=&SearchMethod=1&TocRestrict=n&Toc=&TocEntry=&QField=&QFieldYear=&QFieldMonth=&QFieldDay=&IntQFieldOp=0&ExtQFieldOp=0&XmlQuery=&File=D%3A\zyfiles\Index%20Data\06thru10\Txt\00000033\P100FTEM.txt&User=ANONYMOUS&Password=anonymous&SortMethod=h|-&MaximumDocuments=1&FuzzyDegree=0&ImageQuality=r75g8/r75g8/x150y150g16/i425&Display=p|f&DefSeekPage=x&SearchBack=ZyActionL&Back=ZyActionS&BackDesc=Results%20page&MaximumPages=1&ZyEntry=1&SeekPage=x&ZyPURL

———, "Guidelines for Water Reuse," EPA/600/R-12/618, September 2012. As of November 5, 2015:
http://nepis.epa.gov/Adobe/PDF/P100FS7K.pdf

———, "Hudson River Cleanup," undated. As of November 5, 2015:
http://www.epa.gov/hudson/cleanup.html#quest1

———, "The Importance of Water to the U.S. Economy Part 1: Background Report," September 2012. As of November 5, 2015:
http://water.epa.gov/action/importanceofwater/upload/Background-Report-Public-Review-Draft-2.pdf

———, "List of State and Individual Trading Programs: Excel Spreadsheet," undated.

———, "State and Individual Trading Programs," undated.

———, "Stormwater Management," undated.

———, "Study of the Potential Impacts of Hydraulic Fracturing on Drinking Water Resources: Progress Report," December 2012. As of November 5, 2015:
http://www.epa.gov/hfstudy/pdfs/hf-report20121214.pdf

———, "Water Quality Trading Program Fact Sheets," undated.

———, "What is Sustainability?" undated. As of November 5, 2015:
http://www.epa.gov/sustainability/basicinfo.htm

———, "What Is a Watershed?" March 6, 2012. As of November 5, 2015:
http://water.epa.gov/type/watersheds/whatis.cfm

U.S. Fish and Wildlife Service, "Endangered Species, Section 7 Consultation: A Brief Explanation," undated. As of November 5, 2015:
http://www.fws.gov/midwest/endangered/section7/section7.html

————, "Southwestern Willow Flycatcher," Reno, Nev.: Nevada Fish and Wildlife Office, last updated April 6, 2014. As of November 2, 2015:
http://www.fws.gov/nevada/protected_species/birds/species/swwf.html

U.S. Geological Survey, "Groundwater Depletion," USGS Water Science School, undated. As of November 5, 2015:
http://ga.water.usgs.gov/edu/gwdepletion.html

————, "Ground Water Atlas of the United States Arizona, Colorado, New Mexico, Utah," 1995. As of November 5, 2015:
http://pubs.usgs.gov/ha/ha730/ch_c/C-text3.html

————, "Trends in Water Use in the United States, 1950 to 2005," October 31, 2012. As of November 5, 2015:
http://ga.water.usgs.gov/edu/wateruse-trends.html

U.S. Government Accountability Office, *Defense Infrastructure: Management Issue Requiring Attention in Utility Privatization,* GAO-05-433, May 12, 2005. As of November 2, 2015:
http://www.gao.gov/assets/250/246288.html

————, *Klamath River Basin: Reclamation Met its Water Bank Obligations, But Information Provided to Water Bank Stakeholders Could be Improved*, Washington, D.C., GAO-05-283, 2005. As of November 5, 2015: www.gao.gov/cgi-bin/getrpt?GAO-05-283

U.S. Green Building Council, "LEED 2009 for New Construction and Major Renovations," 2008.

————, "LEED 2009: Technical Advancements to the LEED Rating System," 2011.

U.S. Soil Conservation Service, "Managing Arid and Semi Arid Watersheds: Prescott Active Management Area," undated. As of November 5, 2015:
http://www.fs.fed.us/rm/boise/AWAE/labs/awae_flagstaff/watersheds/highlands/prescott.html

Utah Department of Natural Resources, "Water Reuse in Utah," April 2005. As of November 5, 2015:
http://www.water.utah.gov/WaterReuse/WaterReuseAA.pdf

Walton, Brett, "The Price of Water 2012: 18 Percent Rise Since 2010, 7 Percent Over Last Year in 30 Major U.S. Cities," *Circle of Blue*, May 10, 2012. As of November 5, 2015:
http://www.circleofblue.org/waternews/2012/world/the-price-of-water-2012-18-percent-rise-since-2010-7-percent-over-last-year-in-30-major-u-s-cities/

Wang, Young-Doo, William James Smith, and John Byrne, "Water Conservation-Oriented Rates: Strategies to Extend Supply, Promote Equity, and Meet Minimum Flow Levels," American Water Works Association, 2005. As of November 5, 2015:
http://www.awwa.org/store/productdetail.aspx?productid=6544

Warner, Mildred E., and Amir Hefetz, "Service Characteristics and Contracting: The Importance of Citizen Interest and Competition," *Municipal Year Book 2010*, International City/County Management Association, Washington, D.C., 2010.

Washington State Department of Ecology, "2010 Report to the Legislature: Water Banking in Washington State," Olympia, Wash.: Publication No. 11-11-072, June 2011.

———, "An Overview of Washington State's Watershed Approach to Water Quality Management," last updated April 2008. As of November 5, 2015:
http://www.ecy.wa.gov/programs/wq/watershed/overview.html

———, "Current Status of Water Banking," undated. As of November 5, 2015:
http://www.ecy.wa.gov/programs/wr/instream-flows/wtrbank.html

———, "Pending Water Right Applications in Subbasins of the Yakima River Basin," undated. As of November 5, 2015:
http://www.ecy.wa.gov/programs/wr/cro/yakima-sb31.html

———, "Suncadia Water-rights OK'd for Water-banking Program," February 11, 2010. As of November 5, 2015:
http://www.ecy.wa.gov/news/2010news/2010-025.html

———, "Water Banking and Trust Water Programs: Important Water Management Tools," Olympia, Wash.: Publication No. 09-11-035, December 2009. As of November 5, 2015:
https://fortress.wa.gov/ecy/publications/publications/0911035.pdf

———, "The Watershed Planning Act," undated. As of November 5, 2015:
http://www.ecy.wa.gov/watershed/misc/background.html

Washington State Department of Ecology and WestWater Research, Inc., "Analysis of Water Banks in the Western States," Olympia, Wash.: Publication No. 04-11-011, July 2004.

Washington State Legislature, "Puget Sound Water Quality Protection," 1996. As of November 2, 2015:
http://apps.leg.wa.gov/rcw/default.aspx?cite=90.71&full=true#90.71.21

———, "Watershed Planning," undated. As of November 2, 2015:
http://apps.leg.wa.gov/RCW/default.aspx?cite=90.82&full=true

Water Education Foundation, "Layperson's Guide to California Water," Sacramento, Calif., 2008.

———, "Layperson's Guide to Groundwater," Sacramento, Calif., 2011.

———, "Layperson's Guide to Water Recycling," Sacramento, Calif., 2004.

Water Policy Institute, "Water Wars: Conflicts Over Shared Waters, A Case Study of the Appalachicola-Chattahoochee-Flint and Alabama-Coosa-Tallapoosa River Systems in the Southeastern United States and the Broader Implications of the Conflicts," Water Policy Institute, Hunton & Williams LLP, March 2009. As of November 5, 2015: http://www.huntonfiles.com/files/webupload/WPI_Water_Wars_White_Paper.pdf

Watershed Management Group, "WMG Receives American Planning Association Award for Green Projects," undated.

WaterWired, "Groundwater Depletion in the United States 1900–2008," May 26, 2013. As of November 5, 2015: http://aquadoc.typepad.com/waterwired/2013/05/usgs-publication-groundwater-depletion-in-the-united-states-1900-2008.html

Watson, Reed, and Brandon Scarborough, "Colorado River Water Bank: Making Water Conservation Profitable," Bozeman, Mont.: Property and Environment Research Center, September 28, 2010. As of November 5, 2015: http://perc.org/articles/colorado-river-water-bank-making-water-conservation-profitable

Weiland, Paul, "Federal District Court Strikes Down For Huachuca Biological Opinion," *Endangered Species Law & Policy*, June 5, 2011. As of November 5, 2015: http://www.endangeredspecieslawandpolicy.com/2011/06/articles/court-decisions/federal-district-court-strikes-down-fort-huachuca-biological-opinion/

Western Governors' Association, "Policy Resolutions 10-11: Federal Non-Tribal Fees in General Water Adjudications," undated.

Western Governors' Association and Western States Water Council, *Water Transfers in the West: Projects, Trends and Leading Practices in Voluntary Water Trading*, Denver, Colo., and Murray, Utah, December 2012.

Western Recycled Water Coalition, "Meeting the Water Challenge in the West," undated. As of November 5, 2015: http://www.barwc.org/

———, "Members," 2013.

———, "Projects," 2013.

———, "Western Recycled Water Coalition," fact sheet, March 14, 2013.

Western Resource Advocates, "Arizona Water Meter: A Comparison of Water Conservation Programs in 15 Arizona Communities," October 2010. As of November 5, 2015: www.westernresourceadvocates.org/azmeter/report.pdf

———, "Conservation Measures that Make Cents," 2008.

———, "Filling the Gap: Meeting Future Urban Water Needs in the Arkansas Basin," March 2012. As of November 5, 2015: http://westernresourceadvocates.org/publications/filling-the-gap-arkansas-basin/

———, "Front Range Water Meter: Water Conservation Ratings and Recommendations for 13 Colorado Communities," November 2007.

———, "Smart Water: A Comparative Study of Urban Water Use Across the Southwest," December 2003.

———, "The Reuse Strategy," undated.

———, "Water Rate Structures Structuring Water Rates to Promote Conservation," undated.

Western States Water Council, "Western Water Resources Infrastructure Strategies: Identifying, Prioritizing and Financing Needs," June 2011. As of November 5, 2015: http://www.westernstateswater.org/wp-content/uploads/2012/10/infrastructure-report_final_lowresolution.pdf:

Western States Water Council and Western Governors' Association, "Innovative Water Transfers Overview," 2008.

WestWater Research and Washington State Department of Ecology, "Water Banks in the United States," briefing, undated (data indicate likely 2003–2004).

Wexel, Curt, "Army Utilities Privatization (UP) Program Primer," U.S. Department of the Army, Office of the Assistant Chief of Staff for Installation Management, ACSIM Privatization and Partnerships Division, September 2012.

White House Office of the Press Secretary, "Executive Order 13514—Planning for Federal Sustainability in the Next Decade," October 5, 2009. As of November 2, 2015: http://www.whitehouse.gov/assets/documents/2009fedleader_eo_rel.pdf

———, "Executive Order 13423—Strengthening Federal Environmental, Energy and Transportation Management," January 24, 2007. As of November 2, 2015: http://www.gpo.gov/fdsys/pkg/FR-2007-01-26/pdf/07-374.pdf

———, "Presidential Memorandum—Federal Leadership on Energy Management," December 5, 2013. As of November 7, 2015: https://www.whitehouse.gov/the-press-office/2013/12/05/presidential-memorandum-federal-leadership-energy-management

Whittemore, Donald O., "Water Quality Of The Arkansas River In Southwest Kansas: A Report to the Kansas Water Office," Kansas Water Office, 2000. As of November 5, 2015: http://www.kgs.ku.edu/Hydro/UARC/quality-report.html

Wiener, John D., "Problems with the Arkansas River Basin Water Bank Pilot Program," Boulder, Colo.: University of Colorado paper, undated. As of November 5, 2015: http://www.colorado.edu/ibs/es/wiener/papers/One-pagerArkWBankPilotProgram.pdf

———, "Water Banking as Institutional Adaptation to Climate Variability: The Colorado Experiment," presentation at the AMS 2003 Annual Meeting, Symposium on Impacts of Water Variability: Benefits and Challenges, University of Colorado, Boulder, Colo., 2003.

Wiener, John, "Problems with the Arkansas River Basin Water Bank Pilot Program," One-pager notes, University of Colorado, Boulder, CO, undated. As of May 25, 3012: http://www.colorado.edu/ibs/es/wiener/papers/One-pagerArkWBankPilotProgram.pdf

Wildermuth, John, and John Coté, "Hetch Hetchy Measure Swamped by Voters," *San Francisco Chronicle*, November 7, 2012. As of November 5, 2015: http://www.sfgate.com/politics/article/Hetch-Hetchy-measure-swamped-by-voters-4014676.php

Wildman, Richard A., Jr. and Noelani A. Forde, "Management of Water Shortage in the Colorado River Basin: Evaluating Current Policy and the Viability of Interstate Water Trading," *Journal of the American Water Resources Association (JAWRA)*, Vol. 48, No. 3, June 2012. As of November 5, 2015: http://onlinelibrary.wiley.com/doi/10.1111/j.1752-1688.2012.00665.x/abstract

Willamette Partnership, "In It Together: A How-to Reference for Building Point-Nonpoint Water Quality Trading Programs," Hillsboro, Ore., July 2012.

Williams, Dave, "Georgia Wins Tri-state Water Ruling," *Atlanta Business Chronicle*, June 28, 2011. As of November 5, 2015: http://www.bizjournals.com/atlanta/news/2011/06/28/georgia-wins-tri-state-water-ruling.html

Willoughby, Lynn, "Chattahoochee River," New Georgia Encyclopedia, July 18, 2003. As of October 29, 2015: http://www.georgiaencyclopedia.org/nge/Article.jsp?id=h-950

———, *Flowing Through Time: A History of the Lower Chattahoochee River*, Tuscaloosa, Ala.: University of Alabama Press, 2012.

Winter, Thomas, Judson Harvey, O. Lehn Franke, and William Alley, "Ground Water and Surface Water: A Single Resource," Denver, Colo.: U.S. Geological Survey Circular 1139, 1998. As of November 5, 2015:
http://pubs.usgs.gov/circ/circ1139/

Woodka, Chris, "Blue Mesa Seen as a Water Bank," *Pueblo Chieftain*, February 12, 2012. As of November 5, 2015:
http://www.waterinfo.org/node/5698

Young, Holly, "Live Q&A: Water, Public Good or Private Commodity?" *Guardian Global Development Professionals Network*, September 1, 2014. As of November 9, 2015:
http://www.theguardian.com/global-development-professionals-network/2014/sep/01/water-access-inequality-private-commodity

Zellmer, Sandra, "The Anti-Speculation Doctrine in Water Law: Ghost-busting, Trust-busting, or Ensuring Reasonable, Beneficial Use?" American Bar Association Section of Environment, Energy, and Resources, 26th Annual Water Law Conference, San Diego, Calif., February 21–22, 2008.

Zetland, David, "Using Auctions to Share Scarce Water," *Solutions: For a Sustainable and Desirable Future*, Vol. 3, No. 2, April 2012. As of November 5, 2015:
http://www.thesolutionsjournal.com/node/1075

Zieburtz, Bill, and Rick Giardina (eds.), *Principles of Water Rates, Fees and Charges* (Sixth Edition), Denver, Colo.: American Water Works Association, 2012.

Ziemer, Laura, and Ada Montague, "Can Mitigation Water Banking Play a Role in Montana's Exempt Well Management?" *Trout Unlimited*, September 2011.

Zilberman, David, Ariel Dinar, Neal MacDougall, Madhu Khanna, Cheril Brown, and Frederico Castillo, "Individual and Institutional Responses to the Drought: The Case of California Agriculture," *Journal of Contemporary Water Research and Education*, Vol. 121, 2002. As of November 5, 2015:
http://opensiuc.lib.siu.edu/jcwre/vol121/iss1/3/.

Zito, Kelly, "Suit to Get Kern Water Bank Returned to State," *San Francisco Chronicle*, July 12, 2010. As of November 5, 2015:
http://www.sfgate.com/bayarea/article/Suit-to-get-Kern-Water-Bank-returned-to-state-3181896.php

Zitsch, Katherine H., Rick Brown, and John D. Boyer, "State Water Planning—An Overview of Approaches Used and Lessons Learned Across the U.S.," Proceedings of the 2012 South Carolina Water Resources Conference, October 10–11, 2012.

Zuniga, Jennifer E., "The Central Arizona Project," U.S. Bureau of Reclamation, 2000. As of November 5, 2015:
http://www.usbr.gov/projects/ImageServer?imgName=Doc_1303158888395.pdf